INTRODUCTION TO LINEAR AND NONLINEAR PROGRAMMING

DAVID G. LUENBERGER
Stanford University

ADDISON-WESLEY PUBLISHING COMPANY
Reading, Massachusetts
Menlo Park, California · London · Don Mills, Ontario

Abstract design on cover: "Computer Composition With Lines,"
© A. Michael Noll 1965.

"Computer Composition With Lines," one of the earliest examples of computer-generated art, was inspired by the 1917 painting "Composition With Lines," by the Dutch artist Piet Mondrian. See A. Michael Noll, "Human or Machine: A Subjective Comparison of Piet Mondrian's 'Composition With Lines' (1917) and a Computer-Generated Picture," *The Psychological Record,* 16, 1966, pages 1-10.

 Printed in the United States of America. Published simultaneously in Canada. Library of Congress Catalog Card No. 72–186209.

To Susan, Robert, and Jill

PREFACE

This book is intended as an introduction to a broad range of practical optimization techniques. It is designed for either self-study by professionals or classroom work at the undergraduate or graduate level for students who have a technical background in engineering, mathematics, or science. Over the past four years, the book has served, at various stages of its development, as the basis for several courses in optimization at Stanford University. Like the field of optimization itself which involves many classical disciplines—particularly now with its emphasis on obtaining solutions to real problems—the book should be useful to system analysts, operations researchers, numerical analysts, management scientists, and other specialists from the host of disciplines from which practical optimization applications are drawn. The prerequisites for convenient use of the book are relatively modest; the prime requirement being some familiarity with introductory elements of linear algebra. Certain sections and developments do assume some knowledge of more advanced concepts of linear algebra, such as eigenvector analysis, or some background in sets of real numbers, but the text is structured so that the mainstream of the development can be faithfully pursued without reliance on this more advanced background material.

Although the book is introductory in its level and scope of presentation, it is simultaneously intended to be modern in its approach, reflecting the most recent trends in the foundations of the field. This point is of particular significance since in the last few years these foundations have shifted quite dramatically. A good deal of the modern work in the field, therefore, is concerned not with development of advanced intricate techniques but with development, at an introductory level, of the basic underpinnings of the field. This new work signals the emergence of the field from a collection of isolated theoretical results and heuristic tools into a solidly based discipline in which theory guides the development of techniques. In the past, most of the theory of optimization concentrated on the subject of optimality conditions, and the practical methods of computation were attached to the theoretical body as a somewhat nontheoretical appendage. In the modern setting, theory and practice are better integrated and we find enhanced interplay between them.

A major connection between theory and practice is convergence analysis—the analysis of how particular iterative solution techniques generate points that converge to a solution of a given problem. In the past, convergence analysis, like the appendage of computational methods to which it applied, could at best be viewed as a series of formulae, which had more relation to particular solution techniques than to problem definition and had little coherence or structure. It is somewhat surprising now to find that convergence theory is not only much more coherent than originally suspected, but that it also helps provide a unifying framework for the field. In essence, we have learned that studying the common properties of various solution techniques for a problem is one of the most effective ways to illuminate the fundamental properties of the problem itself.

The material in this book is organized into three separate parts. Part I is a self-contained introduction to linear programming, a key component of optimization theory. The presentation in this part is fairly conventional, covering the main core of both the underlying theory of linear programming and many of the most effective numerical algorithms. It does not, however, cover those specialized areas of linear programming such as network flows or transportation theory that rely on special structural characteristics. Part II, which is independent of Part I, covers the theory of unconstrained optimization, including both derivations of the appropriate optimality conditions and an introduction to basic algorithms. This part of the book explores the general properties of algorithms and defines various notions of convergence. Part III extends the concepts developed in the second part to constrained optimization problems. Except for a few isolated sections, this part is also independent of Part I. It is possible to go directly into Parts II and III omitting Part I, and, in fact, the book has been used in this way at Stanford on several occasions. Each part of the book contains enough material to form the basis of a one-quarter course. In either classroom use or for self-study, it is important not to overlook the suggested exercises at the end of each chapter. The selections generally include exercises of a computational variety designed to test one's understanding of a particular algorithm, a theoretical variety designed to test one's understanding of a given theoretical development, or of the variety that extends the presentation of the chapter to new areas. One should attempt at least four or five exercises from each chapter. In progressing through the book it would be unusual to read straight through from cover to cover. Generally, one will wish to skip around. In order to facilitate this mode, I have indicated sections of a specialized or degressive nature with an asterisk*.

Development of this book would not have been possible without the help of several individuals and institutions. My perception of linear programming was greatly enhanced while working with Adam Shefi, some of whose work I have incorporated into Chapter 5. During the years of writing this book I worked closely with Shumel Oren, Daniel Gabay, Tsuguhiko Tanahashi, and Verne Chant each of whose many suggestions and comments are an integral part of the final product. I wish to thank Harvey Greenberg, Arthur Geoffrion, and Dimitri Bertsekas, who each made several valuable suggestions which were incorporated into the final manuscript. I wish to acknowledge the Department of Engineering-Economic Systems at Stanford University for supplying a stimulating atmosphere and much of the financial assistance. The effort was also partially supported by the National Science Foundation. I wish to express special thanks to Elaine Christensen who unfledgingly typed several drafts of the manuscript. Finally, as with most textbooks of this type, much is owed to the many students who patiently put up with the difficulties associated with an evolving manuscript and unselfishly contributed to its successful completion.

Washington, D.C. D.G.L.
August 1972

CONTENTS

Chapter 12 Penalty and Barrier Methods

Chapter 13 Cutting Plane and Dual Methods

Appendix A Mathematical Review

Appendix B Convex Sets

Appendix C Gaussian Elimination

Chapter 1 INTRODUCTION

1.1 OPTIMIZATION

The concept of optimization is now well-rooted as a principle underlying the analysis of many complex decision or allocation problems. It offers a certain degree of philosophical elegance that is hard to dispute, and it often offers an indispensable degree of operational simplicity. Using this optimization philosophy, one approaches a complex decision problem, involving the selection of values for a number of interrelated variables, by focussing attention on a single objective designed to quantify performance and measure the quality of the decision. This one objective is maximized (or minimized, depending on the formulation) subject to the constraints that may limit the selection of decision variable values. If a suitable single aspect of a problem can be isolated and characterized by an objective, be it profit or loss in a business setting, speed or distance in a physical problem, expected return in the environment of risky investments, or social welfare in the context of government planning, optimization may provide a suitable framework for analysis.

It is, of course, a rare situation in which it is possible to fully represent all the complexities of variable interactions, constraints, and appropriate objectives when faced with a complex decision problem. Thus, as with all quantitative techniques of analysis, a particular optimization formulation should only be regarded as an approximation. Skill in modelling, to capture the essential elements of a problem, and good judgment in the interpretation of results are required to obtain meaningful conclusions. Optimization, then, should be regarded as a tool of conceptualization and analysis rather than as a principle yielding the philosophically correct solution.

Skill and good judgment, with respect to problem formulation and interpretation of results, is enhanced through concrete practical experience and a thorough understanding of relevant theory. Problem formulation itself always involves a tradeoff between the conflicting objectives of building a mathematical model sufficiently complex to accurately capture the problem description and building a model that is tractable. The expert model builder is facile with both aspects of this tradeoff. One aspiring to become such an expert must learn to identify and capture the important issues of a problem mainly through example and experience; he must learn to distinguish tractable models from nontractable ones through a study of available technique and theory and by nurturing the capability to extend existing theory to new situations.

This book is centered around a certain optimization structure—that characteristic of linear and nonlinear programming. Examples of situations leading to this structure are sprinkled throughout the book, and these examples should help to indicate how practical problems can be often fruitfully structured in this form. The book mainly, however, is concerned with the development, analysis, and comparison of algorithms for solving general sub-classes of optimization problems. This is valuable not only for the algorithms themselves, which enable one to solve given problems, but also because identification of the collection of structures they most effectively solve can enhance one's ability to formulate problems.

1.2 TYPES OF PROBLEMS

The content of this book is broken down into three major parts: Linear Programming, Unconstrained Problems, and Constrained Problems. The last two parts together comprise the subject of nonlinear programming.

Linear Programming

Linear programming is without doubt the most natural mechanism for formulating a vast array of problems with modest effort. A linear programming problem is characterized, as the name implies, by linear functions of the unknowns; the objective is linear in the unknowns, and the constraints are linear equalities or linear inequalities in the unknowns. One familiar with other branches of linear mathematics might suspect, initially, that linear programming formulations are popular because the mathematics is nicer, the theory is richer, and the computation simpler for linear problems than for nonlinear ones. But, in fact, these are *not* the primary reasons. In terms of mathematical and computational properties, there are much broader classes of optimization problems than linear programming problems that have elegant and potent theories and for which effective algorithms are available. It seems that the popularity of linear programming lies primarily with the formulation phase of analysis rather than the solution phase—and for good cause. For one thing, a great number of constraints and objectives that arise in practice *are* indisputably linear.

Thus, for example, if one formulates a problem with a budget constraint restricting the total amount of money to be allocated among two different commodities, the budget constraint takes the form $x_1 + x_2 \leqslant B$, where x_i, $i = 1, 2$, is the amount allocated to activity i, and B is the budget. Similarly, if the objective is, for example, maximum weight, then it can be expressed as $w_1 x_1 + w_2 x_2$, where w_i, $i = 1, 2$, is the unit weight of the commodity i. The overall problem would be expressed as

$$\begin{aligned} \text{maximize} \quad & w_1 x_1 + w_2 x_2 \\ \text{subject to} \quad & x_1 + x_2 \leqslant B, \\ & x_1 \geqslant 0, \quad x_2 \geqslant 0 \end{aligned}$$

which is an elementary linear program. The linearity of the budget constraint is extremely natural in this case and does not represent simply an approximation to a more general functional form.

Another reason that linear forms for constraints and objectives are so popular in problem formulation is that they are often the least difficult to define. Thus, even if an objective function is not purely linear by virtue of its inherent definition (as in the above example), it is often far easier to define it as being linear than to decide on some other functional form and convince others that the more complex form is the best possible choice. Linearity, therefore, by virtue of its simplicity, often is selected as the easy way out or, when seeking generality, as the only functional form that will be equally applicable (or nonapplicable) in a class of similar problems.

Of course, the theoretical and computational aspects *do* take on a somewhat special character for linear programming problems—the most significant development being the simplex method. This algorithm is developed in Chapters 2 and 3 and occupies most of the attention that we devote to linear programming.

Unconstrained Problems

It may seem that unconstrained optimization problems are so devoid of structural properties as to preclude their applicability as useful models of meaningful problems. Quite the contrary is true for two reasons. First, it can be argued, quite convincingly, that if the scope of a problem is broadened to the consideration of all relevant decision variables, there may then be no constraints—or put another way, constraints represent artificial delimitations of scope, and when the scope is broadened the constraints vanish. Thus, for example, it may be argued that a budget constraint is not characteristic of a meaningful problem formulation; since by borrowing at some interest rate it is always possible to obtain additional funds, and hence rather than introducing a budget constraint, a term reflecting the cost of funds should be incorporated into the objective. A similar argument applies to constraints describing the availability of other resources which at some cost (however great) could be supplemented.

The second reason that many important problems can be regarded as having no constraints is that constrained problems are sometimes easily converted to unconstrained problems. For instance, the sole effect of equality constraints is simply to limit the degrees of freedom, by essentially making some variables functions of others. These dependencies can sometimes be explicitly characterized, and a new problem having its number of variables equal to the true degree of freedom can be determined. As a simple specific example, a constraint of the form $x_1 + x_2 = B$ can be eliminated by substituting $x_2 = B - x_1$ everywhere else that x_2 appears in the problem.

Aside from representing a significant class of practical problems, the study of unconstrained problems, of course, provides a stepping stone toward the more general case of constrained problems. Many aspects of both theory and algorithms are most naturally motivated and verified for the unconstrained case before progressing to the constrained case.

Constrained Problems

In spite of the arguments given above, many problems met in practice are formulated as constrained problems. This is because in most instances a complex problem such as, for example, the detailed production policy of a giant corporation, the planning of a large government agency, or even the design of a complex device cannot be directly treated in its entirety accounting for all possible choices, but instead must be decomposed into separate subproblems—each subproblem having constraints which are imposed to restrict its scope. Thus, in planning problems, budget constraints are commonly imposed in order to decouple that one problem from a more global one. Therefore, one frequently encounters general nonlinear constrained mathematical programming problems.

The general mathematical programming problem can be stated as

$$\begin{aligned}
&\text{minimize} && f(\mathbf{x}) && \\
&\text{subject to} && h_i(\mathbf{x}) = 0, && i = 1, 2, \ldots, m \\
& && g_j(\mathbf{x}) \leqslant 0, && j = 1, 2, \ldots, r \\
& && \mathbf{x} \in S. &&
\end{aligned}$$

In this formulation, $\mathbf{x}$ is an n-dimensional vector of unknowns, $\mathbf{x} = (x_1, x_2, \ldots, x_n)$, and f, h_i, $i = 1, 2, \ldots, m$, and g_j, $j = 1, 2, \ldots, r$ are real-valued functions of the variables $x_1, x_2, \ldots, x_n$. The set S is a subset of n-dimensional space. The function f is the *objective function* of the problem and the equations, inequalities, and set restrictions are *constraints*.

Generally, in this book, additional assumptions are introduced in order to make the problem smooth in some suitable sense. For example, the functions in the problem are usually required to be continuous, or perhaps to have continuous derivatives. This ensures that small changes in **x** lead to small changes in other values associated with the problem. Also, the set S is not allowed to be arbitrary but usually is required to be a connected region of n-dimensional space, rather than, for example, a set of distinct isolated points. This ensures that small changes in **x** can be made. Indeed, in a majority of problems treated, the set S is taken to be the entire space; there is no set restriction.

In view of these smoothness assumptions, one might characterize the problems treated in this book as *continuous variable programming*, since we generally discuss problems where all variables and function values can be varied continuously. In fact, this assumption forms the basis of many of the algorithms discussed, which operate essentially by making a series of small movements in the unknown **x** vector.

1.3 SIZE OF PROBLEMS

One obvious measure of the complexity of a programming problem is its size, measured in terms of the number of unknown variables or the number of constraints. As might be expected, the size of problems that can be effectively solved has been increasing with advancing computing technology and with advancing theory. Today, with present computing capabilities, however, it is reasonable to distinguish three classes of problems: *small-scale problems* having about five or less unknowns and constraints; *intermediate-scale problems* having from about five to a hundred variables; and *large-scale problems* having on the order of a thousand variables and constraints. This classification is not entirely rigid, but it reflects at least roughly not only size but the basic differences in approach that accompany different size problems.

Much of the early theory associated with optimization, particularly in nonlinear programming, is directed at obtaining necessary and sufficient conditions satisfied by a solution point, rather than at questions of computation. This theory involves mainly the study of Lagrange multipliers, including the Kuhn–Tucker Theorem and its extensions. It tremendously enhances insight into the philosophy of constrained optimization and provides satisfactory basic foundations for other important disciplines, such as the theory of the firm, consumer economics, and optimal control theory. The interpretation of Lagrange multipliers that accompanies this theory is valuable in virtually every optimization setting. As a theoretical basis for computing numerical solutions to optimization, however, this early theory is applicable only to *small-scale* programming problems, because the equations resulting from the necessary conditions can be efficiently solved directly only when the problem is small enough for hand calculation.

If it is acknowledged from the outset that a given problem is too large and too complex to be efficiently solved by hand (and hence it is acknowledged that a computer solution is desirable), then one's theory should be directed toward development of procedures that exploit the efficiencies of computers. In most cases this leads to the abandonment of the idea of solving the set of necessary conditions in favor of the more direct procedure of searching through the space (in an intelligent manner) for ever-improving points. For the development and implementation of such search procedures, the theory of necessary conditions is of little value in itself, but as repeatedly demonstrated throughout this book, it forms the core of a more complete theory that is applicable to such procedures.

Today, search techniques can be effectively applied to more or less general nonlinear programming problems having on the order of 100 to 200 variables, and to linear programming problems having about 400 constraints and 1000 variables. This range of problem that can be solved by a computer with a search method we refer to as *intermediate-scale programming*.

Problems of even greater size, *large-scale programming* problems, can be solved if they possess special structural characteristics that can be exploited by a solution method. The study of large-scale programming consists of the identification of important special structures and the development of techniques that exploit these structures. It is, therefore, a more problem-dependent body of theory than the other theoretical aspects of programming problems.

This book is devoted mainly to the theory of intermediate-scale programming and therefore can be considered as focusing on the aspect of general theory that is most fruitful for computation in the widest class of problems. While necessary and sufficient conditions are examined and their application to small-scale problems is illustrated, our primary interest in such conditions is in their role as the core of a broader theory applicable to the solution of larger problems. At the other extreme, although some instances of structure exploitation are discussed, we focus primarily on the general continuous variable programming problem rather than on special techniques for special structures.

1.4 ITERATIVE ALGORITHMS AND CONVERGENCE

The most important characteristic of a high-speed digital computer is its ability to perform repetitive operations efficiently, and in order to exploit this basic characteristic, most algorithms designed to solve large optimization problems are iterative in nature. Typically, in seeking a vector that solves the programming problem, an initial vector $\mathbf{x}_0$ is selected and the algorithm generates an improved vector $\mathbf{x}_1$. The process is repeated and a still better solution $\mathbf{x}_2$ is found. Continuing in this fashion, a sequence of ever-improving points $\mathbf{x}_0, \mathbf{x}_1, \ldots, \mathbf{x}_k, \ldots,$ is found that approaches a solution point $\mathbf{x}^*$. For linear programming problems, the generated sequence is of finite length, reaching the solution point exactly after a finite (although initially unspecified) number of steps. For nonlinear programming problems, the sequence generally does not ever exactly reach the solution point, but converges toward it. In operation, for nonlinear problems, the process is terminated when a point sufficiently close to the solution point, for practical purposes, is obtained.

The theory of iterative algorithms can be divided into three (somewhat overlapping) aspects. The first is concerned with the creation of the algorithms themselves. Algorithms are not conceived arbitrarily, but are based on a creative examination of the programming problem, its inherent structure, and the efficiencies of digital computers. The second aspect is the verification that a given algorithm will in fact generate a sequence that converges to a solution point. This aspect is referred to as *global convergence analysis*, since it addresses the important question of whether the algorithm, when initiated far from the solution point, will eventually converge to it. The third aspect is referred to as *local convergence analysis* and is concerned with the rate at which the generated sequence of points converges to the solution. One cannot regard a problem as solved simply because an algorithm is known which will converge to the solution, since it may require an exorbitant amount of time to reduce the error to an acceptable tolerance. It is essential when prescribing algorithms that some estimate of the time required be available. It is the convergence-rate aspect of the theory that allows some quantitative evaluation and comparison of different algorithms, and at least crudely, assigns a measure of tractability to a problem, as discussed in Section 1.1.

A modern-day technical version of Confucius' most famous saying, and one which represents an underlying philosophy of this book, might be, "One good theory is worth a thousand computer runs". Thus the convergence properties of an iterative algorithm can be estimated with confidence either by performing numerous computer experiments on different problems or by a simple well-directed theoretical analysis. A simple theory, of course, provides invaluable insight as well as the desired estimate.

It is perhaps somewhat surprising that there does not yet exist a useful convergence theory for the simplex method of linear programming, one of the oldest and most important optimization techniques. This seems to be due to the fact that, since convergence occurs in a finite number of steps, an estimate of the total number of steps, rather than a rate of convergence, must be found. There has, however, been accumulated a vast body of consistent experimental data from which a heuristic rule has been extracted, enabling the time required for a new problem to be estimated with confidence.

For nonlinear programming problems, a large number of different algorithms are in fairly common use, but the limited computational experience associated with each of these methods is in itself not vast enough to allow definitive comparisons to be made. The convergence properties of a large class of iterative algorithms for nonlinear programming problems, however, can be deduced analytically by fairly simple means, and this analysis is substantiated by the limited computational experience available. Presentation of convergence analysis, which seems to be really the natural focal point for a theory directed at obtaining specific answers, is a unique feature of this book. Much of the theory appears here for the first time.

The convergence rate theory presented has two somewhat surprising but definitely pleasing aspects. First, the theory is, for the most part, extremely simple in nature. Although initially one might fear that a theory aimed at predicting the speed of convergence of a complex algorithm might itself be doubly complex, in fact the associated convergence analysis often turns out to be exceedingly elementary, requiring only a line or two of calculation. Second, a large class of seemingly distinct algorithms turn out to converge at the same rate. Indeed, as emphasized in the later chapters of the book, there is a *canonical rate* associated with a given programming problem that seems to govern the speed of convergence of many algorithms when applied to that problem. It is this fact that underlies the potency of the theory, allowing definitive comparisons among algorithms to be made even without detailed knowledge of the problems to which they will be applied. Together these two properties, simplicity and potency, assure convergence analysis a permanent position of major importance in mathematical programming theory.

PART I
LINEAR PROGRAMMING

Chapter 2 BASIC PROPERTIES OF LINEAR PROGRAMS

2.1 INTRODUCTION

A linear programming problem is a mathematical program in which the objective function is linear in the unknowns and the constraints consist of linear equalities and linear inequalities. The exact form of these constraints may differ from one problem to another, but as shown below, any linear program can be transformed into the following *standard form*:

$$\begin{aligned}
&\text{minimize} && c_1x_1 + c_2x_2 + \cdots + c_nx_n \\
&\text{subject to} && a_{11}x_1 + a_{12}x_2 + \cdots + a_{1n}x_n = b_1 \\
& && a_{21}x_1 + a_{22}x_2 + \cdots + a_{2n}x_n = b_2 \\
& && \quad\vdots \qquad\qquad\qquad\qquad\quad \vdots \\
& && a_{m1}x_1 + a_{m2}x_2 + \cdots + a_{mn}x_n = b_m \\
&\text{and} && x_1 \geqslant 0,\ x_2 \geqslant 0,\ \ldots,\ x_n \geqslant 0,
\end{aligned} \tag{1}$$

where the b_i's, c_i's and a_{ij}'s are fixed real constants, and the x_i's are real numbers to be determined. We always assume that each equation has been multiplied by minus unity, if necessary, so that each $b_i \geqslant 0$.

In more compact vector notation,† this standard problem becomes

$$\begin{aligned}
&\text{minimize} && \mathbf{cx} \\
&\text{subject to} && \mathbf{Ax} = \mathbf{b} \quad \text{and} \quad \mathbf{x} \geqslant \mathbf{0}.
\end{aligned} \tag{2}$$

Here $\mathbf{x}$ is an n-dimensional column vector, $\mathbf{c}$ is an n-dimensional row vector,

† See Appendix A for a description of the vector notation used throughout this book.

A is an $m \times n$ matrix and **b** is an m-dimensional column vector. The vector inequality $\mathbf{x} \geqslant \mathbf{0}$ means that each component of **x** is nonnegative.

Before giving some examples of areas in which linear programming problems arise naturally, we indicate how various other forms of linear programs can be converted to the standard form.

Example 1 (Slack variables). Consider the problem

$$\begin{array}{ll} \text{minimize} & c_1x_1 + c_2x_2 + \cdots + c_nx_n, \\ \text{subject to} & a_{11}x_1 + a_{12}x_2 + \cdots + a_{1n}x_n \leqslant b_1 \\ & a_{21}x_1 + a_{22}x_2 + \cdots + a_{2n}x_n \leqslant b_2 \\ & \vdots \\ & a_{m1}x_1 + a_{m2}x_2 + \cdots + a_{mn}x_n \leqslant b_m \\ \text{and} & x_1 \geqslant 0,\ x_2 \geqslant 0,\ \ldots,\ x_n \geqslant 0. \end{array}$$

In this case the constraint set is determined entirely by linear inequalities. The problem may be alternatively expressed as

$$\begin{array}{ll} \text{minimize} & c_1x_1 + c_2x_2 + \cdots + c_nx_n \\ \text{subject to} & a_{11}x_1 + a_{12}x_2 + \cdots + a_{1n}x_n + y_1 \qquad\qquad = b_1 \\ & a_{21}x_1 + a_{22}x_2 + \cdots + a_{2n}x_n \qquad + y_2 \qquad = b_2 \\ & \vdots \\ & a_{m1}x_1 + a_{m2}x_2 + \cdots + a_{mn}x_n \qquad\qquad + y_m = b_m \\ \text{and} & x_1 \geqslant 0,\ x_2 \geqslant 0,\ \ldots,\ x_n \geqslant 0, \\ \text{and} & y_1 \geqslant 0,\ y_2 \geqslant 0,\ \ldots,\ y_m \geqslant 0. \end{array}$$

The new positive variables y_i introduced to convert the inequalities to equalities are called *slack variables* (or more loosely, *slacks*). By considering the problem as one having $n + m$ unknowns $x_1, x_2, \ldots, x_n, y_1, y_2, \ldots, y_m$, the problem takes the standard form. The $m \times (n + m)$ matrix that now describes the linear equality constraints is of the special form $[\mathbf{A}, \mathbf{I}]$, (that is, its columns can be partitioned into two sets; the first n columns make up the original **A** matrix and the last m columns make up an $m \times m$ identity matrix).

Example 2 (Surplus variables). If the linear inequalities of Example 1 are reversed so that a typical inequality is

$$a_{i1}x_1 + a_{i2}x_2 + \cdots + a_{in}x_n \geqslant b_i,$$

it is clear that this is equivalent to

$$a_{i1}x_1 + a_{i2}x_2 + \cdots + a_{in}x_n - y_i = b_i$$

with $y_i \geqslant 0$. Variables, such as y_i, adjoined in this fashion to convert a "greater than or equal to" inequality to equality are called *surplus variables.*

It should be clear that by suitably multiplying by minus unity, and adjoining slack and surplus variables, any set of linear inequalities can be converted to standard form if the unknown variables are restricted to be nonnegative.

Example 3 (Free variables—first method). If a linear program is given in standard form except that one or more of the unknown variables is not required to be nonnegative, the problem can be transformed to standard form by either of two simple techniques.

To describe the first technique, suppose in (1), for example, that the restriction $x_1 \geqslant 0$ is not present and hence x_1 is free to take on either positive or negative values. We then write

$$x_1 = u_1 - v_1, \tag{3}$$

where we require $u_1 \geqslant 0$ and $v_1 \geqslant 0$. If we substitute $u_1 - v_1$ for x_1 everywhere in (1), the linearity of the constraints is preserved and all variables are now required to be nonnegative. The problem is then expressed in terms of the $n + 1$ variables $u_1, v_1, x_2, x_3, \ldots, x_n$.

There is obviously a certain degree of redundancy introduced by this technique, however, since a constant added to u_1 and v_1 does not change x_1 (that is, the representation of a given value x_1 is not unique). Nevertheless, this does not hinder the simplex method of solution.

Example 4 (Free variables—second method). A second approach for converting to standard form when x_1 is unconstrained in sign is to eliminate x_1 together with one of the constraint equations. Take any one of the m equations in (1) which has a nonzero coefficient for x_1. Say, for example,

$$a_{i1}x_1 + a_{i2}x_2 + \cdots + a_{in}x_n = b_i, \tag{4}$$

where $a_{i1} \neq 0$. Then x_1 can be expressed as a linear combination of the other variables plus a constant. If this expression is substituted for x_1 everywhere in (1), we are led to a new problem of exactly the same form but expressed in terms of the variables $x_2, x_3, \ldots, x_n$ only. Furthermore, the ith equation, used to determine x_1, is now identically zero and it too can be eliminated. This substitution scheme is valid since any combination of nonnegative variables $x_2, x_3, \ldots, x_n$ leads to a feasible x_1 from (4), since the sign of x_1 is unrestricted. As a result of this simplification, we obtain a standard linear program having $n-1$ váriables and $m-1$ constraint equations. The value of the variable x_1 can be determined after solution through (4).

Example 5 (Specific case). As a specific instance of the above technique consider the problem

$$\begin{array}{ll} \text{minimize} & x_1 + 3x_2 + 4x_3 \\ \text{subject to} & x_1 + 2x_2 + x_3 = 5 \\ & 2x_1 + 3x_2 + x_3 = 6 \\ & x_2 \geqslant 0, \qquad x_3 \geqslant 0. \end{array}$$

Since x_1 is free, we solve for it from the first constraint, obtaining

$$x_1 = 5 - 2x_2 - x_3. \tag{5}$$

Substituting this into the objective and the second constraint, we obtain the equivalent problem (substracting five from the objective)

$$\begin{array}{ll} \text{minimize} & x_2 + 3x_3 \\ \text{subject to} & x_2 + x_3 = 4 \\ & x_2 \geqslant 0, \qquad x_3 \geqslant 0, \end{array}$$

which is a problem in standard form. After the smaller problem is solved (the answer is $x_2 = 4$, $x_3 = 0$) the value for x_1 ($x_1 = -3$) can be found from (5).

2.2 EXAMPLES OF LINEAR PROGRAMMING PROBLEMS

Linear programming has long proved its merit as a significant model of numerous allocation problems and economic phenomena. The continuously expanding literature of applications repeatedly demonstrates the importance of linear programming as a general framework for problem formulation. In this section we present some classic examples of situations that have natural formulations.

Example 1 (The diet problem). How can we determine the most economical diet that satisfies the basic minimum nutritional requirements for good health? Such a problem might, for example, be faced by the dietician of a large army. We assume that there are available at the market n different foods and that the ith food sells at a price c_i per unit. In addition there are m basic nutritional ingredients and, to achieve a balanced diet, each individual must receive at least b_j units of the jth nutrient per day. Finally we assume that each unit of food i contains a_{ji} units of the jth nutrient.

If we denote by x_i the number of units of food i in the diet, the problem then is to select the x_i's to minimize the total cost

$$c_1x_1 + c_2x_2 + \cdots + c_nx_n$$

subject to the nutritional constraints

$$\begin{aligned}
a_{11}x_1 + a_{12}x_2 + \cdots + a_{1n}x_n &\geqslant b_1 \\
a_{21}x_1 + a_{22}x_2 + \cdots + a_{2n}x_n &\geqslant b_2 \\
\vdots \qquad\qquad & \quad \vdots \\
a_{m1}x_1 + a_{m2}x_2 + \cdots + a_{mn}x_n &\geqslant b_m
\end{aligned}$$

and the nonnegativity constraints

$$x_1 \geqslant 0, \; x_2 \geqslant 0, \; \ldots, \; x_n \geqslant 0$$

on the food quantities.

This problem can be converted to standard form by substracting a nonnegative surplus variable from the left side of each of the m linear inequalities. The diet problem is discussed further in Chapter 4.

Example 2 (The transportation problem). Quantities $a_1, a_2, \ldots, a_m$, respectively, of a certain product are to be shipped from each of m locations and received in amounts $b_1, b_2, ., b_n$, respectively, at each of n destinations. Associated with the shipping of a unit of product from origin i to destination j is a unit shipping cost c_{ij}. It is desired to determine the amounts x_{ij} to be shipped between each origin–destination pair $i = 1, 2, \ldots, m$; $j = 1, 2, \ldots, n$; so as to satisfy the shipping requirements and minimize the total cost of transportation.

To formulate this problem as a linear programming problem, we set up the array shown below:

$$\begin{array}{|cccc|c}
x_{11} & x_{12} & \cdots & x_{1n} & a_1 \\
x_{21} & x_{22} & \cdots & x_{2n} & a_2 \\
\vdots & & & \vdots & \vdots \\
x_{m1} & x_{m2} & \cdots & x_{mn} & a_m \\
\hline
b_1 & b_2 & \cdots & b_n &
\end{array}$$

The ith row in this array defines the variables associated with the ith origin, while the jth column in this array defines the variables associated with the jth destination. The problem is to place nonnegative variables x_{ij} in this array so that the sum across the ith row is a_i, the sum down the jth column is b_j, and the weighted sum $\sum_{j=1}^{n} \sum_{i=1}^{m} c_{ij}x_{ij}$, representing the transportation cost, is minimized.

Thus we have the linear programming problem:

$$\begin{aligned} &\text{minimize} && \sum_{ij} c_{ij}x_{ij} && \\ &\text{subject to} && \sum_{j=1}^{n} x_{ij} = a_i \quad \text{for} \quad i = 1, 2, \ldots, m && (6) \\ & && \sum_{i=1}^{m} x_{ij} = b_j \quad \text{for} \quad j = 1, 2, \ldots, n && (7) \\ & && x_{ij} \geqslant 0 \quad \text{for} \quad i = 1, 2, \ldots, m; && \\ & && j = 1, 2, \ldots, n. && \end{aligned}$$

In order that the constraints (6), (7) be consistent, we must, of course, assume that $\sum_{i=1}^{m} a_i = \sum_{j=1}^{n} b_j$ which corresponds to assuming that the total amount shipped is equal to the total amount received.

The transportation problem is now clearly seen to be a linear programming problem in mn variables. The equations (6), (7) can be combined and expressed in matrix form in the usual manner and this results in an $(m + n) \times (mn)$ coefficient matrix consisting of zeros and ones only.

Example 3 (Manufacturing problem). Suppose we own a facility that is capable of engaging in n different production activities, each of which produces various amounts of m commodities. Each activity can be operated at any level $x_i \geqslant 0$ but when operated at the unity level the ith activity costs c_i dollars and yields a_{ji} units of the jth commodity. Assuming linearity of the production facility, if we are given a set of m numbers $b_1, b_2, \ldots, b_m$ describing the output requirements of the m commodities, and we wish to produce these at minimum cost, ours is the linear program (1).

2.3 BASIC SOLUTIONS

Consider the system of equalities

$$\mathbf{Ax} = \mathbf{b} \tag{8}$$

where $\mathbf{x}$ is an n-vector, $\mathbf{b}$ an m-vector and $\mathbf{A}$ is an $m \times n$ matrix. Suppose that from the n columns of $\mathbf{A}$ we select a set of m linearly independent columns (such a set exists if the rank of $\mathbf{A}$ is m). For notational simplicity assume that we select the first m columns of $\mathbf{A}$ and denote the $m \times m$ matrix determined by these columns by $\mathbf{B}$. The matrix $\mathbf{B}$ is then nonsingular and we may uniquely solve the equation

$$\mathbf{Bx_B} = \mathbf{b} \tag{9}$$

for the m-vector $\mathbf{x_B}$. By putting $\mathbf{x} = (\mathbf{x_B}, \mathbf{0})$ (that is, setting the first m

components of $\mathbf{x}$ equal to those of $\mathbf{x_B}$ and the remaining components equal to zero), we obtain a solution to $\mathbf{Ax} = \mathbf{b}$. This leads to the following definition.

Definition. Given the set of m simultaneous linear equations in n unknowns (8), let $\mathbf{B}$ by any nonsingular $m \times m$ submatrix made up of columns of $\mathbf{A}$. Then, if all $n - m$ components of $\mathbf{x}$ not associated with columns of $\mathbf{B}$ are set equal to zero, the solution to the resulting set of equations is said to be a *basic solution* to (8) with respect to the basis $\mathbf{B}$. The components of $\mathbf{x}$ associated with columns of $\mathbf{B}$ are called *basic variables.*

In the above definition we refer to $\mathbf{B}$ as a basis, since $\mathbf{B}$ consists of m linearly independent columns that can be regarded as a basis for the space E^m. The basic solution corresponds to an expression for the vector $\mathbf{b}$ as a linear combination of these basis vectors. This interpretation is discussed further in the next section.

In general, of course, Eq. (8) may have no basic solutions. However, we may avoid trivialities and difficulties of a nonessential nature by making certain elementary assumptions regarding the structure of the matrix $\mathbf{A}$. First, we usually assume that $n > m$, that is, the number of variables x_i exceeds the number of equality constraints. Second, we usually assume that the rows of $\mathbf{A}$ are linearly independent, corresponding to linear independence of the m equations. A linear dependency among the rows of $\mathbf{A}$ would lead either to contradictory constraints and hence no solutions to (8), or to a redundancy that could be eliminated. For most purposes, then, we may assume that $\mathbf{A}$ has rank m and therefore that there is at least one basic solution to (8).

Definition. If one or more of the basic variables in a basic solution has value zero, that solution is said to be a *degenerate basic solution.*

We note that in a non-degenerate basic solution the basic variables, and hence the basis $\mathbf{B}$, can be immediately identified from the positive components of the solution. There is ambiguity associated with a degenerate basic solution, however, since the zero-valued basic and nonbasic variables can be interchanged.

So far in the discussion of basic solutions we have treated only the equality constraint (8) and have made no reference to positivity constraints on the variables. Similar definitions apply when these constraints are also considered. Thus, consider now the system of constraints

$$\begin{aligned} \mathbf{Ax} &= \mathbf{b}, \\ \mathbf{x} &\geqslant \mathbf{0} \end{aligned} \tag{10}$$

which represent the constraints of a linear program in standard form.

Definition. A vector $\mathbf{x}$ satisfying (10) is said to be *feasible* for these constraints. A feasible solution to the constraints (10) that is also basic is said to be a *basic feasible solution*; if this solution is also a degenerate basic solution, it is called a *degenerate basic feasible solution.*

2.4 THE FUNDAMENTAL THEOREM OF LINEAR PROGRAMMING

In this section, through the fundamental theorem of linear programming, we establish the primary importance of basic feasible solutions in solving linear programming problems. The method of proof of the theorem is in many respects as important as the result itself, since it represents the beginning of the development of the simplex method. The theorem itself shows that it is necessary only to consider basic feasible solutions when seeking an optimal solution to a linear program because the optimal value is always achieved at such a solution.

Corresponding to a linear program in standard form

$$\begin{aligned} &\text{minimize} && \mathbf{cx} \\ &\text{subject to} && \mathbf{Ax} = \mathbf{b}, \\ & && \mathbf{x} \geqslant \mathbf{0} \end{aligned} \tag{11}$$

a feasible solution to the constraints that achieves the minimum value of the objective function subject to those constraints is said to be an *optimal feasible solution.* If this solution is basic, it is an *optimal basic feasible solution.*

Fundamental theorem of linear programming. *Given a linear program in standard form* (11) *where* $\mathbf{A}$ *is an* $m \times n$ *matrix of rank* m,

i) *if there is a feasible solution, there is a basic feasible solution*;

ii) *if there is an optimal feasible solution, there is an optimal basic feasible solution.*

Proof of (i). Denote the columns of $\mathbf{A}$ by $\mathbf{a}_1, \mathbf{a}_2, \ldots, \mathbf{a}_n$. Suppose $\mathbf{x} = (x_1, x_2, \ldots, x_n)$ is a feasible solution. Then, in terms of the columns of $\mathbf{A}$, this solution satisfies:

$$x_1\mathbf{a}_1 + x_2\mathbf{a}_2 + \cdots + x_n\mathbf{a}_n = \mathbf{b}.$$

Assume that exactly p of the variables x_i are greater than zero, and for convenience, that they are the first p variables. Thus

$$x_1\mathbf{a}_1 + x_2\mathbf{a}_2 + \cdots + x_p\mathbf{a}_p = \mathbf{b}. \tag{12}$$

There are now two cases, corresponding as to whether the set $\mathbf{a}_1, \mathbf{a}_2, \ldots, \mathbf{a}_p$ is linearly independent or linearly dependent.

CASE 1: Assume $\mathbf{a}_1, \mathbf{a}_2, \ldots, \mathbf{a}_p$ are linearly independent. Then clearly, $p \leqslant m$. If $p = m$, the solution is basic and the proof is complete. If $p < m$, then, since $\mathbf{A}$ has rank m, $m - p$ vectors can be found from the remaining $n - p$ vectors so that the resulting set of m vectors is linearly independent. (See Exercise 4.) Assigning the value zero to the corresponding $m - p$ variables yields a (degenerate) basic feasible solution.

CASE 2: Assume $\mathbf{a}_1, \mathbf{a}_2, \ldots, \mathbf{a}_p$ are linearly dependent. Then there is a non-trivial linear combination of these vectors that is zero. Thus there are constants $y_1, y_2, \ldots, y_p$, at least one of which can be assumed to be positive, such that

$$y_1\mathbf{a}_1 + y_2\mathbf{a}_2 + \cdots + y_p\mathbf{a}_p = \mathbf{0}. \tag{13}$$

Multiplying this equation by a scalar ε and subtracting it from (12), we obtain

$$(x_1 - \varepsilon y_1)\mathbf{a}_1 + (x_2 - \varepsilon y_2)\mathbf{a}_2 + \cdots + (x_p - \varepsilon y_p)\mathbf{a}_p = \mathbf{b}. \tag{14}$$

This equation holds for every ε, and for each ε the components $x_i - \varepsilon y_i$ correspond to a solution of the linear equalities—although they may violate $x_i - \varepsilon y_i \geqslant 0$. Denoting $\mathbf{y} = (y_1, y_2, \ldots, y_p, 0, 0, \ldots, 0)$, we see that for any ε

$$\mathbf{x} - \varepsilon\mathbf{y} \tag{15}$$

is a solution to the equalities. For $\varepsilon = 0$, this reduces to the original feasible solution. As ε is increased from zero, the various components increase, decrease, or remain constant, depending upon whether the corresponding y_i is positive, negative, or zero. Since we assume at least one y_i is positive, at least one component will decrease as ε is increased. Increase ε to the first point where one or more components becomes zero. Specifically, set

$$\varepsilon = \min\,\{x_i/y_i : y_i > 0\}.$$

For this value of ε the solution given by (15) is feasible and has at most $p - 1$ positive variables. Repeating this process if necessary, we can eliminate positive variables until we have a feasible solution with corresponding columns that are linearly independent. At that point case 1 applies. ∎

Proof of (ii). Let $\mathbf{x} = (x_1, x_2, \ldots, x_n)$ be an optimal feasible solution and, as in the proof of (i) above, suppose there are exactly p positive variables $x_1, x_2, \ldots, x_p$. Again there are two cases; and case 1, corresponding to linear independence, is exactly the same as before.

Case 2 also goes exactly the same as before, but it must be shown that for any ε the solution (15) is optimal. To show this, note that the value of the solution $\mathbf{x} - \varepsilon\mathbf{y}$ is

$$\mathbf{c}\mathbf{x} - \varepsilon\mathbf{c}\mathbf{y}. \tag{16}$$

For ε sufficiently small in magnitude, $\mathbf{x} - \varepsilon\mathbf{y}$ is a feasible solution for positive or negative values of ε. Thus we conclude that $\mathbf{cy} = 0$. For, if $\mathbf{cy} \neq 0$, an ε of small magnitude and proper sign could be determined so as to render (16) smaller than $\mathbf{cx}$ while maintaining feasibility. This would violate the assumption of optimality of $\mathbf{x}$ and hence we must have $\mathbf{cy} = 0$.

Having established that the new feasible solution with fewer positive components is also optimal, the remainder of the proof may be completed exactly as in part (i). ▮

This theorem reduces the task of solving a linear programming problem to that of searching over basic feasible solutions. Since for a problem having n variables and m constraints there are at most

$$\binom{n}{m} = \frac{n!}{m!(n-m)!}$$

basic solutions (corresponding to the number of ways of selecting m of n columns), there are only a finite number of possibilities. Thus the fundamental theorem yields an obvious, but terribly inefficient, finite search technique. By expanding upon the technique of proof as well as the statement of the fundamental theorem, the efficient simplex procedure is derived.

It should be noted that the proof of the fundamental theorem given above is of a simple algebraic character. In the next section the geometric interpretation of this theorem is explored in terms of the general theory of convex sets. Although the geometric interpretation is aesthetically pleasing and theoretically important, the reader should bear in mind, lest he be diverted by the somewhat more advanced arguments employed, the underlying elementary level of the fundamental theorem.

2.5 RELATIONS TO CONVEXITY

Our development to this point and, in particular, the proof of the fundamental theorem has been based only on elementary properties of systems of linear equations. These results, however, have interesting interpretations in terms of the theory of convex sets that can lead not only to an alternative derivation of the fundamental theorem, but also to a clearer geometric understanding of the result. The main link between the algebraic and geometric theories is the formal relation between basic feasible solutions of linear inequalities in standard form and extreme points of polytopes. We now establish this correspondence. The reader is referred to Appendix B for a summary of basic concepts related to convexity.

Theorem (Equivalence of extreme points and basic solutions). *Let* $\mathbf{A}$ *be an* $m \times n$ *matrix and* $\mathbf{b}$ *an m-vector. Let K be the convex polytope consisting of all n-vectors* $\mathbf{x}$ *satisfying,*

$$\begin{aligned} \mathbf{Ax} &= \mathbf{b}, \\ \mathbf{x} &\geqslant \mathbf{0}. \end{aligned} \tag{17}$$

A vector $\mathbf{x}$ *is an extreme point of K if and only if* $\mathbf{x}$ *is a basic feasible solution to* (17).

Proof. Suppose first that $\mathbf{x} = (x_1, x_2, \ldots, x_m, 0, 0, \ldots, 0)$ is a basic feasible solution to (17). Then

$$x_1\mathbf{a}_1 + x_2\mathbf{a}_2 + \cdots + x_m\mathbf{a}_m = \mathbf{b}$$

where $\mathbf{a}_1, \mathbf{a}_2, \ldots, \mathbf{a}_m$, the first m columns of $\mathbf{A}$, are linearly independent. Suppose that $\mathbf{x}$ could be expressed as a convex combination of two other points in K; say, $\mathbf{x} = \alpha\mathbf{y} + (1-\alpha)\mathbf{z}$, $0 < \alpha < 1$, $\mathbf{y} \neq \mathbf{z}$. Since all components of $\mathbf{x}$, $\mathbf{y}$, $\mathbf{z}$ are nonnegative and since $0 < \alpha < 1$, it follows immediately that the last $n - m$ components of $\mathbf{y}$ and $\mathbf{z}$ are zero. Thus, in particular, we have

$$y_1\mathbf{a}_1 + y_2\mathbf{a}_2 + \cdots + y_m\mathbf{a}_m = \mathbf{b}$$

and

$$z_1\mathbf{a}_1 + z_2\mathbf{a}_2 + \cdots + z_m\mathbf{a}_m = \mathbf{b}.$$

Since the vectors $\mathbf{a}_1, \mathbf{a}_2, \ldots, \mathbf{a}_m$ are linearly independent, however, it follows that $\mathbf{x} = \mathbf{y} = \mathbf{z}$ and hence $\mathbf{x}$ is an extreme point of K.

Conversely, assume that $\mathbf{x}$ is an extreme point of K. Let us assume that the nonzero components of $\mathbf{x}$ are the first k components. Then

$$x_1\mathbf{a}_1 + x_2\mathbf{a}_2 + \cdots + x_k\mathbf{a}_k = \mathbf{b}$$

with $x_i > 0, i = 1, 2, \ldots, k$. To show that $\mathbf{x}$ is a basic feasible solution it must be shown that the vectors $\mathbf{a}_1, \mathbf{a}_2, \ldots, \mathbf{a}_k$ are linearly independent. We do this by contradiction. Suppose $\mathbf{a}_1, \mathbf{a}_2, \ldots, \mathbf{a}_k$ are linearly dependent. Then there is a nontrivial linear combination that is zero:

$$y_1\mathbf{a}_1 + y_2\mathbf{a}_2 + \cdots + y_k\mathbf{a}_k = \mathbf{0}.$$

Define the n-vector $\mathbf{y} = (y_1, y_2, \ldots y_k, 0, 0, \ldots, 0)$. Since $x_i > 0, 1 \leqslant i \leqslant k$, it is possible to select ε such that

$$\mathbf{x} + \varepsilon\mathbf{y} \geqslant \mathbf{0}, \qquad \mathbf{x} - \varepsilon\mathbf{y} \geqslant \mathbf{0}.$$

We then have $\mathbf{x} = \frac{1}{2}(\mathbf{x} + \varepsilon\mathbf{y}) + \frac{1}{2}(\mathbf{x} - \varepsilon\mathbf{y})$ which expresses $\mathbf{x}$ as a convex combination of two distinct vectors in K. This cannot occur, since $\mathbf{x}$ is an extreme point of K. Thus $\mathbf{a}_1, \mathbf{a}_2, \ldots, \mathbf{a}_k$ are linearly independent and $\mathbf{x}$ is a basic feasible solution. ▌

This correspondence between extreme points and basic feasible solutions enables us to prove certain geometric properties of the convex polytope K defining the constraint set of a linear programming problem.

Corollary 1. *If the convex set K corresponding to* (17) *is non-empty, it has at least one extreme point.*

Proof. This follows from the first part of the Fundamental Theorem and the Equivalence Theorem above. ▌

Corollary 2. *If there is a finite optimal solution to a linear programming problem, there is a finite optimal solution which is an extreme point of the constraint set.*

Corollary 3. *The constraint set K corresponding to* (17) *possesses at most a finite number of extreme points.*

Proof. There are obviously only a finite number of basic solutions obtained by selecting m basis vectors from the n columns of $\mathbf{A}$. The extreme points of K are a subset of these basic solutions. ▌

Finally, we come to the special case which occurs most frequently in practice and which in some sense is characteristic of well-formulated linear programs—the case where the constraint set K is nonempty and bounded. In this case we combine the results of the Equivalence Theorem and Corollary 3 above to obtain the following corollary.

Corollary 4. *If the convex polytope K corresponding to* (17) *is bounded, then K is a convex polyhedron, that is, K consists of points that are convex combinations of a finite number of points.*

In order to partially close the circle on the string of results derived above, we now present a direct simple proof that, in the case where K is a convex polyhedron, the objective of a linear program achieves a minimum at an extreme point.

Proposition. *A linear objective function* $\mathbf{cx}$ *achieves its minimum over a convex polyhedron K at an extreme point of K.*

Proof. Let $\mathbf{x}_1, \mathbf{x}_2, \ldots, \mathbf{x}_k$ be the extreme points of K. Then any point $\mathbf{x} \in K$ can be expressed in the form

$$\mathbf{x} = \alpha_1\mathbf{x}_1 + \alpha_2\mathbf{x}_2 + \cdots + \alpha_k\mathbf{x}_k \tag{18}$$

where

$$\alpha_i \geqslant 0, \qquad i = 1, 2, \ldots, k$$

$$\alpha_1 + \alpha_2 + \cdots + \alpha_k = 1.$$

Then

$$\mathbf{cx} = \alpha_1(\mathbf{cx}_1) + \alpha_2(\mathbf{cx}_2) + \cdots + \alpha_k(\mathbf{cx}_k).$$

Now let $z_0 = \min\{\mathbf{c}\mathbf{x}_i : i = 1, 2, \ldots, k\}$. Then from (18) we have

$$\mathbf{c}\mathbf{x} \geqslant (\alpha_1 + \alpha_2 + \cdots + \alpha_k)z_0 = z_0,$$

and hence the minimum of $\mathbf{c}\mathbf{x}$ over K is z_0. ▌

The proof given above possesses a certain degree of philosophical superiority over the proof of the Fundamental Theorem given in Section 2.3, since it is based directly on extreme points, which are intrinsically defined by the constraint set, rather than basic solutions, which are definable only in terms of a particular standard form representation of the constraint set. Although the two concepts are intimately related, one is coordinate independent and the other is not. Furthermore, transformation of a linear program to standard form often requires augmentation of additional variables and hence a change in the dimension of the space in which the problem is defined. This means that to obtain the extreme point theorem with respect to the original variables, a correspondence between extreme points of the two representations must be derived. All of this can, of course, be satisfactorily worked out. (See Exercises 8 and 9.)

Another approach, which is perhaps the most artistic, but which is significantly more technical than the development given here, is to first prove all of the extreme point properties in a coordinate-free fashion. This approach is discussed in Exercises 6 and 7.

Some of these considerations are illustrated by the following examples:

Example 1. Consider the constraint set in E^3 defined by

$$x_1 + x_2 + x_3 = 1,$$
$$x_1 \geqslant 0, \quad x_2 \geqslant 0, \quad x_3 \geqslant 0.$$

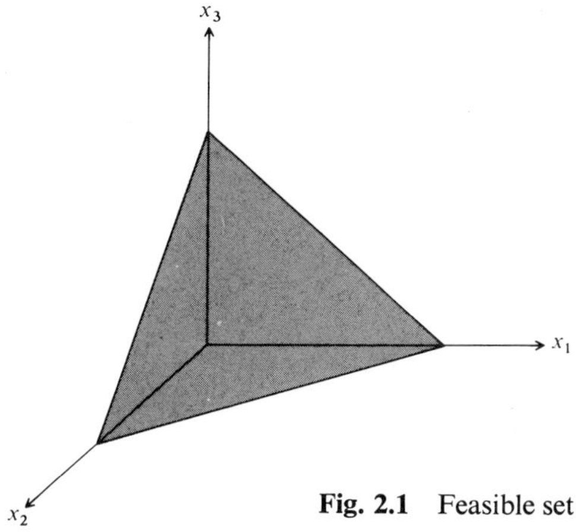

Fig. 2.1 Feasible set for Example 1

This set is illustrated in Fig. 2.1. It has three extreme points, corresponding to the three basic solutions to $x_1 + x_2 + x_3 = 1$.

Example 2. Consider the constraint set in E^3 defined by

$$\begin{aligned} x_1 + \ \ x_2 + x_3 &= 1 \\ 2x_1 + 3x_2 \qquad &= 1 \\ x_1 \geqslant 0, \quad x_2 \geqslant 0, \quad x_3 &\geqslant 0. \end{aligned}$$

This set is illustrated in Fig. 2.2. It has two extreme points, corresponding to the two basic feasible solutions. Note that the system of equations itself has three basic solutions, one of which is not feasible.

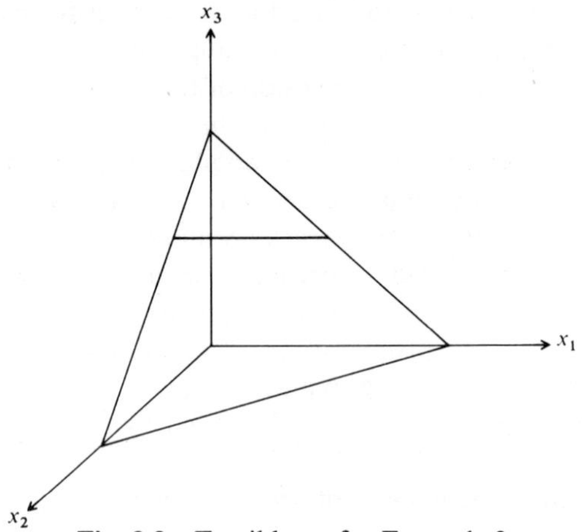

Fig. 2.2 Feasible set for Example 2

Example 3. Consider the constraint set in E^2 defined in terms of the inequalities

$$\begin{aligned} x_1 + \tfrac{8}{3}x_2 &\leqslant 4 \\ x_1 + \ \ x_2 &\leqslant 2 \\ 2x_1 \qquad &\leqslant 3 \\ x_1 \geqslant 0, \quad x_2 &\geqslant 0. \end{aligned}$$

This set is illustrated in Fig. 2.3. We see by inspection that this set has five extreme points. In order to compare this example with our general results we must introduce slack variables to yield the equivalent set in E^5

$$\begin{aligned} x_1 + \tfrac{8}{3}x_2 + x_3 \qquad\qquad &= 4 \\ x_1 + \ \ x_2 \qquad + x_4 \qquad &= 2 \\ 2x_1 \qquad\qquad\qquad + x_5 &= 3 \\ x_1 \geqslant 0, \quad x_2 \geqslant 0, \quad x_3 \geqslant 0, \quad x_4 \geqslant 0, \quad x_5 &\geqslant 0. \end{aligned}$$

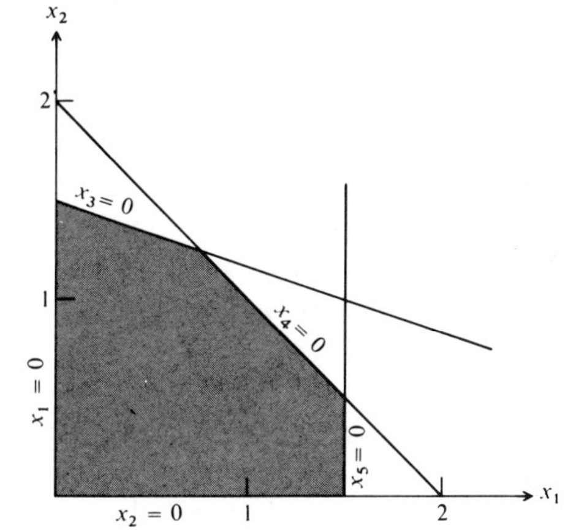

Fig. 2.3 Feasible set for Example 3

A basic solution for this system is obtained by setting any two variables to zero and solving for the remaining three. As indicated in Fig. 2.3, each edge of the figure corresponds to one variable being zero, and the extreme points are the points where two variables are zero.

2.6 EXERCISES

1. Convert the following problem to a linear program in standard form:

$$\begin{aligned} \text{minimize} \quad & |x| + |y| + |z| \\ \text{subject to} \quad & x + y \leqslant 1 \\ & 2x + z = 3. \end{aligned}$$

2. Convert the following problem to standard form and solve:

$$\begin{aligned} \text{maximize} \quad & x_1 + 4x_2 + x_3 \\ \text{subject to} \quad & 2x_1 - 2x_2 + x_3 = 4 \\ & x_1 \qquad\quad - x_3 = 1 \\ & x_2 \geqslant 0, \qquad x_3 \geqslant 0. \end{aligned}$$

3. Discuss the situation of a linear program that has one or more columns of the $\mathbf{A}$ matrix equal to zero. Consider both the case where the corresponding variables are required to be nonnegative and the case where some are free.

4. Suppose that the matrix $\mathbf{A} = (\mathbf{a}_1, \mathbf{a}_2, \ldots, \mathbf{a}_n)$ has rank m, and that for some $p < m$, $\mathbf{a}_1, \mathbf{a}_2, \ldots, \mathbf{a}_p$ are linearly independent. Show that $m - p$ vectors from the remaining $n - p$ vectors can be adjoined to form a set of m linearly independent vectors.

5. Suppose that $\mathbf{x}$ is a feasible solution to the linear program (11), with $\mathbf{A}$ an $m \times n$ matrix of rank m. Show that there is a feasible solution $\mathbf{y}$ having the same value (that is, $\mathbf{cy} = \mathbf{cx}$) and having at most $m + 1$ positive components.
6. Prove the following directly, using the results in Appendix B.

   Let C be a closed convex set contained in the positive orthant of E^n and let T be the subset of C where the linear function $f(\mathbf{x}) = \mathbf{cx}$ attains its maximum. If T is nonempty it contains an extreme point.
7. Using the result of Exercise 6 show that any linear program, having both linear inequalities and equalities as constraints but having no free variables, has an extreme point optimal solution if it has any optimal solution. Give an example that shows the result may not be true if there are free variables.
8. Let S be a convex set in E^n and S^* a convex set in E^m. Suppose $\mathbf{T}$ is an $m \times n$ matrix that establishes a one-to-one correspondence between S and S^*, i.e. for every $\mathbf{s} \in S$ there is $\mathbf{s}^* \in S^*$ such that $\mathbf{Ts} = \mathbf{s}^*$, and for every $\mathbf{s}^* \in S^*$ there is a single $\mathbf{s} \in S$ such that $\mathbf{Ts} = \mathbf{s}^*$. Show that there is a one-to-one correspondence between extreme points of S and S^*.
9. Consider the two linear programming problems in Example 1, Section 2.1; one in E^n and the other in E^{n+m}. Show that there is a one-to-one correspondence between extreme points of these two problems.
10. Find the three basic solutions of Example 2, Section 2.5.
11. What are the basic solutions of Example 3, Section 2.5?

REFERENCES

2.1–2.4. The approach taken in this chapter, which is continued in the next, is the more or less standard approach to linear programming as presented in, for example, Dantzig [D5], Hadley [H1], Gass [G2], Simonnard [S3], and Gale [G1].

2.5. An excellent discussion of this type can be found in Simonnard [S3].

Chapter 3 THE SIMPLEX METHOD

The idea of the simplex method is to proceed from one basic feasible solution (that is, one extreme point) of the constraint set of a problem in standard form to another, in such a way as to continually decrease the value of the objective function until a minimum is reached. The results of Chapter 2 assure us that it is sufficient to only consider basic feasible solutions in our search for an optimal feasible solution. This chapter demonstrates that an efficient method for moving among basic solutions to the minimum can be constructed.

In the first five sections of this chapter the simplex machinery is developed from a careful examination of the system of linear equations that defines the constraints and the basic feasible solutions of the system. This approach, which focuses on individual variables and their relation to the system, is probably the simplest, but unfortunately is not easily expressed in compact form. In the last few sections of the chapter, the simplex method is viewed from a matrix theoretic approach, which focuses on all variables together. This more sophisticated viewpoint leads to a compact notational representation, increased insight into the simplex process, and to alternative methods for implementation.

3.1 PIVOTS

To obtain a firm grasp of the simplex procedure, it is essential that one first understands the process of pivoting in a set of simultaneous linear equations. There are two dual interpretations of the pivot procedure.

First Interpretation

Consider the set of simultaneous linear equations

$$
\begin{aligned}
a_{11}x_1 + a_{12}x_2 + \cdots + a_{1n}x_n &= b_1 \\
a_{21}x_1 + a_{22}x_2 + \cdots + a_{2n}x_n &= b_2 \\
\cdot \qquad\qquad\qquad\qquad & \quad \cdot \\
\cdot \qquad\qquad\qquad\qquad & \quad \cdot \\
\cdot \qquad\qquad\qquad\qquad & \quad \cdot \\
a_{m1}x_1 + a_{m2}x_2 + \cdots + a_{mn}x_n &= b_m,
\end{aligned}
\tag{1}
$$

where $m \leqslant n$. In matrix form we write this as

$$\mathbf{Ax} = \mathbf{b}. \tag{2}$$

In the space E^n we interpret this as a collection of m linear relations that must be satisfied by a vector $\mathbf{x}$. Thus denoting by $\mathbf{a}^i$ the ith row of $\mathbf{A}$ we may express (1) as:

$$\begin{aligned} \mathbf{a}^1\mathbf{x} &= b_1 \\ \mathbf{a}^2\mathbf{x} &= b_2 \\ &\ \ \cdot \\ &\ \ \cdot \\ &\ \ \cdot \\ \mathbf{a}^m\mathbf{x} &= b_m. \end{aligned} \tag{3}$$

This corresponds to the most natural interpretation of (1) as a set of m equations.

If $m < n$ and the equations are linearly independent, then there is not a unique solution but a whole linear variety of solutions (see Appendix B). A unique solution results, however, if $n - m$ additional independent linear equations are adjoined. For example, we might specify $n - m$ equations of the form $\mathbf{e}^k\mathbf{x} = 0$, where $\mathbf{e}^k$ is the kth unit vector, (which is equivalent to $x_k = 0$) in which case we obtain a basic solution to (1). Different basic solutions are obtained by imposing different additional equations of this special form.

If the equations (3) are linearly independent, we may replace a given equation by any nonzero multiple of itself plus any linear combination of the other equations in the system. This leads to the well-known Gaussian reduction schemes, whereby multiples of equations are systematically subtracted from one another to yield either a triangular or canonical form. It is well known, and easily proved, that if the first m columns of $\mathbf{A}$ are linearly independent, the system (1) can, by a sequence of such multiplications and subtractions, be converted to the following *canonical form*:

$$\begin{array}{lllll} x_1 & + y_{1,m+1}x_{m+1} + y_{1,m+2}x_{m+2} + \cdots + y_{1,n}x_n = y_{10} \\ \quad x_2 & + y_{2,m+1}x_{m+1} + y_{2,m+2}x_{m+2} + \cdots + y_{2,n}x_n = y_{20} \\ & \cdot \\ & \cdot \\ & \cdot \\ \qquad x_m & + y_{m,m+1}x_{m+1} + \quad \cdots \quad + y_{m,n}x_n = y_{m0}. \end{array} \tag{4}$$

Corresponding to this canonical representation of the system, the variables $x_1, x_2, \ldots, x_m$ are called *basic* and the other variables are *nonbasic*. The corresponding basic solution is then:

$$x_1 = y_{10}, \quad x_2 = y_{20}, \ldots, \quad x_m = y_{m0}, \quad x_{m+1} = 0, \ldots, \quad x_n = 0,$$

or in vector form: $\mathbf{x} = (\mathbf{y}_0, \mathbf{0})$ where $\mathbf{y}_0$ is m-dimensional and $\mathbf{0}$ is the $n - m$ dimensional zero vector.

Actually, we relax our definition somewhat and consider a system to be in *canonical form* if, among the n variables, there are m basic ones with the property that each appears in only one equation, its coefficient in that equation is unity, and if no two of these m variables appear in any one equation. This is equivalent to saying that a system is in canonical form if by some reordering of the equations and the variables it takes the form (4).

Also it is customary, from the dictates of economy, to represent the system (4) by its corresponding array of coefficients or *tableau*:

$$\begin{array}{cccccccc}
1 & 0 \cdot \cdot \cdot 0 & y_{1,m+1} & y_{1,m+2} & \cdot \cdot \cdot & y_{1n} & y_{10} \\
0 & 1 \cdot \cdot \cdot 0 & y_{2,m+1} & y_{2,m+2} & \cdot \cdot \cdot & y_{2n} & y_{20} \\
0 & 0 \cdot \cdot \cdot 0 & \cdot & \cdot & & \cdot & \cdot \\
\cdot & \cdot \quad \cdot \quad \cdot & \cdot & \cdot & & \cdot & \cdot \\
\cdot & \cdot \quad \cdot \quad \cdot & \cdot & \cdot & & \cdot & \cdot \\
\cdot & \cdot \quad \cdot \quad \cdot & \cdot & \cdot & & \cdot & \cdot \\
0 & 0 \cdot \cdot \cdot 1 & y_{m,m+1} & y_{m,m+2} & \cdot \cdot \cdot & y_{mn} & y_{m0}.
\end{array}$$

The question solved by pivoting is this: given a system in canonical form, suppose a basic variable is to be made nonbasic and a nonbasic variable is to be made basic; what is the new canonical form corresponding to the new set of basic variables? The procedure is quite simple. Suppose in the canonical system (4) we wish to replace the basic variable x_p, $1 \leqslant p \leqslant m$, by the nonbasic variable x_q. This can be done if and only if y_{pq} is nonzero; it is accomplished by dividing row p by y_{pq} to get a unit coefficient for x_q in the pth equation, and then subtracting suitable multiples of row p from each of the other rows in order to get a zero coefficient for x_q in all other equations. This transforms the qth column of the tableau so that it is zero except in its pth entry (which is unity) and does not affect the columns of the other basic variables. Denoting the coefficients of the new system in canonical form by y'_{ij}, we have explicitly

$$\begin{cases} y'_{ij} = y_{ij} - \dfrac{y_{pj}}{y_{pq}} y_{iq}, & i \neq p \\[2ex] y'_{pj} = \dfrac{y_{pj}}{y_{pq}}. \end{cases} \tag{5}$$

Equations (5) are the pivot equations that arise frequently in linear programming. The element y_{pq} in the original system is said to be the *pivot element*.

Example 1. Consider the system in canonical form:

$$\begin{aligned} x_1 \quad\quad\quad\quad &+ x_4 + x_5 - x_6 = 5 \\ x_2 \quad\quad &+ 2x_4 - 3x_5 + x_6 = 3 \\ x_3 &- x_4 + 2x_5 - x_6 = -1. \end{aligned}$$

Let us find the basic solution having basic variables x_4, x_5, x_6. We set up the coefficient array below:

$$\begin{array}{ccccccc} x_1 & x_2 & x_3 & x_4 & x_5 & x_6 & \\ 1 & 0 & 0 & \textcircled{1} & 1 & -1 & 5 \\ 0 & 1 & 0 & 2 & -3 & 1 & 3 \\ 0 & 0 & 1 & -1 & 2 & -1 & -1. \end{array}$$

The circle indicated is our first pivot element and corresponds to the replacement of x_1 by x_4 as a basic variable. After pivoting we obtain the array

$$\begin{array}{ccccccc} x_1 & x_2 & x_3 & x_4 & x_5 & x_6 & \\ 1 & 0 & 0 & 1 & 1 & -1 & 5 \\ -2 & 1 & 0 & 0 & \textcircled{-5} & 3 & -7 \\ 1 & 0 & 1 & 0 & 3 & -2 & 4 \end{array}$$

and again we have circled the next pivot element indicating our intention to replace x_2 by x_5. We then obtain

$$\begin{array}{ccccccc} x_1 & x_2 & x_3 & x_4 & x_5 & x_6 & \\ 3/5 & 1/5 & 0 & 1 & 0 & -2/5 & 18/5 \\ 2/5 & -1/5 & 0 & 0 & 1 & -3/5 & 7/5 \\ -1/5 & 3/5 & 1 & 0 & 0 & \textcircled{-1/5} & -1/5 \end{array}$$

Continuing, there results

$$\begin{array}{ccccccc} x_1 & x_2 & x_3 & x_4 & x_5 & x_6 & \\ 1 & 1 & -2 & 1 & 0 & 0 & 4 \\ 1 & -2 & -3 & 0 & 1 & 0 & 2 \\ 1 & -3 & -5 & 0 & 0 & 1 & 1. \end{array}$$

From this last canonical form we obtain the new basic solution

$$x_4 = 4, \qquad x_5 = 2, \qquad x_6 = 1.$$

Second Interpretation

The set of simultaneous equations represented by (1) and (2) can be interpreted in E^m as a vector equation. Denoting the columns of $\mathbf{A}$ by $\mathbf{a}_1, \mathbf{a}_2, \ldots, \mathbf{a}_n$ we write (1) as

$$x_1\mathbf{a}_1 + x_2\mathbf{a}_2 + \cdots + x_n\mathbf{a}_n = \mathbf{b}. \tag{6}$$

In this interpretation we seek to express $\mathbf{b}$ as a linear combination of the $\mathbf{a}_i$'s.

If $m < n$ and the vectors $\mathbf{a}_i$ span E^m then there is not a unique solution but a whole family of solutions. The vector $\mathbf{b}$ has a unique representation, however, as a linear combination of a given linearly independent subset of these vectors. The corresponding solution with $n - m$ x_i variables set equal to zero is a basic solution to (1).

Suppose now that we start with a system in the canonical form corresponding to the tableau

$$\begin{array}{cccccccc}
1 & 0 & \ldots & 0 & y_{1,m+1} & y_{1,m+2} & \cdots & y_{1n} & y_{10} \\
0 & 1 & \cdot & 0 & y_{2,m+1} & y_{2,m+2} & \cdots & y_{2n} & y_{20} \\
0 & 0 & \cdot & 0 & \cdot & \cdot & \cdot & \cdot & \cdot \\
\cdot & \cdot & \cdot & \cdot & \cdot & \cdot & \cdot & \cdot & \cdot \\
\cdot & \cdot & \cdot & \cdot & \cdot & \cdot & \cdot & \cdot & \cdot \\
\cdot & \cdot & \cdot & \cdot & \cdot & \cdot & \cdot & \cdot & \cdot \\
0 & 0 & \cdot & 1 & y_{m,m+1} & y_{m,m+2} & \cdot & y_{mn} & y_{m0}.
\end{array} \tag{7}$$

In this case the first m vectors form a basis. Furthermore, every other vector represented in the tableau can be expressed as a linear combination of these basis vectors by simply reading the coefficients down the corresponding column. Thus,

$$\mathbf{a}_j = y_{1j}\mathbf{a}_1 + y_{2j}\mathbf{a}_2 + \cdots + y_{mj}\mathbf{a}_m. \tag{8}$$

The tableau can be interpreted as giving the representations of the vectors $\mathbf{a}_i$ in terms of the basis; the ith column of the tableau is the representation for the vector $\mathbf{a}_i$. In particular, the expression for $\mathbf{b}$ in terms of the basis is given in the last column.

We now consider the operation of replacing one member of the basis by another vector not already in the basis. Suppose for example we wish to replace the basis vector $\mathbf{a}_p$, $1 \leqslant p \leqslant m$, by the vector $\mathbf{a}_q$. Provided that the first m vectors with $\mathbf{a}_p$ replaced by $\mathbf{a}_q$ are linearly independent these vectors constitute a basis and every vector can be expressed as a linear combination of this new basis. To find the new representations of the vectors we must update the tableau. The linear independence condition holds if and only if $y_{pq} \neq 0$.

Any vector $\mathbf{a}_j$ can be expressed in terms of the old array through (8). For $\mathbf{a}_q$ we have

$$\mathbf{a}_q = \sum_{\substack{i=1 \\ i \neq p}}^{m} y_{iq}\mathbf{a}_i + y_{pq}\mathbf{a}_p$$

from which we may solve for $\mathbf{a}_p$,

$$\mathbf{a}_p = \frac{1}{y_{pq}}\mathbf{a}_q - \sum_{\substack{i=1 \\ i \neq p}}^{m} \frac{y_{iq}}{y_{pq}}\mathbf{a}_i. \tag{9}$$

Substituting (9) into (8) we obtain:

$$\mathbf{a}_j = \sum_{\substack{i=1 \\ i \neq p}}^{m} \left(y_{ij} - \frac{y_{iq}}{y_{pq}} y_{pj} \right) \mathbf{a}_i + \frac{y_{pj}}{y_{pq}}\mathbf{a}_q. \tag{10}$$

Denoting the coefficients of the new tableau, which gives the linear combinations, by y'_{ij} we obtain immediately from (10)

$$\begin{cases} y'_{ij} = y_{ij} - \dfrac{y_{iq}}{y_{pq}} y_{pj}, & i \neq p \\[2ex] y'_{pj} = \dfrac{y_{pj}}{y_{pq}}. \end{cases} \tag{11}$$

These formulae are identical to (5).

If a system of equations is not originally given in canonical form, we may put it into canonical form by adjoining the m unit vectors to the tableau and, starting with these vectors as the basis, successively replace each of them with columns of A using the pivot operation.

Example 2. Suppose we wish to solve the simultaneous equations

$$\begin{aligned} x_1 + x_2 - x_3 &= 5 \\ 2x_1 - 3x_2 + x_3 &= 3 \\ -x_1 + 2x_2 - x_3 &= -1 \end{aligned}$$

To obtain an original basis, we form the augmented tableau

$\mathbf{e}_1$	$\mathbf{e}_2$	$\mathbf{e}_3$	$\mathbf{a}_1$	$\mathbf{a}_2$	$\mathbf{a}_3$	$\mathbf{b}$
1	0	0	1	1	−1	5
0	1	0	2	−3	1	3
0	0	1	−1	2	−1	−1

and replace $\mathbf{e}_1$ by $\mathbf{a}_1$, $\mathbf{e}_2$ by $\mathbf{a}_2$, and $\mathbf{e}_3$ by $\mathbf{a}_3$. The required operations are identical to those of Example 1.

3.2 ADJACENT EXTREME POINTS

In Chapter 2 it was discovered that it is only necessary to consider basic feasible solutions to the system

$$\begin{aligned} \mathbf{A}\mathbf{x} &= \mathbf{b} \\ \mathbf{x} &\geqslant \mathbf{0}, \end{aligned} \tag{12}$$

when solving a linear program, and in the previous section it was demonstrated that the pivot operation can generate a new basic solution from an old one by replacing one basic variable by a nonbasic variable. It is clear, however, that although the pivot operation takes one basic solution into another, the nonnegativity of the solution will not in general be preserved. Special conditions must be satisfied in order that a pivot operation maintain feasibility. In this section we show how it is possible to select pivots so that we may transfer from one basic feasible solution to another.

We show that although it is not possible to arbitrarily specify the pair of variables whose roles are to be interchanged and expect to maintain the nonnegativity condition, it is possible to arbitrarily specify which nonbasic variable is to become basic and then determine which basic variable should become nonbasic. As is conventional, we base our derivation on the vector interpretation of the linear equations although the dual interpretation could alternatively be used.

Nondegeneracy Assumption

Many arguments in linear programming are substantially simplified upon the introduction of the following.

Nondegeneracy assumption: Every basic feasible solution of (12) is a nondegenerate basic feasible solution.

This assumption is invoked throughout our development of the simplex method, since when it does not hold the simplex method can break down if it is not suitably amended. The assumption, however, should be regarded as one made for convenience only, since all arguments can be extended to include degeneracy, and the simplex method itself can be easily modified to account for it.

Determination of Vector to Leave Basis

Suppose we have the basic feasible solution $\mathbf{x} = (x_1, x_2, \ldots, x_m, 0, 0, \ldots, 0)$

or, equivalently, the representation

$$x_1\mathbf{a}_1 + x_2\mathbf{a}_2 + \cdots + x_m\mathbf{a}_m = \mathbf{b}. \tag{13}$$

Under the nondegeneracy assumption, $x_i > 0$, $i = 1, 2, \ldots, m$. Suppose also that we have decided to bring into the representation the vector $\mathbf{a}_k$, $k > m$. We have available a representation of $\mathbf{a}_k$ in terms of the current basis

$$\mathbf{a}_k = y_{1k}\mathbf{a}_1 + y_{2k}\mathbf{a}_2 + \cdots + y_{mk}\mathbf{a}_m. \tag{14}$$

Multiplying (14) by a variable $\varepsilon \geqslant 0$ and subtracting from (13), we have

$$(x_1 - \varepsilon y_{1k})\mathbf{a}_1 + (x_2 - \varepsilon y_{2k})\mathbf{a}_2 + \cdots + (x_m - \varepsilon y_{mk})\mathbf{a}_m + \varepsilon\mathbf{a}_k = \mathbf{b}. \tag{15}$$

Thus, for any $\varepsilon \geqslant 0$ (15) gives $\mathbf{b}$ as a linear combination of at most $m + 1$ vectors. For $\varepsilon = 0$ we have the old basic feasible solution. As ε is increased from zero, the coefficient of $\mathbf{a}_k$ increases, and it is clear that for small enough ε (15) gives a feasible but nonbasic solution. The coefficients of the other vectors will either increase or decrease linearly as ε is increased. If any decrease, we may set ε equal to the value corresponding to the first place where one (or more) of the coefficients vanishes. That is

$$\varepsilon = \min_i \{x_i/y_{ik} : y_{ij} > 0\}. \tag{16}$$

In this case we have a new basic feasible solution, with the vector $\mathbf{a}_k$ replacing the vector $\mathbf{a}_i$ where i corresponds to the minimum in (16). If the minimum in (16) is achieved by more than a single index i, then the new solution is degenerate and any of the vectors with zero component can be regarded as the one which left the basis.

If none of the y_{ik}'s are positive, then all coefficients in the representation (15) increase (or remain constant) as ε is increased, and no new basic feasible solution is obtained. We observe, however, that in this case, where none of the y_{ik}'s are positive, there are feasible solutions to (12) having arbitrarily large coefficients. This means that the set K of feasible solutions to (12) is unbounded, and this special case, as we shall see, is of special significance in the simplex procedure.

In summary, we have deduced that given a basic feasible solution and an arbitrary vector $\mathbf{a}_k$, there is either a new basic feasible solution having $\mathbf{a}_k$ in its basis and one of the original vectors removed, or the set of basic feasible solutions is unbounded.

Let us consider how the calculation of this section can be displayed in our tableau. We assume that corresponding to the constraints

$$\mathbf{A}\mathbf{x} = \mathbf{b}$$
$$\mathbf{x} \geqslant \mathbf{0},$$

we have a tableau of the form

$$
\begin{array}{cccccccccc}
\mathbf{a}_1 & \mathbf{a}_2 & \mathbf{a}_3 & \cdots & \mathbf{a}_m & \mathbf{a}_{m+1} & \mathbf{a}_{m+2} & \cdots & \mathbf{a}_n & \mathbf{b} \\
1 & 0 & 0 & \cdots & 0 & y_{1,m+1} & y_{1,m+2} & \cdots & y_{1n} & y_{10} \\
0 & 1 & 0 & & 0 & y_{2,m+1} & y_{2,m+2} & & \cdot & y_{20} \\
0 & 0 & 1 & & \cdot & \cdot & \cdot & & \cdot & \cdot \\
\cdot & \cdot & \cdot & & \cdot & \cdot & \cdot & & \cdot & \cdot \\
\cdot & \cdot & \cdot & & \cdot & \cdot & \cdot & & \cdot & \cdot \\
\cdot & \cdot & \cdot & & \cdot & \cdot & \cdot & & \cdot & \cdot \\
0 & 0 & \cdot & & 1 & y_{m,m+1} & y_{m,m+2} & \cdots & y_{mn} & y_{m0}.
\end{array}
\tag{17}
$$

This tableau may be the result of several pivot operations applied to the original tableau, but in any event, it represents a solution with basis $\mathbf{a}_1, \mathbf{a}_2, \ldots, \mathbf{a}_m$. We assume that $y_{10}, y_{20}, \ldots, y_{m0}$ are nonnegative, so that the corresponding basic solution $x_1 = y_{10}, x_2 = y_{20}, \ldots, x_m = y_{m0}$ is feasible. We wish to bring into the basis the vector $\mathbf{a}_k$, $k > m$, and maintain feasibility. In order to determine which element in the kth column to use as pivot (and hence which vector in the basis will leave), we use (16) and compute the ratios $x_i/y_{ik} = y_{i0}/y_{ik}$, $i = 1, 2, \ldots, m$, select the smallest nonnegative ratio, and pivot on the corresponding y_{ik}.

Example. Consider the system

$\mathbf{a}_1$	$\mathbf{a}_2$	$\mathbf{a}_3$	$\mathbf{a}_4$	$\mathbf{a}_5$	$\mathbf{a}_6$	$\mathbf{b}$
1	0	0	2	4	6	4
0	1	0	1	2	3	3
0	0	1	−1	2	1	1

which has basis $\mathbf{a}_1, \mathbf{a}_2, \mathbf{a}_3$ yielding a basic feasible solution $\mathbf{x} = (4, 3, 1, 0, 0, 0)$. Suppose we elect to bring $\mathbf{a}_4$ into the basis. To determine which element in the fourth column is the appropriate pivot, we compute the three ratios:

$$4/2 = 2, \qquad 3/1 = 3, \qquad 1/-1 = -1$$

and select the smallest nonnegative one. This gives 2 as the pivot element. The new tableau is

$\mathbf{a}_1$	$\mathbf{a}_2$	$\mathbf{a}_3$	$\mathbf{a}_4$	$\mathbf{a}_5$	$\mathbf{a}_6$	$\mathbf{b}$
1/2	0	0	1	2	3	2
−1/2	1	0	0	0	0	1
1/2	0	1	0	4	4	3

with corresponding basic feasible solution $\mathbf{x} = (0, 1, 3, 2, 0, 0)$.

Our derivation of the method for selecting the pivot in a given column that will yield a new feasible solution has been based on the vector interpretation of the equation $\mathbf{Ax} = \mathbf{b}$. An alternative derivation can be constructed by considering the dual approach that is based on the rows of the tableau rather than the columns. Briefly, the argument runs like this: if we pivot on y_{ik}, then we first divide the ith row by y_{ik} to get a unity in the ikth element. In order for the new $y_{i0} = x_i$ to remain positive, it is clear that we must have $y_{ik} > 0$. Next, we subtract, from the jth row, y_{jk}/y_{ik} times the ith row to obtain zeros in the remaining elements of the kth column. This gives a new

$$x_j' = x_j - \frac{y_{jk}}{y_{ik}} x_i$$

which to be nonnegative requires $x_j/y_{jk} \geqslant x_i/y_{ik}$, and hence again we are led to the conclusion that we select i to minimize x_i/y_{ik}.

3.3 DETERMINING A MINIMUM FEASIBLE SOLUTION

In the last section we showed how it is possible to pivot from one basic feasible solution to another (or determine that the solution set is unbounded) by arbitrarily selecting a column to pivot on and then appropriately selecting the pivot in that column. The idea of the simplex method is to select the column so that the resulting new basic feasible solution will yield a lower value to the objective function than the previous one. This then provides the final link in the simplex procedure. By an elementary calculation, which is derived below, it is possible to determine which vector should enter the basis so that the objective value is reduced, and by another simple calculation, derived in the previous section, it is possible to then determine which vector should leave in order to maintain feasibility.

Suppose we have a basic feasible solution

$$(\mathbf{x_B}, \mathbf{0}) = (y_{10}, y_{20}, \ldots, y_{m0}, 0, 0, \ldots, 0)$$

together with a tableau having an identity matrix appearing in the first m columns as shown below:

$$\begin{array}{cccccc} \mathbf{a}_1 & \mathbf{a}_2 \;\cdots & \mathbf{a}_m & \mathbf{a}_{m+1} & \cdots\; \mathbf{a}_n & \mathbf{b} \\ 1 & 0 & 0 & y_{1,m+1} & \cdots\; y_{1n} & y_{10} \\ 0 & 1 & 0 & y_{2,m+1} & \cdots\; y_{2n} & y_{20} \\ \cdot & \cdot & \cdot & \cdot & \cdot & \cdot \\ \cdot & \cdot & \cdot & \cdot & \cdot & \cdot \\ \cdot & \cdot & \cdot & \cdot & \cdot & \cdot \\ 0 & 0 & 1 & y_{m,m+1} & \cdots\; y_{mn} & y_{m0}. \end{array} \tag{18}$$

The value of the objective function corresponding to any solution $\mathbf{x}$ is

$$z = c_1x_1 + c_2x_2 + \cdots + c_nx_n \tag{19}$$

and hence for the basic solution, the corresponding value is

$$z_0 = \mathbf{c_B x_B} \tag{20}$$

where $\mathbf{c_B} = [c_1, c_2, \ldots, c_m]$.

Although it is natural to use the basic solution $(\mathbf{x_B}, \mathbf{0})$ when we have the tableau (18), it is clear that if arbitrary values are assigned to $x_{m+1}, x_{m+2}, \ldots, x_n$, we can easily solve for the remaining variables as

$$\begin{aligned} x_1 &= y_{10} - \sum_{j=m+1}^{n} y_{1j}x_j \\ x_2 &= y_{20} - \sum_{j=m+1}^{n} y_{2j}x_j \\ &\;\;\vdots \\ x_m &= y_{m0} - \sum_{j=m+1}^{n} y_{mj}x_j. \end{aligned} \tag{21}$$

Using (21) we may eliminate $x_1, x_2, \ldots, x_m$ from the general formula (19). Doing this we obtain

$$z = \mathbf{cx} = z_0 + (c_{m+1} - z_{m+1})x_{m+1} + (c_{m+2} - z_{m+2})x_{m+2} + \cdots + (c_n - z_n)x_n \tag{22}$$

where

$$z_j = y_{1j}c_1 + y_{2j}c_2 + \cdots + y_{mj}c_m, \qquad m+1 \leqslant j \leqslant n \tag{23}$$

which is the fundamental relation required to determine the pivot column. The important point is that this equation gives the values of the objective function z for any solution of $\mathbf{Ax} = \mathbf{b}$ in terms of the variables $x_{m+1}, \ldots, x_n$. From it we can determine if there is any advantage in changing the basic solution by introducing one of the nonbasic variables. For example, if $c_j - z_j$ is negative for some j, $m + 1 \leqslant j \leqslant n$, then increasing x_j from zero to some positive value would decrease the total cost, and therefore would yield a better solution. The formulae (22) and (23) automatically take into account the changes that would be required in the values of the basic variables $x_1, x_2, \ldots, x_m$ to accommodate the change in x_j.

Let us derive these relations from a different viewpoint. Let $\mathbf{y}_i$ be the ith column of the tableau. Then any solution satisfies

$$x_1\mathbf{e}_1 + x_2\mathbf{e}_2 + \cdots + x_m\mathbf{e}_m = \mathbf{y}_0 - x_{m+1}\mathbf{y}_{m+1} - x_{m+2}\mathbf{y}_{m+2} - \cdots - x_n\mathbf{y}_n.$$

Taking the inner product of this vector equation with $\mathbf{c_B}$, we have

$$\sum_{i=1}^{m} c_i x_i = \mathbf{c_B y}_0 - \sum_{j=m+1}^{n} z_j x_j$$

where $z_j = \mathbf{c_B y}_j$.

Thus, adding $\sum_{j=m+1}^{n} c_j x_j$ to both sides,

$$\mathbf{cx} = z_0 + \sum_{j=m+1}^{n} (c_j - z_j) x_j \tag{24}$$

as before.

Finally, let us view this relation from the dual point of view of a system of equations. We regard z as an additional variable and

$$c_1 x_1 + c_2 x_2 + \cdots + c_n x_n - z = 0 \tag{25}$$

as another linear equation. We pivot the entire set of $m + 1$ equations so that the $m + 1$ basic variables are $x_1, x_2, \ldots, x_m, z$. The last row of the tableau for this system then gives z in terms of the nonbasic variables. The elements in this row are of course the $c_i - z_i$'s.

We now state the condition for improvement, which follows easily from the above observation, as a theorem.

Theorem (Improvement of basic feasible solution). *Given a nondegenerate basic feasible solution with corresponding objective value z_0, suppose that for some j there holds $c_j - z_j < 0$. Then there is a feasible solution with objective value $z < z_0$. If the column $\mathbf{a}_j$ can be substituted for some vector in the original basis to yield a new basic feasible solution, this new solution will have $z < z_0$. If $\mathbf{a}_j$ cannot be substituted to yield a basic feasible solution, then the solution set K is unbounded and the objective function can be made arbitrarily small (toward minus infinity).*

Proof. The result is an immediate consequence of the previous discussion. Let $(x_1, x_2, \ldots, x_m, 0, 0, \ldots, 0)$ be the basic feasible solution with objective value z_0 and suppose $c_{m+1} - z_{m+1} < 0$. Then, in any case, new feasible solutions can be constructed of the form $(x_1', x_2', \ldots, x_m', x_{m+1}', 0, 0, \ldots, 0)$ with $x_{m+1}' > 0$. Substituting this solution in (22) we have

$$z - z_0 = (c_{m+1} - z_{m+1}) x_{m+1}' < 0,$$

and hence $z < z_0$ for any such solution. It is clear that we desire to make x_{m+1}' as large as possible. As x_{m+1}' is increased, the other components increase, remain constant, or decrease. Thus x_{m+1}' can be increased until one $x_i' = 0, i \leqslant m$, in which case we obtain a new basic feasible solution, or if none of the x_i''s decrease, x_{m+1}' can be increased without bound indicating an unbounded solution set and an objective value without lower bound. ▌

We see that if at any stage $c_j - z_j < 0$ for some j, it is possible to make x_j positive and decrease the objective function. The final question remaining is whether $c_j - z_j \geqslant 0$ for all j implies optimality.

Optimality Condition Theorem. *If for some basic feasible solution $c_j - z_j \geqslant 0$ for all j, then that solution is optimal.*

Proof. This follows immediately from (22), since any other feasible solution must have $x_i \geqslant 0$ for all i, and hence the value z of the objective will satisfy $z - z_0 \geqslant 0$. ∎

Since the constants $c_j - z_j$ play such a central role in the development of the simplex method, it is convenient to introduce the somewhat abbreviated notation $r_j = c_j - z_j$ and refer to the r_j's as the *relative cost coefficients* or, alternatively, the *reduced cost coefficients* (both terms occur in common usage). These coefficients measure the cost of a variable relative to a given basis.

We conclude this section by giving an economic interpretation of the relative cost coefficients. Let us agree to interpret the linear programming problem:

$$\begin{aligned} \text{minimize} \quad & \mathbf{cx} \\ \text{subject to} \quad & \mathbf{Ax} = \mathbf{b} \\ & \mathbf{x} \geqslant \mathbf{0} \end{aligned}$$

as a diet problem (see Section 2.2) where the nutritional requirements must be met exactly. A column of **A** gives the nutritional equivalent of a unit of a particular food. With a given basis consisting of, say, the first m columns of **A**, the corresponding simplex tableau shows how any food (or more precisely, the nutritional content of any food) can be constructed as a combination of foods in the basis. For instance, if carrots are not in the basis we can, using the description given by the tableau, construct a *synthetic* carrot which is nutritionally equivalent to a carrot, by an appropriate combination of the foods in the basis.

In considering whether or not the solution represented by the current basis is optimal, we consider a certain food not in the basis—say carrots—and determine if it would be advantageous to bring it into the basis. This is very easily determined by examining the cost of carrots as compared with the cost of synthetic carrots. If carrots are food j, then the unit cost of carrots is c_j. The cost of a unit of synthetic carrots is, on the other hand,

$$z_j = \sum_{i=1}^{m} c_i y_{ij}.$$

If $r_j = c_j - z_j < 0$, it is advantageous to use real carrots in place of synthetic carrots, and carrots should be brought into the basis.

In general the z_j's can be thought of as the price of a unit of the column $\mathbf{a}_j$ when constructed from the current basis. The difference between this synthetic price and the direct price of that column determines whether that column should enter the basis.

3.4 COMPUTATIONAL PROCEDURE—SIMPLEX METHOD

In previous sections the theory, and indeed much of the technique, necessary for the detailed development of the simplex method has been established. It is only necessary to put it all together and illustrate it with examples.

In this section we assume that we begin with a basic feasible solution and that the tableau corresponding to $\mathbf{Ax} = \mathbf{b}$ is in the canonical form for this solution. Methods for obtaining this first basic feasible solution, when one is not obvious, are described in the next section.

In addition to beginning with the array $\mathbf{Ax} = \mathbf{b}$ expressed in canonical form corresponding to a basic feasible solution, we transform the cost function $z = \mathbf{cx}$ into the form given in the last section by calculating the r_j's or, equivalently, by adding appropriate linear combinations of rows of the $\mathbf{Ax} = \mathbf{b}$ array to a row corresponding to $\mathbf{cx} - z = 0$. We define a *simplex tableau* consisting of the array corresponding to $\mathbf{Ax} = \mathbf{b}$ augmented by the transformed row expressing z. Thus if we assume the basic variables are (in order) $x_1, x_2, \ldots, x_m$, the simplex tableau takes the initial form shown below.

$\mathbf{a}_1$	$\mathbf{a}_2$	$\ldots$	$\mathbf{a}_m$	$\mathbf{a}_{m+1}$	$\mathbf{a}_{m+2}$	$\ldots$	$\mathbf{a}_j$	$\ldots$	$\mathbf{a}_n$	$\mathbf{b}$
1	0	$\ldots$	0	$y_{1,m+1}$	$y_{1,m+2}$	$\cdots$	y_{1j}	$\cdots$	y_{1n}	y_{10}
0	1		·	·	·		·		·	·
·	·		·	·	·		·		·	·
·	·		·	·	·		·		·	·
0	0		·	$y_{i,m+1}$	$y_{i,m+2}$	$\cdots$	y_{ij}	$\cdots$	y_{in}	y_{i0}
·	·		·	·	·		·		·	·
·	·		·	·	·		·		·	·
0	0		1	$y_{m,m+1}$	$y_{m,m+2}$	$\cdots$	y_{mj}	$\cdots$	y_{mn}	$y_{m+1,0}$
0	0	$\ldots$	0	r_{m+1}	r_{m+2}	$\cdots$	r_j	$\cdots$	r_n	$-z_0$

Fig. 3.1 Canonical simplex tableau

The basic solution corresponding to this tableau is

$$x_i = \begin{cases} y_{i0} & 0 \leqslant i \leqslant m \\ 0 & m+1 \leqslant i \leqslant n \end{cases}$$

which we have assumed is feasible, that is, $y_{i0} \geqslant 0, i = 1, 2, \ldots, m$. The corresponding value of the objective function is z_0.

The relative cost coefficients r_j indicate whether the value of the objective will increase or decrease if x_j is pivoted into the solution. If these coefficients are all nonnegative, then the indicated solution is optimal. If some of them are negative, an improvement can be made (assuming nondegeneracy) by bringing the corresponding component into the solution. When more than one of the relative cost coefficients is negative, any one of them may be selected to determine in which column to pivot. Common practice is to select the most negative value. (See Exercise 15 for further discussion of this point.)

After a column j is selected in which to pivot, the final selection of the pivot element is made by computing the ratio y_{i0}/y_{ij} for the positive elements y_{ij}, $i = 1, 2, \ldots, m$, of the jth column and selecting the element i yielding the minimum ratio. Pivoting on this element will maintain feasibility as well as (assuming nondegeneracy) decrease the value of the objective function. If there are ties, any element yielding the minimum can be used. If there are no nonnegative elements in the column, the problem is unbounded. After updating the entire tableau with y_{ij} as pivot and transforming the last row in the same manner as all other rows (except row j), we obtain a new tableau in canonical form. The new value of the objective function again appears in the lower right hand corner of the tableau.

Example 3. Maximize $3x_1 + x_2 + 3x_3$ subject to

$$\begin{aligned} 2x_1 + x_2 + x_3 &\leqslant 2 \\ x_1 + 2x_2 + 3x_3 &\leqslant 5 \\ 2x_1 + 2x_2 + x_3 &\leqslant 6 \\ x_1 \geqslant 0, \quad x_2 \geqslant 0, \quad x_3 &\geqslant 0. \end{aligned}$$

To transform the problem into standard form so that the simplex procedure can be applied, we change the maximization to minimization by multiplying the objective function by minus one, and introduce three nonnegative slack variables x_4, x_5, x_6. We then have the initial tableau

$\mathbf{a}_1$	$\mathbf{a}_2$	$\mathbf{a}_3$	$\mathbf{a}_4$	$\mathbf{a}_5$	$\mathbf{a}_6$	$\mathbf{b}$
2	(1)	1	1	0	0	2
1	2	3	0	1	0	5
2	2	1	0	0	1	6
−3	−1	−3	0	0	0	0

First tableau

The problem is already in canonical form with the three slack variables serving as the basic variables. We have at this point $r_j = c_j - z_j = c_j$, since the costs of the slacks are zero. Application of the criterion for selecting a column in which to pivot shows that any of the first three columns would yield an improved solution. In each of these columns the appropriate pivot element is determined by computing the ratios y_{i0}/y_{ij} and selecting the smallest positive one. The three allowable pivots are all circled on the tableau. It is only necessary to determine one allowable pivot, and normally we would not bother to calculate them all. For hand calculation on problems of this size, however, we may wish to examine the allowable pivots and select one that will minimize (at least in the short run) the amount of division required. Thus for this example we select (1).

2	1	1	1	0	0	2
−3	0	(1)	−2	1	0	1
−2	0	−1	−2	0	1	2
−1	0	−2	1	0	0	2

Second tableau

We note that the objective function—we are using the negative of the original one—has decreased from zero to minus two. Again we pivot on (1).

(5)	1	0	3	−1	0	1
−3	0	1	−2	1	0	1
−5	0	0	−4	1	1	3
−7	0	0	−3	2	0	4

Third tableau

The value of the objective function has now decreased to minus four and we may pivot in either the first or fourth column. We select (5).

1	1/5	0	3/5	−1/5	0	1/5
0	3/5	1	−1/5	2/5	0	8/5
0	1	0	−1	0	1	4
0	7/5	0	6/5	3/5	0	27/5

Fourth tableau

Since the last row has no negative elements, we conclude that the solution corresponding to the fourth tableau is optimal. Thus $x_1 = \frac{1}{5}$, $x_2 = 0$, $x_3 = \frac{8}{5}$, $x_4 = 0$, $x_5 = 0$, $x_6 = 4$ is the optimal solution with a corresponding value of the (negative) objective of $-\frac{27}{5}$.

3.5 ARTIFICIAL VARIABLES

A basic feasible solution is sometimes immediately available for linear programs. For example, in problems with constraints of the form

$$\begin{aligned} \mathbf{Ax} &\leqslant \mathbf{b}, \\ \mathbf{x} &\geqslant \mathbf{0} \end{aligned} \tag{26}$$

with $\mathbf{b} \geqslant \mathbf{0}$, a basic feasible solution to the corresponding standard form of the problem is provided by the slack variables. This provides a means for initiating the simplex procedure. The example in the last section was of this type. An initial basic feasible solution is not always apparent for other types of linear programs, however, and it is necessary to develop a means for determining one so that the simplex method can be initiated. Interestingly (and fortunately), an auxiliary linear program and corresponding application of the simplex method can be used to determine the required initial solution.

By elementary straightforward operations the constraints of a linear programming problem can always be expressed in the form

$$\begin{aligned} \mathbf{Ax} &= \mathbf{b}, \\ \mathbf{x} &\geqslant \mathbf{0} \end{aligned} \tag{27}$$

with $\mathbf{b} \geqslant \mathbf{0}$. In order to find a solution to (27) consider the (artificial) minimization problem

$$\begin{aligned} \text{minimize} \quad & \sum_{i=1}^{m} y_i, \\ \text{subject to} \quad & \mathbf{Ax} + \mathbf{y} = \mathbf{b}, \\ & \mathbf{x} \geqslant \mathbf{0} \\ & \mathbf{y} \geqslant \mathbf{0} \end{aligned} \tag{28}$$

where $\mathbf{y} = (y_1, y_2, \ldots, y_m)$ is a vector of artificial variables. If there is a feasible solution to (27), then it is clear that (28) has a minimum value of zero with $\mathbf{y} = \mathbf{0}$. If (27) has no feasible solution, then the minimum value of (28) is greater than zero.

Now problem (28) is itself a linear programming problem in the variables $\mathbf{x}$, $\mathbf{y}$, and the system is already in canonical form with basic feasible solution $\mathbf{y} = \mathbf{b}$. If (28) is solved using the simplex technique, a basic feasible solution is obtained at each step. If the minimum value of (28) is zero, then the final basic solution will have all $y_i = 0$, and hence barring degeneracy, the final solution will have no y_i variables basic. If in the final solution some y_i are both zero and basic, indicating a degenerate solution, these basic variables can be exchanged for nonbasic x_i variables (again at zero level) to yield a basic feasible solution involving x variables only.

Example 1. Find a basic feasible solution to

$$\begin{aligned} 2x_1 + x_2 + 2x_3 &= 4 \\ 3x_1 + 3x_2 + x_3 &= 3 \\ x_1 \geqslant 0, \quad x_2 \geqslant 0, \quad x_3 &\geqslant 0. \end{aligned}$$

We introduce artificial variables $x_4 \geqslant 0$, $x_5 \geqslant 0$ and an objective function $x_4 + x_5$. The initial tableau is

x_1	x_2	x_3	x_4	x_5	**b**
2	1	2	1	0	4
3	3	1	0	1	3
0	0	0	1	1	0

Initial tableau

A basic feasible solution to the expanded system is given by the artificial variables. To initiate the simplex procedure we must update the last row so that it has zero components under the basic variables. This yields:

2	1	2	1	0	4
(3)	3	1	0	1	3
−5	−4	−3	0	0	−7.

First tableau

(The pivot entry 3 is circled.)

Pivoting in the column having the most negative bottom row component as indicated, we obtain:

0	−1	(4/3)	1	−2/3	2
1	1	1/3	0	1/3	1
0	1	−4/3	0	5/3	−2.

Second tableau

(The pivot entry 4/3 is circled.)

In the second tableau there is only one choice for pivot, and it leads to the final tableau shown.

0	−3/4	1	3/4	−1/2	3/2
1	5/4	0	−1/4	1/2	1/2
0	0	0	1	1	0 .

Final tableau

Both of the artificial variables have been driven out of the basis, thus reducing the value of the objective function to zero and leading to the basic feasible solution to the original problem

$$x_1 = 1/2, \qquad x_2 = 0, \qquad x_3 = 3/2.$$

Using artificial variables, we attack a general linear programming problem by use of the *two-phase method.* This method consists simply of a *phase I* in which artificial variables are introduced as above and a basic feasible solution is found (or it is determined that no feasible solutions exist); and a *phase II* in which, using the basic feasible solution resulting from phase I, the original objective function is minimized. During phase II the artificial variables and the objective function of phase I are omitted.

Example 2. Consider the problem

$$\begin{aligned} \text{minimize} \quad & 4x_1 + x_2 + x_3 \\ \text{subject to} \quad & 2x_1 + x_2 + 2x_3 = 4 \\ & 3x_1 + 3x_2 + x_3 = 3 \\ & x_1 \geqslant 0, \quad x_2 \geqslant 0, \quad x_3 \geqslant 0. \end{aligned}$$

There is no basic feasible solution apparent, so we use the two-phase method. The first phase was done in Example 1 for these constraints, so we shall not repeat it here. We give only the final tableau with the columns corresponding to the artificial variables deleted, since they are not used in phase II. We use the new cost function in place of the old one. Temporarily writing **c** in the bottom row we have

x_1	x_2	x_3	**b**
0	−3/4	1	3/2
1	5/4	0	1/2
4	1	1	0

Initial tableau

Transforming the last row so that zeros appear in the basic columns, we have

0	$-3/4$	0	$3/2$
0	(5/4)	0	$1/2$
0	$-13/4$	0	$-7/2$

First tableau

$3/5$	0	1	$9/5$
$4/5$	1	0	$2/5$
$13/5$	0	0	$-11/5$

Second tableau

and hence the optimal solution is $x_1 = 0$, $x_2 = 2/5$, $x_3 = 9/5$.

Example 3. (A free variable problem).

$$\begin{aligned} \text{minimize} \quad & -2x_1 + 4x_2 + 7x_3 + x_4 + 5x_5 \\ \text{subject to} \quad & -x_1 + x_2 + 2x_3 + x_4 + 2x_5 = 7 \\ & -x_1 + 2x_2 + 3x_3 + x_4 + x_5 = 6 \\ & -x_1 + x_2 + x_3 + 2x_4 + x_5 = 4 \end{aligned}$$

$$x_1 \text{ free}, \quad x_2 \geqslant 0, \quad x_3 \geqslant 0, \quad x_4 \geqslant 0, \quad x_5 \geqslant 0.$$

Since x_1 is free, it can be eliminated, as described in Chapter 2, by solving for x_1 in terms of the other variables from the first equation and substituting everywhere else. This can all be done with the simplex tableau as follows:

x_1	x_2	x_3	x_4	x_5	**b**
(−1)	1	2	1	2	7
−1	2	3	1	1	6
−1	1	1	2	1	4
−2	4	7	1	5	0

Initial tableau

We select any nonzero element in the first column to pivot on—this will eliminate x_1.

1	−1	−2	−1	−2	−7
0	1	1	0	−1	−1
0	0	−1	1	−1	−3
0	2	3	−1	1	−14

Equivalent problem

We now save the first row for future reference, but our linear program only involves the sub-tableau indicated. There is no obvious basic feasible solution for this problem, so we introduce artificial variables x_6 and x_7.

x_2	x_3	x_4	x_5	x_6	x_7	**b**
−1	−1	0	1	1	0	1
0	1	−1	1	0	1	3
0	0	0	0	1	1	0

Initial tableau for phase I

Transforming the last row appropriately we obtain

x_2	x_3	x_4	x_5	x_6	x_7	**b**
−1	−1	0	(1)	1	0	1
0	1	−1	1	0	1	3
1	0	1	−2	0	0	−4

First tableau—phase I

−1	−1	0	1	1	0	1
(1)	2	−1	0	−1	1	2
−1	−2	1	0	2	0	−2

Second tableau—phase I

0	1	−1	1	0	1	3
1	2	−1	0	−1	1	2
0	0	0	0	1	1	0

Final tableau—phase I

Now we go back to the equivalent reduced problem

x_2	x_3	x_4	x_5	**b**
0	1	-1	1	3
1	2	-1	0	2
2	3	-1	1	-14

Initial tableau—phase II

Transforming the last row appropriately we proceed with:

0	1	-1	1	3
1	(2)	-1	0	2
0	-2	2	0	-21

First tableau—phase II

$-1/2$	0	$-1/2$	1	2
$1/2$	1	$-1/2$	0	1
1	0	1	0	-19

Final tableau—phase II

The solution $x_3 = 1$, $x_5 = 2$ can be inserted in the expression for x_1 giving

$$x_1 = -7 + 2\cdot 1 + 2\cdot 2 = -1,$$

thus the final solution is

$$x_1 = -1, \qquad x_2 = 0, \qquad x_3 = 1, \qquad x_4 = 0, \qquad x_5 = 2.$$

*3.6 VARIABLES WITH UPPER BOUNDS

Many linear programs arising from practical situations involve variables that are subject to both lower and upper bounds. Thus the level of a production activity may be limited by available facilities, the amount of material transported between two points may be subject to a capacity constraint, or the variable voltages in a large electric power network might be required to lie within prescribed bounds. Hence, a typical variable x_i in a linear programming problem will be subject to bounds of the form $g_i \leqslant x_i \leqslant h_i$ for some values of g_i and h_i.

If the variable x_i is subject to no finite bounds, then that variable is free, and as explained in Section 2.1, it can be eliminated from consideration. If the variable is subject to a single bound, that bound can, by possibly changing sign and translating by a constant, always be assumed to take the standard form $x_i \geqslant 0$. If x_i has finite upper and lower bounds, they can similarly be assumed to take the form $0 \leqslant x_i \leqslant h_i$. In this section, therefore, we consider the linear program with upper bounds in the standard form

$$\begin{aligned} \text{minimize} \quad & \mathbf{cx} \\ \text{subject to} \quad & \mathbf{Ax} = \mathbf{b}, \\ & \mathbf{0} \leqslant \mathbf{x} \leqslant \mathbf{h}. \end{aligned} \tag{29}$$

For simplicity of notation we assume that each x_i is subject to a finite upper bound. In practice, if some variables in the program are not subject to upper bounds, either an extremely large bound can be artificially introduced, or, more simply, the technique we describe can be trivially modified.

The reader will surely notice that the problem with upper bounds (29) can be easily converted to the usual standard form by the introduction of slack variables y_i for each of the upper bound constraints. This conversion yields

$$\begin{aligned} \text{minimize} \quad & \mathbf{cx} \\ \text{subject to} \quad & \mathbf{Ax} = \mathbf{b} \\ & \mathbf{x} + \mathbf{y} = \mathbf{h} \\ & \mathbf{x} \geqslant \mathbf{0}, \qquad \mathbf{y} \geqslant \mathbf{0}. \end{aligned} \tag{30}$$

Although the usual simplex procedure can be applied to this form of the problem, there is a high price to be paid in terms of computing and storage requirements. If the matrix $\mathbf{A}$ were $m \times n$, the matrix associated with the problem in the form (30) would be $(m + n) \times 2n$, which clearly shows that the addition of upper bounds greatly increases the dimensionality of the standard form. In this section we describe an alternative approach that, by slightly generalizing the simplex method, allows computations and storage, for problems with upper bounds, to be treated without explicitly increasing the dimension of the problem.

To describe the method we introduce a single new definition.

Definition. An *extended basic feasible solution* corresponding to problem (29) is a feasible solution for which $n - m$ variables are equal to either their lower (zero) or their upper bound; and the remaining m (basic) variables correspond to linearly independent columns of $\mathbf{A}$.

In our development, we assume that every extended basic feasible solution is nondegenerate, which means that the m basic variables take values not equal to either of their bounds.

The idea underlying the method is quite simple. Suppose we start with an extended basic feasible solution to the problem. We examine the non-basic variables (the variables that are on one of their bounds) as possible candidates for change so as to obtain an improved solution. A variable at its lower bound can only be increased, and an increase will be beneficial if the corresponding relative cost coefficient is negative. A variable at its upper bound can only be decreased, and a decrease will be beneficial if the corresponding relative cost coefficient is positive. Suppose a certain nonbasic variable x_i is selected for change in this way. As its value is changed continuously from one bound toward the other, the associated cost will continuously decrease. At the same time the values of the m basic variables will change in such a way that the solution continues to satisfy the linear equalities. The value of the nonbasic variable can be continuously changed until either (i) the value of a basic variable becomes equal to one of its bounds or (ii) the nonbasic variable being changed reaches its opposite bound. If (i) occurs first, then that corresponding basic variable is declared nonbasic and the nonbasic variable that was changed is declared basic. If (ii) occurs first, then the basis is not changed. The simultaneous occurrence of (i) and (ii) results in a degenerate solution and, as before, we ignore this possibility. This procedure is carried out step by step until no further change leading to an improvement is possible.

The justification of the procedure described above rests on the following theorem.

Upper Bound Optimality Theorem. *An extended basic feasible solution is optimal for* (29) *if for the nonbasic variables* x_j

$$r_j \geqslant 0 \quad \textit{if} \quad x_j = 0$$

$$r_j \leqslant 0 \quad \textit{if} \quad x_j = h_j.$$

Proof. This follows directly from (22). ▌

The technique discussed above leads to an improved solution if the conditions of the Upper Bound Optimality Theorem do not hold. The computations themselves proceed analogously to that of the usual simplex method, except that the choice of pivot must be modified slightly.

To derive appropriate simplex operations, it is convenient to introduce the notation $x_i^+ = x_i$, $x_i^- = h_i - x_i$. As the method progresses we change back and forth from x_i^+ to x_i^-, depending on whether the variable x_i has most recently been at its lower or upper bound, respectively. At any stage, the system of equalities and cost function are represented in the *extended canonical form* by the tableau shown below.

$$
\begin{array}{ccccccccc}
1 & 0 & \cdots & 0 & y_{1,m+1} & y_{1,m+2} & \cdots & y_{1n} & y_{10} \\
0 & 1 & & 0 & y_{2,m+1} & y_{2,m+2} & & y_{2n} & y_{20} \\
0 & 0 & & 0 & \cdot & \cdot & & \cdot & \cdot \\
\cdot & \cdot & & \cdot & \cdot & \cdot & & \cdot & \cdot \\
\cdot & \cdot & & \cdot & \cdot & \cdot & & \cdot & \cdot \\
0 & 0 & & 1 & y_{m,m+1} & y_{m,m+2} & & y_{mn} & y_{m0} \\
0 & 0 & & 0 & r_{m+1} & r_{m+2} & & r_n & -z_0 \\
e_1 & e_2 & \cdots & e_m & e_{m+1} & e_{m+2} & & e_n &
\end{array}
$$

Fig. 3.2 Extended tableau

Below each column we place an indicator $e_i = \pm$ to indicate if in the current solution that column corresponds to $x_i^+ = x_i$ $(e_i = +)$ or to $x_i^- = h_i - x_i$ $(e_i = -)$. The tableau is interpreted as representing the equations

$$\sum_{j=1}^{n} y_{ij} x_j^{e_j} = y_{i0}, \qquad i = 1, 2, \ldots, m.$$

With this notation the procedure for transforming from one extended basic feasible solution to another, following the strategy outlined above the Upper Bound Optimality Theorem and analogous to the simplex method, can be easily implemented. The procedure follows.

Step 1. Determine a nonbasic variable x_j for which $r_j < 0$. If no such variable exists, stop; the current solution is optimal.

Step 2. Evaluate the three numbers

a) h_j

b) $\min\limits_{y_{ij}>0} y_{i0}/y_{ij}$

c) $\min\limits_{y_{ij}<0} (y_{i0} - h_i)/y_{ij}$,

where h_i is the upper bound associated with the ith basic variable.

Step 3. According as to which number in step 2 is smallest, update the extended tableau as follows:

a) The variable x_j goes to its opposite bound. Substract h_j times column j from column 0. Multiply column j by minus unity (including a change in sign of e_j). The basis does not change and no pivot is required.

b) Suppose i is the minimizing index in (b) of step 2. Then the ith basic variable returns to its old bound. Pivot on the ijth element.

c) Suppose i is the minimizing index in (c) of step 2. Then the ith basic variable goes to its opposite bound. Subtract h_i from y_{i0}, change the signs of y_{ii} and e_i, and pivot on the ijth element.

Return to step 1.

Example.

$$\begin{aligned} \text{minimize} \quad & 2x_1 + x_2 + 3x_3 - 2x_4 + 10x_5 \\ \text{subject to} \quad & x_1 \quad\; + x_3 - \;\, x_4 + 2x_5 = 5 \\ & \quad\; x_2 + 2x_3 + 2x_4 + \;\, x_5 = 9 \end{aligned}$$

$$0 \leqslant x_1 \leqslant 7, \quad 0 \leqslant x_2 \leqslant 10, \quad 0 \leqslant x_3 \leqslant 1, \quad 0 \leqslant x_4 \leqslant 5, \quad 0 \leqslant x_5 \leqslant 3.$$

The original tableau for this problem, after calculating the relative cost coefficients is

x_1	x_2	x_3	x_4	x_5	**b**
1	0	1	−1	2	5
0	1	2	2	1	9
0	0	−1	−2	5	−19
+	+	+	+	+	+

First tableau

which represents a basic feasible solution. We decide that column 4 should enter. Making the required calculations we find the numbers

(a) 5
(b) 9/2
(c) 2

and hence case (c) applies. Before pivoting we modify the tableau to

−1	0	1	(−1)	2	−2
0	1	2	2	1	9
0	0	−1	−2	5	−19
−	+	+	+	+	+

Modified first tableau

Pivoting as dictated by case (c) we obtain

1	0	−1	1	−2	2
−2	1	4	0	5	5
2	0	−3	0	1	−15
−	+	+	+	+	

Second tableau

Next we decide to enter column 3. This time we find

(a) 1
(b) 5/4
(c) 3

so case (a) applies. Thus no pivot is necessary, but we obtain

1	0	1	1	−2	3
−2	1	−4	0	5	1
2	0	3	0	1	−12
−	+	−	+	+	

Final tableau

Since all reduced cost coefficients are nonnegative this tableau represents an optimal solution. The solution is

$$x_1 = 7, \qquad x_2 = 1, \qquad x_3 = 1, \qquad x_4 = 3, \qquad x_5 = 0.$$

3.7 MATRIX FORM OF THE SIMPLEX METHOD

Although the elementary pivot transformations associated with the simplex method are in many respects most easily discernible in the tableau format, with attention focused on the individual elements, there is much insight to be gained by studying a matrix interpretation of the procedure. The vector–matrix relationships that exist between the various rows and columns of the tableau leads, however, not only to increased understanding but also, in a rather direct way, to the *revised simplex procedure* which in many cases can result in considerable computational advantage. The matrix formulation is also a natural setting for the discussion of dual linear programs and other topics related to linear programming.

A preliminary observation in the development is that the tableau at any point in the simplex procedure can be determined solely by a knowledge of which variables are basic. As before we denote by $\mathbf{B}$ the submatrix of the original $\mathbf{A}$ matrix consisting of the m columns of $\mathbf{A}$ corresponding to the basic variables. These columns are linearly independent and hence the columns of $\mathbf{B}$ form a basis for E^m. We refer to $\mathbf{B}$ as the basis matrix.

As usual, let us assume that $\mathbf{B}$ consists of the first m columns of $\mathbf{A}$. Then by partitioning $\mathbf{A}$, $\mathbf{x}$, and $\mathbf{c}$ as

$$\mathbf{A} = [\mathbf{B}, \mathbf{D}]$$
$$\mathbf{x} = (\mathbf{x_B}, \mathbf{x_D}), \qquad \mathbf{c} = [\mathbf{c_B}, \mathbf{c_D}]$$

the standard linear programming problem becomes

$$\begin{aligned} \text{minimize} \quad & \mathbf{c_B x_B} + \mathbf{c_D x_D} \\ \text{subject to} \quad & \mathbf{B x_B} + \mathbf{D x_D} = \mathbf{b} \\ & \mathbf{x_B} \geqslant \mathbf{0}, \quad \mathbf{x_D} \geqslant \mathbf{0}. \end{aligned} \tag{31}$$

The basic solution, which we assume is also feasible, corresponding to the basis $\mathbf{B}$ is $\mathbf{x} = (\mathbf{x_B}, \mathbf{0})$ where $\mathbf{x_B} = \mathbf{B}^{-1}\mathbf{b}$. The basic solution results from setting $\mathbf{x_D} = \mathbf{0}$. However, for any value of $\mathbf{x_D}$ the necessary value of $\mathbf{x_B}$ can be computed from (31) as

$$\mathbf{x_B} = \mathbf{B}^{-1}\mathbf{b} - \mathbf{B}^{-1}\mathbf{D}\mathbf{x_D} \tag{32}$$

and this general expression when substituted in the cost function yields

$$\begin{aligned} z &= \mathbf{c_B}(\mathbf{B}^{-1}\mathbf{b} - \mathbf{B}^{-1}\mathbf{D}\mathbf{x_D}) + \mathbf{c_D x_D} \\ &= \mathbf{c_B B}^{-1}\mathbf{b} + (\mathbf{c_D} - \mathbf{c_B B}^{-1}\mathbf{D})\mathbf{x_D} \end{aligned} \tag{33}$$

which expresses the cost of any solution to (31) in terms of $\mathbf{x_D}$. Thus

$$\mathbf{r} = \mathbf{c_D} - \mathbf{c_B B}^{-1}\mathbf{D} \tag{34}$$

is the relative cost vector. It is the components of this vector that are used to determine which vector to bring into the basis.

Having derived the vector expression for the relative cost is it now possible to write the simplex tableau in matrix form. The initial tableau takes the form

$$\left[\begin{array}{c|c} \mathbf{A} & \mathbf{b} \\ \hline \mathbf{c} & 0 \end{array}\right] = \left[\begin{array}{c|c|c} \mathbf{B} & \mathbf{D} & \mathbf{b} \\ \hline \mathbf{c_B} & \mathbf{c_D} & 0 \end{array}\right] \tag{35}$$

which is not in general in canonical form and does not correspond to a point in the simplex procedure. If the matrix $\mathbf{B}$ is used as a basis, then the corresponding tableau becomes

$$\mathbf{T} = \left[\begin{array}{c|c|c} \mathbf{I} & \mathbf{B}^{-1}\mathbf{D} & \mathbf{B}^{-1}\mathbf{b} \\ \hline \mathbf{0} & \mathbf{c}_D - \mathbf{c}_B\mathbf{B}^{-1}\mathbf{D} & -\mathbf{c}_B\mathbf{B}^{-1}\mathbf{b} \end{array}\right] \tag{36}$$

which is the matrix form we desire.

3.8 THE REVISED SIMPLEX METHOD

Extensive experience with the simplex procedure applied to problems from various fields, and having various values of n and m, has indicated that the method can be expected to converge to an optimum solution in about m, or perhaps $3m/2$, pivot operations. Thus, particularly if m is much smaller than n, that is, if the matrix $\mathbf{A}$ has far fewer rows than columns, pivots will occur in only a small fraction of the columns during the course of optimization. Since the other columns are not explicitly used, it appears that the work expended in calculating the elements in these columns after each pivot, is, in some sense, wasted effort. The revised simplex method is a scheme for ordering the computations required of the simplex method so that unnecessary calculations are avoided.

The revised form of the simplex method is this: Given the inverse $\mathbf{B}^{-1}$ of a current basis, and the current solution $\mathbf{x}_B = \mathbf{y}_0 = \mathbf{B}^{-1}\mathbf{b}$,

Step 1. Calculate the current relative cost coefficients $\mathbf{r} = \mathbf{c}_D - \mathbf{c}_B\mathbf{B}^{-1}\mathbf{D}$. This can best be done by first calculating $\boldsymbol{\lambda} = \mathbf{c}_B\mathbf{B}^{-1}$ and then the relative cost vector $\mathbf{c}_D - \boldsymbol{\lambda}\mathbf{D}$. If $\mathbf{r} \geqslant \mathbf{0}$ stop; the current solution is optimal.

Step 2. Determine which vector $\mathbf{a}_j$ is to enter the basis by selecting the most negative cost coefficient; and calculate $\mathbf{y}_j = \mathbf{B}^{-1}\mathbf{a}_j$ which gives the vector $\mathbf{a}_j$ expressed in terms of the current basis.

Step 3. Calculate the ratios y_{i0}/y_{ij} to determine which vector is to leave the basis.

Step 4. Update $\mathbf{B}^{-1}$ and the current solution $\mathbf{B}^{-1}\mathbf{b}$. Return to step 1.

Updating of $\mathbf{B}^{-1}$ is accomplished by the usual pivot operations applied to an array consisting of $\mathbf{B}^{-1}$ and $\mathbf{y}_j$, where the pivot is the appropriate element in $\mathbf{y}_j$. Of course $\mathbf{B}^{-1}\mathbf{b}$ may be updated at the same time by adjoining it as another column.

To begin the procedure one requires, as always, an initial basic feasible solution and, in this case, the inverse of the initial basis. In most problems the initial basis (and hence also its inverse) is an identity matrix, resulting either from slack or surplus variables or from artificial variables.

The inverse of any initial basis can, however, be explicitly calculated in order to initiate the revised simplex procedure.

To illustrate the method and to indicate how the computations and storage can be handled, we consider an example.

Example 1. We solve again Example 1 of Section 3.4. The vectors are listed once for reference

$\mathbf{a}_1$	$\mathbf{a}_2$	$\mathbf{a}_3$	$\mathbf{a}_4$	$\mathbf{a}_5$	$\mathbf{a}_6$	$\mathbf{b}$
2	1	1	1	0	0	2
1	2	3	0	1	0	5
2	2	1	0	0	1	6

and the objective function is determined by

$$\mathbf{c} = [-3, -1, -3, 0, 0, 0].$$

We start with an initial basic feasible solution and corresponding $\mathbf{B}^{-1}$ as shown in the tableau below

Variable	$\mathbf{B}^{-1}$			$\mathbf{x_B}$
4	1	0	0	2
5	0	1	0	5
6	0	0	1	6 .

We compute

$$\boldsymbol{\lambda} = [0, 0, 0]\,\mathbf{B}^{-1} = [0, 0, 0]$$

and then

$$\mathbf{c_D} - \boldsymbol{\lambda}\mathbf{D} = [-3, -1, -3].$$

We decide to bring $\mathbf{a}_2$ into the basis (violating the rule of selecting the most negative relative cost in order to simplify the hand calculation). Its current representation is found by multiplying by $\mathbf{B}^{-1}$; thus we have

Variable	$\mathbf{B}^{-1}$			$\mathbf{x_B}$	$\mathbf{y}_2$
4	1	0	0	2	(1)
5	0	1	0	5	2
6	0	0	1	6	2

After computing the ratios in the usual manner, we select the pivot indicated. The updated tableau becomes

Variable	$\mathbf{B}^{-1}$			$\mathbf{x_B}$
2	1	0	0	2
5	−2	1	0	1
6	−2	0	1	2

then

$$\lambda = [-1, 0, 0]\,\mathbf{B}^{-1} = [-1, 0, 0]$$

$$c_1 - z_1 = -1, \qquad c_3 - z_3 = -2, \qquad c_4 - z_4 = 1.$$

We select $\mathbf{a}_3$ to enter. We have the tableau

Variable	$\mathbf{B}^{-1}$			$\mathbf{x_B}$	$\mathbf{y}_3$
2	1	0	0	2	1
5	−2	1	0	1	(1)
6	−2	0	1	2	−1

Using the pivot indicated we obtain

Variable	$\mathbf{B}^{-1}$			$\mathbf{x_B}$
2	3	−1	0	1
3	−2	1	0	1
6	−4	1	1	3

Now

$$\lambda = [-1, -3, 0]\,\mathbf{B}^{-1} = [3, -2, 0]$$

and

$$c_1 - z_1 = -7, \qquad c_4 - z_4 = -3, \qquad c_5 - z_5 = 2.$$

We select $\mathbf{a}_1$ to enter the basis. We have the tableau

Variable	$\mathbf{B}^{-1}$			$\mathbf{x_B}$	$\mathbf{y}_1$
2	3	−1	0	1	(5)
3	−2	1	0	1	−3
6	−4	1	1	3	−5

Using the pivot indicated we obtain

Variable	$\mathbf{B}^{-1}$			$\mathbf{x_B}$
1	3/5	−1/5	0	1/5
3	−1/5	2/5	0	8/5
6	−1	0	1	4

Now

$$\boldsymbol{\lambda} = [-3, -3, 0]\,\mathbf{B}^{-1} = [-6/5, -3/5, 0]$$

and

$$c_2 - z_2 = 7/5, \qquad c_4 - z_4 = 6/5, \qquad c_5 - z_5 = 3/5.$$

Since the $c_i - z_i$'s are all nonnegative, we conclude that the solution $\mathbf{x} = (1/5, 0, 8/5, 0, 0, 4)$ is optimal.

The Product Form of the Inverse

A variant of the revised simplex procedure is based on a product representation for the inverse of the basis matrix. This variant often requires fewer computations than other methods, but its primary advantage is its minimal high speed core storage requirements. For this reason it is used extensively for large linear programs involving, say, over a hundred constraints.

Suppose a tableau $\mathbf{T}$ is transformed by a pivot operation where the pivot column is

$$\mathbf{y}_j = (y_{1j}, y_{2j}, \ldots, y_{mj})$$

with pivot element y_{pj}. Then the result of the pivot operation is the matrix

$$\mathbf{ET} \tag{37}$$

where $\mathbf{E}$ is the *elementary matrix*

$$\mathbf{E} = \begin{bmatrix} 1 & 0 \ldots 0 & v_1 & 0 \ldots 0 \\ 0 & 1 \quad\; 0 & v_2 & 0 \quad\; 0 \\ \cdot & 0 \quad\;\; \cdot & \cdot & \cdot \quad\;\; \cdot \\ \cdot & \cdot \quad\;\; \cdot & \cdot & \cdot \quad\;\; \cdot \\ \cdot & \cdot \quad\;\; 1 & \cdot & \cdot \quad\;\; \cdot \\ \cdot & \cdot \quad\;\; 0 & v_p & 0 \quad\;\; \cdot \\ \cdot & \cdot \quad\;\; \cdot & \cdot & 1 \quad\;\; \cdot \\ \cdot & \cdot \quad\;\; \cdot & \cdot & 0 \quad\;\; \cdot \\ \cdot & \cdot \quad\;\; \cdot & \cdot & \cdot \quad\;\; \cdot \\ 0 & 0 \ldots 0 & v_m & 0 \ldots 1 \end{bmatrix} \tag{38}$$

and where
$$v_i = -y_{ij}/y_{pj}, \qquad i \neq p$$
$$v_p = 1/y_{pj} \tag{39}$$

The elementary matrix **E** is determined entirely by the elements of the pivot column. Indeed a little reflection on the nature of pivoting will quickly lead the reader first to the observation that any column can be transformed upon knowing the components of that column and of the pivot column; and next to the verification of (37), (38) and (39).

If we initiate a simplex procedure with a basis that is the identity matrix, then after the kth pivot the inverse of the basis matrix is

$$\mathbf{B}^{-1} = \mathbf{E}_k\mathbf{E}_{k-1} \ldots \mathbf{E}_2\mathbf{E}_1 \tag{40}$$

where $\mathbf{E}_i$ is the elementary matrix corresponding to the ith pivot operation. We now apply this representation of the inverse to obtain an alternate revised simplex method.

Suppose after the kth pivot we have stored $\mathbf{E}_1, \mathbf{E}_2, \ldots, \mathbf{E}_k$.

Step 1. Calculate the current basic solution by the recursive formula

$$\mathbf{x}_B = \Big(\mathbf{E}_k(\mathbf{E}_{k-1} \ldots (\mathbf{E}_1\mathbf{b}))\Big).$$

Step 2. Calculate the current relative cost coefficients

$$\mathbf{r} = \mathbf{c}_\mathbf{D} - \mathbf{c}_\mathbf{B}\mathbf{B}^{-1}\mathbf{D}$$

by using the form

$$\mathbf{r} = \mathbf{c}_\mathbf{D} - \boldsymbol{\lambda}\mathbf{D} \qquad \text{where} \qquad \boldsymbol{\lambda} = \mathbf{c}_\mathbf{B}\mathbf{B}^{-1}.$$

The row vector $\boldsymbol{\lambda}$ is first calculated by the recursive formula

$$\boldsymbol{\lambda} = \Big(((\mathbf{c}_\mathbf{B}\mathbf{E}_k)\mathbf{E}_{k-1}) \ldots \mathbf{E}_1\Big).$$

If $\mathbf{r} \geqslant \mathbf{0}$, stop; the current solution is optimal.

Step 3. Select the most negative relative cost coefficient and calculate the corresponding vector

$$\mathbf{y}_j = \Big(\mathbf{E}_k(\mathbf{E}_{k-1} \ldots (\mathbf{E}_1\mathbf{a}_j))\Big).$$

Step 4. Calculate the ratios y_{i0}/y_{ij} to determine which vector is to leave the basis—thus determining which component is the next pivot, and hence defining $\mathbf{E}_{k+1}$. Return to step 1.

It would not be sensible to store each elementary matrix in full form, since most of its elements are zero. It is necessary only to store the $m + 1$ numbers $(p, v_1, v_2, \ldots, v_m)$ in order to reconstitute **E**. In fact there is no need to reconstitute **E** explicitly, since all that is required are the formulae for multiplying a vector by **E**. In steps 1 and 3, this is multiplication of a vector by **E** on its left, and this is equivalent to a standard pivot transformation. In step 2 a vector is multiplied by **E** on the right, and this operation leaves all components of the vector, except one, unchanged.

When solving large problems, almost all data can be stored in slow access memory, such as magnetic tape, drum or disk, when using this method. Columns of **A** can be brought into the high speed memory individually, and they can be multiplied by $\mathbf{B}^{-1}$ by successively bringing into the high speed memory the $m + 1$ dimensional vectors defining the elementary matrices.

Reinversion

As the simplex method progresses from one basis to another, it often happens, particularly in large problems, that the inevitable small round-off errors introduced at each stage may accumulate to the point where the current stored value of $\mathbf{B}^{-1}$ is highly inaccurate. If such a condition occurs (it can be detected by evaluation $\mathbf{B}^{-1}\mathbf{B}$), it is advisable to directly invert the current basis **B** as accurately as possible, and proceed from that point with a fresh start. This operation, referred to as *reinversion*, is an essential component of any major library program designed for general use. Any of several methods can be employed to obtain the new inverse, but commonly it is done by pivoting in the matrix **B**; the order of pivot selection being determined solely on the grounds of accuracy.

When using the product form of the inverse, reinversion implies that the current, perhaps long, string of elementary matrices is replaced by a string of m such matrices. Thus in this case, reinversion serves to control storage requirements as well as accuracy.

*3.9 THE SIMPLEX METHOD AND LU DECOMPOSITION

We may go one step further in the matrix interpretation of the simplex method and note that execution of a single simplex cycle is not explicitly dependent on having $\mathbf{B}^{-1}$ but rather on the ability to solve linear systems with **B** as the coefficient matrix. Thus, the simplex method stated at the beginning of Section 3.8 can be restated as: Given the current basis **B**,

Step 1. Calculate the current solution $\mathbf{x_B} = \mathbf{y}_0$ satisfying $\mathbf{B}\mathbf{y}_0 = \mathbf{b}$.

Step 2. Solve $\boldsymbol{\lambda}\mathbf{B} = \mathbf{c_B}$, and set $\mathbf{r} = \mathbf{c_D} - \boldsymbol{\lambda}\mathbf{D}$. If $\mathbf{r} \geqslant \mathbf{0}$, stop; the current solution is optimal.

Step 3. Determine which vector $\mathbf{a}_j$ is to enter the basis by selecting the most negative relative cost coefficient, and solve $\mathbf{B}\mathbf{y}_j = \mathbf{a}_j$.

Step 4. Calculate the ratios y_{i0}/y_{ij} and select the smallest nonnegative one to determine which vector is to leave the basis.

Step 5. Update **B**. Return to step 1.

In this form it is apparent that there is no explicit need for having $\mathbf{B}^{-1}$, but rather it is only necessary to solve three systems of equations, two involving the matrix $\mathbf{B}$ and one (the one for λ) involving $\mathbf{B}'$. In previous sections these three equations were solved, as the method progressed, by the pivoting operations. From the viewpoints of efficiency and numerical stability, however, this pivoting procedure is not as effective as the method of Gaussian elimination for general systems of linear equations (see Appendix C), and it, therefore, seems appropriate to investigate the possibility of adapting the numerically superior method of Gaussian elimination to the simplex method. The result is a version of the revised simplex method that possesses better numerical stability than other methods, and which for large-scale problems can offer tremendous storage advantages.

We concentrate on the problem of solving the linear systems

$$\mathbf{B}\mathbf{y}_0 = \mathbf{b}, \qquad \boldsymbol{\lambda}\mathbf{B} = \mathbf{c}_{\mathbf{B}}, \qquad \mathbf{B}\mathbf{y}_j = \mathbf{a}_j \tag{41}$$

that are required by a single step of the simplex method. Suppose $\mathbf{B}$ has been decomposed into the form $\mathbf{B} = \mathbf{L}\mathbf{U}$ where $\mathbf{L}$ is a lower triangular matrix and $\mathbf{U}$ is an upper triangular matrix.† Then each of the linear systems (41) can be solved by solving two triangular systems. Since solving in this fashion is simple, knowledge of $\mathbf{L}$ and $\mathbf{U}$ is almost as good as knowledge of $\mathbf{B}^{-1}$.

Next, we show how the **LU** decomposition of $\mathbf{B}$ can be updated when a single basis vector is changed. At the beginning of the simplex cycle suppose $\mathbf{B}$ has the form

$$\mathbf{B} = [\mathbf{a}_1, \mathbf{a}_2, \ldots, \mathbf{a}_m].$$

At the end of the cycle we have the new basis

$$\overline{\mathbf{B}} = [\mathbf{a}_1, \mathbf{a}_2, \ldots, \mathbf{a}_{k-1}, \mathbf{a}_{k+1}, \ldots, \mathbf{a}_m, \mathbf{a}_j]$$

where it should be noted that when $\mathbf{a}_k$ is dropped all subsequent vectors are shifted to the left, and the new vector $\mathbf{a}_j$ is appended on the right. This procedure leads to a fairly simple updating technique.

We have

$$\begin{aligned}\mathbf{L}^{-1}\overline{\mathbf{B}} &= [\mathbf{L}^{-1}\mathbf{a}_1, \mathbf{L}^{-1}\mathbf{a}_2, \ldots, \mathbf{L}^{-1}\mathbf{a}_{k-1}, \mathbf{L}^{-1}\mathbf{a}_{k+1}, \ldots, \mathbf{L}^{-1}\mathbf{a}_m, \mathbf{L}^{-1}\mathbf{a}_j] \\ &= [\mathbf{u}_1, \mathbf{u}_2, \ldots, \mathbf{u}_{k-1}, \mathbf{u}_{k+1}, \ldots, \mathbf{u}_m, \mathbf{L}^{-1}\mathbf{a}_j] = \overline{\mathbf{H}}\end{aligned}$$

---

† For simplicity, we are assuming that no row interchanges are required to produce the **LU** decomposition. This assumption can be relaxed with an attendant complexification of notation.

where the $\mathbf{u}_i$'s are the columns of $\mathbf{U}$. The matrix $\overline{\mathbf{H}}$ takes the form

$\overline{\mathbf{H}} =$

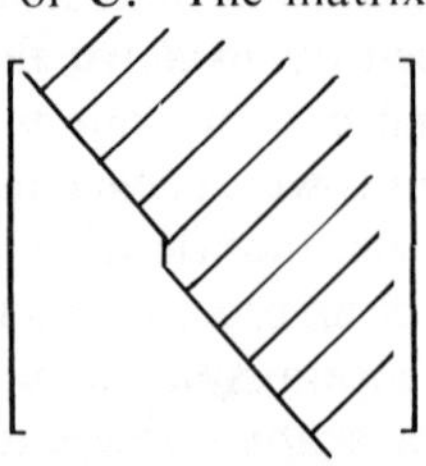

with zeros below the main diagonal in the first $k - 1$ columns, and zeros below the element immediately under the diagonal in all other columns. The matrix $\overline{\mathbf{H}}$ itself can be constructed without additional computation, since the $\mathbf{u}_i$'s are known and $\mathbf{L}^{-1}\mathbf{a}_j$ is a by-product in the computation of $\mathbf{y}_j$.

$\overline{\mathbf{H}}$ can be reduced to upper triangular form by using Gaussian elimination to zero out the subdiagonal elements. Thus the upper triangular matrix $\overline{\mathbf{U}}$ can be obtained from $\overline{\mathbf{H}}$ by application of a series of transformations, each having the form

$$\mathbf{M}_i = \begin{bmatrix} 1 & & & & & & \\ & 1 & & & & & \\ & & \cdot & & & & \\ & & & \cdot & & & \\ & & & & 1 & & \\ & & & & m_i & 1 & \\ & & & & & & \cdot \\ & & & & & & & \cdot \\ & & & & & & & & 1 \end{bmatrix} \tag{42}$$

for $i = k, k + 1, \ldots, m - 1$. The matrix $\overline{\mathbf{U}}$ becomes

$$\overline{\mathbf{U}} = \mathbf{M}_{m-1}\mathbf{M}_{m-2} \ldots \mathbf{M}_k\overline{\mathbf{H}}. \tag{43}$$

We then have

$$\overline{\mathbf{B}} = \mathbf{L}\overline{\mathbf{H}} = \mathbf{L}\mathbf{M}_k^{-1}\mathbf{M}_{k+1}^{-1} \ldots \mathbf{M}_{m-1}^{-1}\overline{\mathbf{U}} \tag{44}$$

and thus evaluating

$$\overline{\mathbf{L}} = \mathbf{L}\mathbf{M}_k^{-1} \ldots \mathbf{M}_{m-1}^{-1} \tag{45}$$

we obtain the decomposition

$$\overline{\mathbf{B}} = \overline{\mathbf{L}}\overline{\mathbf{U}}. \tag{46}$$

Since $\mathbf{M}_i^{-1}$ is simply $\mathbf{M}_i$ with the sign of the off-diagonal term reversed, evaluation of $\overline{\mathbf{L}}$ is straightforward.

A major advantage of this **LU** decomposition method, in addition to its numerical accuracy, is its effectiveness in handling sparse matrices (that is, matrices with relatively few nonzero elements). This is because sparcity is preserved by the **LU** decomposition while it is not preserved in the inverse. Thus, the **LU** decomposition provides a more economical representation of the essential problem structure than storage of the inverse.

There are numerous variations of this basic idea. The elementary transformations (42) can be carried (much as in the product form of the inverse) rather than explicitly evaluating **L**, the **LU** decomposition can be periodically re-evaluated (much as in re-inversion), and row and column interchanges can be handled in such a way as to maximize stability or minimize the density of the decomposition. This area is a fruitful one for further investigation.

3.10 SUMMARY

The simplex method is founded on the fact that the optimal value of a linear program, if finite, is always attained at a basic feasible solution. Using this foundation there are two ways in which to visualize the simplex process. The first is to view the process as one of continuous change. One starts with a basic feasible solution and imagines that some nonbasic variable is increased slowly from zero. As the value of this variable is increased, the values of the current basic variables are continuously adjusted so that the overall vector continues to satisfy the system of linear equality constraints. The change in the objective function due to a unit change in this nonbasic variable, taking into account the corresponding required changes in the values of the basic variables, is the relative cost coefficient associated with the nonbasic variable. If this coefficient is negative, then the objective value will be continuously improved as the value of this nonbasic variable is increased, and therefore one increases the variable as far as possible; to the point where further increase would violate feasibility. At this point the value of one of the basic variables is zero, and that variable is declared nonbasic; while the nonbasic variable that was increased is declared basic.

The other viewpoint is more discrete in nature. Realizing that only basic feasible solutions need be considered, various bases are selected and the corresponding basic solutions are calculated by solving the associated set of linear equations. The logic for the systematic selection of new bases again involves the relative cost coefficients and, of course, is derived largely from the first, continuous, viewpoint.

3.11 EXERCISES

1. Show that if the column vectors $\mathbf{a}_1, \mathbf{a}_2, \ldots, \mathbf{a}_m$ are linearly independent, then the system of equations (1) can be put in canonical form (4) by forming linear combinations of rows.

2. Show that if the vectors $\mathbf{a}_1, \mathbf{a}_2, \ldots, \mathbf{a}_m$ are a basis in E^m, the vectors $\mathbf{a}_1, \mathbf{a}_2, \ldots, \mathbf{a}_{p-1}, \mathbf{a}_q, \mathbf{a}_{p+1}, \ldots, \mathbf{a}_m$ also are a basis if and only if $y_{pq} \neq 0$, where y_{pq} is defined by the tableau (7).

3. Suppose $\mathbf{B}$ is an $m \times m$ square nonsingular matrix, and let the tableau $\mathbf{T}$ be constructed, $\mathbf{T} = [\mathbf{I}, \mathbf{B}]$ where $\mathbf{I}$ is the $m \times m$ identity matrix. Suppose that pivot operations are performed on this tableau so that it takes the form $[\mathbf{C}, \mathbf{I}]$. Show that $\mathbf{C} = \mathbf{B}^{-1}$.

4. Using pivoting solve the simultaneous equations

$$\begin{aligned} 3x_1 + 2x_2 &= 5 \\ 5x_1 + x_2 &= 9 \end{aligned}$$

5. Using pivoting solve the simultaneous equations

$$\begin{aligned} x_1 + 2x_2 + x_3 &= 7 \\ 2x_1 - x_2 + 2x_3 &= 6 \\ x_1 + x_2 + 3x_3 &= 12. \end{aligned}$$

6. Solve the equations in Exercise 5 by Gaussian elimination as described in Appendix C.

7. Discuss the various possibilities that arise when trying to find a basic feasible solution adjacent to a given degenerate basic feasible solution.

8. Show that a degenerate basic feasible solution may be optimal without satisfying $r_j \geqslant 0$ for all j.

9. Show that the vector $\mathbf{r}$ of relative cost coefficients can be obtained by adjoining a row $\mathbf{c}$ to the bottom of the tableau of equations in canonical form and zeroing out the components of this row that correspond to basic variables by adding appropriate multiples of the other rows.

10. *Degeneracy.* If a basic feasible solution is degenerate, then there is the possibility that if $\mathbf{a}_k$ is to be brought into the basis, and $\mathbf{a}_i$ is to go out, we have $y_{i0}/y_{ik} = 0$. This means that the new basic feasible solution will also be degenerate, and the new value of the objective function will be identical to the old value. It is then theoretically possible that a sequence of degenerate basic feasible solutions will be generated that endlessly cycles without making progress. It is the purpose of this exercise and the next two to develop a technique that can be applied to the simplex method to avoid this *cycling.*

    Corresponding to the linear system $\mathbf{Ax} = \mathbf{b}$ where $\mathbf{A} = [\mathbf{a}_1, \mathbf{a}_2, \ldots, \mathbf{a}_n]$ define the perturbed system $\mathbf{Ax} = \mathbf{b}(\varepsilon)$ where $\mathbf{b}(\varepsilon) = \mathbf{b} + \varepsilon\mathbf{a}_1 + \varepsilon^2\mathbf{a}_2 + \cdots + \varepsilon^n\mathbf{a}_n$, $\varepsilon > 0$. Show that if there is a basic feasible solution (possibly degenerate) to the unperturbed system with basis $\mathbf{B} = [\mathbf{a}_1, \mathbf{a}, \ldots, \mathbf{a}_m]$, then corresponding to the same basis, there is a nondegenerate basic feasible solution to the perturbed system for some range of $\varepsilon > 0$.

11. Show that corresponding to any basic feasible solution to the perturbed system of Exercise 10, which is nondegenerate for some range of $\varepsilon > 0$, and to a vector $\mathbf{a}_k$ not in the basis, there is a unique vector $\mathbf{a}_i$ in the basis which when replaced by $\mathbf{a}_k$ leads to a basic feasible solution; and that solution is nondegenerate for a range of $\varepsilon > 0$.

12. Show that the tableau associated with a basic feasible solution of the perturbed system of Exercise 10, and which is nondegenerate for a range of $\varepsilon > 0$, is identical with that of the unperturbed system except in the column under $\mathbf{b}(\varepsilon)$. Show how the proper pivot in a given column to preserve feasibility of the perturbed system can be determined from the tableau of the unperturbed system. Conclude that the simplex method will avoid cycling if whenever there is a choice in the pivot element of a column k, arising from a tie in the minimum of y_{i0}/y_{ik} among the elements $i \in I_0$, the tie is resolved by finding the minimum of y_{i1}/y_{ik}, $i \in I_0$. If there still remain ties among elements $i \in I$, the process is repeated with y_{i2}/y_{ik}, etc., until there is a unique element.

13. Using the simplex procedure solve

$$\begin{array}{ll} \text{maximize} & 2x_1 + 4x_2 + x_3 + x_4 \\ \text{subject to} & x_1 + 3x_2 \qquad\qquad + x_4 \leqslant 4 \\ & 2x_1 + x_2 \qquad\qquad\qquad \leqslant 3 \\ & \qquad x_2 + 4x_3 + x_4 \leqslant 3 \\ & x_i \geqslant 0, \quad i = 1, 2, 3, 4. \end{array}$$

14. Using the two-phase simplex procedure solve

a)
$$\begin{array}{ll} \text{minimize} & -3x_1 + x_2 + 3x_3 - x_4 \\ \text{subject to} & x_1 + 2x_2 - x_3 + x_4 = 0 \\ & 2x_1 - 2x_2 + 3x_3 + 3x_4 = 9 \\ & x_1 - x_2 + 2x_3 - x_4 = 6 \\ & x_i \geqslant 0, \quad i = 1, 2, 3, 4. \end{array}$$

b)
$$\begin{array}{ll} \text{minimize} & x_1 + 6x_2 - 7x_3 + x_4 + 5x_5 \\ \text{subject to} & 5x_1 - 4x_2 + 13x_3 - 2x_4 + x_5 = 20 \\ & x_1 - x_2 + 5x_3 - x_4 + x_5 = 8 \\ & x_i \geqslant 0, \quad i = 1, 2, 3, 4, 5. \end{array}$$

15. Rather than select the variable corresponding to the most negative relative cost coefficient as the variable to enter the basis, it has been suggested that a better criterion would be to select that variable which, when pivoted in, will produce the greatest improvement in the objective function. Show that this criterion leads to selecting the variable x_k corresponding to the index k minimizing $\max\limits_{i, y_{ik} > 0} r_k y_{i0}/y_{ik}$.

16. Show that in the phase I procedure of a problem that has feasible solutions, if an artificial variable becomes nonbasic, it need never again be made basic. Thus, when an artificial variable becomes nonbasic its column can be eliminated from future tableaus.

17. For the linear program of Exercise 13

    a) How much can the elements of $\mathbf{b} = (4, 3, 3)$ be changed without changing the optimal basis?

    b) How much can the elements of $\mathbf{c} = [2, 3, 1, 1]$ be changed without changing the optimal basis?

    c) What happens to the optimal cost for small changes in $\mathbf{b}$?

    d) What happens to the optimal cost for small changes in $\mathbf{c}$?

18. In the ordinary simplex method one new vector is brought into the basis and one removed at every step. Consider the possibility of bringing two new vectors into the basis and removing two at each stage. Develop a complete procedure that operates in this fashion.

19. If $r_j > 0$ for every j corresponding to a variable x_j that is not basic, show that the corresponding basic feasible solution is the unique optimal solution.

20. Consider the system of linear inequalities $\mathbf{Ax} \geqslant \mathbf{b}$, $\mathbf{x} \geqslant \mathbf{0}$ with $\mathbf{b} \geqslant \mathbf{0}$. This system can be transformed to standard form by the introduction of m surplus variables so that it becomes $\mathbf{Ax} - \mathbf{y} = \mathbf{b}$, $\mathbf{x} \geqslant \mathbf{0}$, $\mathbf{y} \geqslant \mathbf{0}$. Let $b_k = \max_i b_i$ and consider the new system in standard form obtained by adding the kth row to the negative of every other row. Show that the new system requires the addition of only a single artificial variable to obtain an initial basic feasible solution.

    Use this technique to find a basic feasible solution to the system.

$$
\begin{aligned}
x_1 + 2x_2 + x_3 &\geqslant 4 \\
2x_1 + x_2 + x_3 &\geqslant 5 \\
2x_1 + 3x_2 + 2x_3 &\geqslant 6 \\
x_i \geqslant 0, \qquad i = 1, 2, 3.&
\end{aligned}
$$

21. What happens to the phase I procedure applied to $\mathbf{Ax} = \mathbf{b}$, $\mathbf{x} \geqslant \mathbf{0}$ if there is a linear dependency (redundancy) among the original equations?

22. Find a basic feasible solution to

$$
\begin{aligned}
x_1 + 2x_2 - x_3 + x_4 &= 3 \\
2x_1 + 4x_2 + x_3 + 2x_4 &= 12 \\
x_1 + 4x_2 + 2x_3 + x_4 &= 9 \\
x_i \geqslant 0, \qquad i = 1, 2, 3, 4.&
\end{aligned}
$$

23. Explain how the procedure for handling problems with upper bounds can be modified to handle problems in which only some variables are subject to upper bounds.

24. Using the method of Section 3.6 solve

$$\begin{aligned} \text{minimize} \quad & 3x_1 + x_3 \\ \text{subject to} \quad & x_1 + 2x_2 + x_3 + x_4 = 10 \\ & x_1 - 2x_2 + 2x_3 = 6 \end{aligned}$$

$$0 \leqslant x_1 \leqslant 4, \quad 0 \leqslant x_2 \leqslant 4, \quad 0 \leqslant x_3 \leqslant 4, \quad 0 \leqslant x_4 \leqslant 12.$$

25. Modify the revised simplex method to accommodate variables with upper bounds.

26. Using the revised simplex method find a basic feasible solution to

$$\begin{aligned} x_1 + 2x_2 - x_3 + x_4 &= 3 \\ 2x_1 + 4x_2 + x_3 + 2x_4 &= 12 \\ x_1 + 4x_2 + 2x_3 + x_4 &= 9 \\ x_i \geqslant 0, \quad i &= 1, 2, 3, 4. \end{aligned}$$

27. Work out an extension of **LU** decomposition, as described in Appendix C, when row interchanges are introduced.

28. Work out the details of **LU** decomposition applied to the simplex method when row interchanges are required.

REFERENCES

3.1–3.8. All of this is now standard material contained in most courses in linear programming. See the references cited at the end of Chapter 2. For the original work in this area, see Dantzig [D2] for development of the simplex method; Orden [O2] for the artificial basis technique; Dantzig, Orden and Wolfe [D7], Orchard-Hays [O1], and Dantzig [D3] for the revised simplex method; and Charnes and Lemke [C3] and Dantzig [D4] for upper bounds. The synthetic carrot interpretation is due to Gale [G1].

3.9. The idea of using **LU** decomposition for the simplex method is due to Bartels [B1]. See also Bartels and Golub [B2] and Gill and Murray [G3]. For a nice simple introduction to Gaussian elimination see Forsythe and Moler [F9].

3.11. The degeneracy technique discussed in Exercises 10–12 is due to Charnes [C2].

Chapter 4 DUALITY

Associated with every linear programming problem, and intimately related to it, is a corresponding dual linear programming problem. Both problems are constructed from the same underlying cost and constraint coefficients but in such a way that if one of these problems is one of minimization the other is one of maximization, and the optimal values of the corresponding objective functions, if finite, are equal. The variables of the dual problem can be interpreted as prices associated with the constraints of the original (primal) problem, and through this association it is possible to give an economically meaningful characterization to the dual whenever there is such a characterization for the primal.

The variables of the dual problem are also intimately related to the calculation of the relative cost coefficients in the simplex method. Thus, a study of duality sharpens our understanding of the simplex procedure and motivates certain alternative solution methods. Indeed, the simultaneous consideration of a problem from both the primal and dual viewpoints often provides significant computational advantage as well as economic insight.

4.1 DUAL LINEAR PROGRAMS

In this section we define the dual program that is associated with a given linear program. Initially, we depart from our usual strategy of considering problems in standard form, since the duality relationship is most symmetric for problems expressed solely in terms of inequalities. Specifically then, we define duality through the pair of problems displayed below.

	Primal		*Dual*	
minimize	$\mathbf{cx}$	maximize	$\boldsymbol{\lambda}\mathbf{b}$	
subject to	$\mathbf{Ax} \geqslant \mathbf{b}$	subject to	$\boldsymbol{\lambda}\mathbf{A} \leqslant \mathbf{c}$	(1)
	$\mathbf{x} \geqslant \mathbf{0}$		$\boldsymbol{\lambda} \geqslant \mathbf{0}$	

If $\mathbf{A}$ is an $m \times n$ matrix, then $\mathbf{x}$ is an n-dimensional column vector, $\mathbf{b}$ is an m-dimensional column vector, $\mathbf{c}$ is an n-dimensional row vector, and $\boldsymbol{\lambda}$ is an m-dimensional row vector. The vector $\mathbf{x}$ is the variable of the primal problem, and $\boldsymbol{\lambda}$ is the variable of the dual problem.

The pair of problems (1) is called the *symmetric form* of duality and, as explained below, can be used to define the dual of any linear program. It is important to note that the role of primal and dual can be reversed. Thus, studying in detail the process by which the dual is obtained from the primal: interchange of cost and constraint vectors, transposition of coefficient matrix, reversal of constraint inequalities, and change of minimization to maximization; we see that this same process applied to the dual yields the primal. Put another way, if the dual is transformed, by multiplying the objective and the constraints by minus unity, so that it has the structure of the primal (but is still expressed in terms of $\boldsymbol{\lambda}$), its corresponding dual will be equivalent to the original primal.

The dual of any linear programming problem can be found by converting the problem to the form of the primal problem shown above. For example, given a linear program in standard form

$$\begin{array}{ll} \text{minimize} & \mathbf{cx} \\ \text{subject to} & \mathbf{Ax} = \mathbf{b} \\ & \mathbf{x} \geqslant \mathbf{0} \end{array}$$

we write it in the equivalent form

$$\begin{array}{ll} \text{minimize} & \mathbf{cx} \\ \text{subject to} & \mathbf{Ax} \geqslant \mathbf{b} \\ & -\mathbf{Ax} \geqslant -\mathbf{b} \\ & \mathbf{x} \geqslant \mathbf{0} \end{array}$$

which is in the form of the primal of (1) but with coefficient matrix $\left[\begin{array}{c} \mathbf{A} \\ \hline -\mathbf{A} \end{array}\right]$. Using a dual vector partitioned as $[\mathbf{u}, \mathbf{v}]$, the corresponding dual is

$$\begin{array}{ll} \text{maximize} & \mathbf{ub} - \mathbf{vb} \\ \text{subject to} & \mathbf{uA} - \mathbf{vA} \leqslant \mathbf{c}, \\ & \mathbf{u} \geqslant \mathbf{0}, \\ & \mathbf{v} \geqslant \mathbf{0}. \end{array}$$

Letting $\boldsymbol{\lambda} = \mathbf{u} - \mathbf{v}$ we may simplify the representation of the dual problem so that we obtain the pair of problems displayed below

$$\begin{array}{llll} & \textit{Primal} & & \textit{Dual} \\ \text{minimize} & \mathbf{cx} & \text{maximize} & \boldsymbol{\lambda}\mathbf{b} \\ \text{subject to} & \mathbf{Ax} = \mathbf{b}, & \text{subject to} & \boldsymbol{\lambda}\mathbf{A} \leqslant \mathbf{c}. \\ & \mathbf{x} \geqslant \mathbf{0} & & \end{array} \tag{2}$$

This is the *unsymmetric form* of the duality relation. In this form the dual vector $\boldsymbol{\lambda}$ (which is really a composite of $\mathbf{u}$ and $\mathbf{v}$) is not restricted to be nonnegative.

Similar transformations can be worked out for any linear programming problem to first get the primal in the form (1), calculate the dual, and then simplify the dual to account for special structure.

In general, if some of the linear inequalities in the primal problem of (1) are changed to equality, the corresponding components of $\boldsymbol{\lambda}$ in the dual problem become free variables. If some of the components of $\mathbf{x}$ in the primal problem are free variables, then the corresponding inequalities in $\boldsymbol{\lambda}\mathbf{A} \leqslant \mathbf{c}$ are changed to equality in the dual. We mention again that these are not arbitrary rules but are direct consequences of the original definition and the equivalence of various forms of linear programming problems.

Example 1. (Dual of the diet problem). The diet problem, Example 1, Section 2.2, was the problem faced by a dietician trying to select a combination of foods to meet certain nutritional requirements at minimum cost. This problem has the form

$$\begin{aligned} &\text{minimize} && \mathbf{cx} \\ &\text{subject to} && \mathbf{Ax} \geqslant \mathbf{b}, \\ & && \mathbf{x} \geqslant \mathbf{0} \end{aligned}$$

and hence can be regarded as the primal problem of the symmetric pair above. We describe an interpretation of the dual problem.

Imagine a pharmaceutical company that produces in pill form each of the nutrients considered important by the dietician. The pharmaceutical company tries to convince the dietician to buy pills, and thereby supply the nutrients directly rather than through purchase of various foods. The problem faced by the drug company is that of determining positive unit prices $\lambda_1, \lambda_2, \ldots, \lambda_m$ for the nutrients so as to maximize revenue while at the same time being competitive with real food. To be competitive with real food, the cost of a unit of food i made synthetically from pure nutrients bought from the druggist must be no greater than c_i, the market price of the food. Thus, denoting by $\mathbf{a}_i$ the ith food, the company must satisfy $\boldsymbol{\lambda}\mathbf{a}_i \leqslant c_i$ for each i. In matrix form this is equivalent to $\boldsymbol{\lambda}\mathbf{A} \leqslant \mathbf{c}$. Since b_j units of the jth nutrient will be purchased, the problem of the druggist is

$$\begin{aligned} &\text{maximize} && \boldsymbol{\lambda}\mathbf{b} \\ &\text{subject to} && \boldsymbol{\lambda}\mathbf{A} \leqslant \mathbf{c} \\ & && \boldsymbol{\lambda} \geqslant \mathbf{0} \end{aligned}$$

which is the dual problem.

Example 2. (Dual of the transportation problem). The transportation problem, Example 2, Section 2.2, is the problem, faced by a manufacturer, of selecting the pattern of product shipments between several fixed origins and destinations so as to minimize transportation cost while satisfying demand. To interpret the dual problem, we imagine an entrepreneur who, feeling that he can ship more efficiently, comes to the manufacturer with the offer to buy his product at the plant sites (origins) and sell it at the warehouses (destinations). The product price that is to be used in these transactions varies from point to point, and is determined by the entrepreneur in advance. He must choose these prices, of course, so that his offer will be attractive to the manufacturer.

The entrepreneur, then, must select prices $\lambda_1, \lambda_2, \ldots, \lambda_m$ for the m origins and $\mu_1, \mu_2, \ldots, \mu_n$ for the n destinations. To be competitive with usual transportation modes, his prices must satisfy $\mu_j - \lambda_i \leqslant c_{ij}$ for all i, j, since $\mu_j - \lambda_i$ represents the net amount the manufacturer must pay to sell a unit of product at origin i and buy it back again at destination j. Subject to this constraint, the entrepreneur will adjust his prices to maximize his revenue. Thus, his problem is to

$$\begin{aligned} &\text{maximize} \quad \sum_{i=1}^{m} \lambda_i a_i - \sum_{j=1}^{n} \mu_j b_j \\ &\text{subject to} \quad \mu_j - \lambda_i \leqslant c_{ij}, \qquad i = 1, 2, \ldots, m, \qquad j = 1, 2, \ldots, n. \end{aligned}$$

This is the dual of the transportation problem.

4.2 THE DUALITY THEOREM

To this point the relation between the primal and dual problems has been simply a formal one based on what might appear as an arbitrary definition. In this section, however, the deeper connection between a problem and its dual, as expressed by the Duality Theorem, is derived and the significance of this result serves, after the fact, as motivation for the definition.

The proof of the Duality Theorem given in this section relies on the Separating Hyperplane Theorem (Appendix B) and is therefore somewhat more advanced than previous arguments. An alternative result, which is established by a simpler technique but which is somewhat less general, is given in the next section.

Throughout this section we consider the primal problem in standard form

$$\begin{aligned} &\text{minimize} \quad \mathbf{cx} \\ &\text{subject to} \quad \mathbf{Ax} = \mathbf{b} \\ &\qquad\qquad\quad \mathbf{x} \geqslant \mathbf{0} \end{aligned} \tag{3}$$

and its corresponding dual

$$\begin{aligned} &\text{maximize} \quad \boldsymbol{\lambda}\mathbf{b} \\ &\text{subject to} \quad \boldsymbol{\lambda}\mathbf{A} \leqslant \mathbf{c} \end{aligned} \tag{4}$$

The following lemma is easily established and gives us an important relation between the two problems.

Lemma 1. *If* $\mathbf{x}$ *and* $\boldsymbol{\lambda}$ *are feasible for* (3) *and* (4), *respectively, then* $\mathbf{cx} \geqslant \boldsymbol{\lambda}\mathbf{b}$.

Proof. We have

$$\boldsymbol{\lambda}\mathbf{b} = \boldsymbol{\lambda}\mathbf{A}\mathbf{x} \leqslant \mathbf{cx},$$

the last inequality being valid since $\mathbf{x} \geqslant \mathbf{0}$ and $\boldsymbol{\lambda}\mathbf{A} \leqslant \mathbf{c}$. ∎

This lemma shows that a feasible vector to either problem yields a bound on the value of the other problem. From this we have an important corollary.

Corollary. *If* $\mathbf{x}^0$ *and* $\boldsymbol{\lambda}^0$ *are feasible for* (3) *and* (4), *respectively, and if* $\mathbf{cx}^0 = \boldsymbol{\lambda}^0\mathbf{b}$, *then* $\mathbf{x}^0$ *and* $\boldsymbol{\lambda}^0$ *are optimal for their respective problems.*

The above corollary shows that if a pair of feasible vectors can be found to the primal and dual problems with equal objective values, then these are both optimal. The Duality Theorem of linear programming states that the converse is also true.

Duality Theorem of Linear Programming. *If either of the problems* (3) *or* (4) *has a finite optimal solution, so does the other, and the corresponding values of the objective functions are equal. If either problem has an unbounded objective, the other problem has no feasible solution.*

Proof. We note first that the second statement is an immediate consequence of Lemma 1. For if the primal is unbounded and $\boldsymbol{\lambda}$ is feasible for the dual, we must have $\boldsymbol{\lambda}\mathbf{b} \leqslant -M$ for arbitrarily large M; which is clearly impossible.

Second we note that although the primal and dual are not stated in symmetric form it is sufficient, in proving the first statement, to assume that the primal has a finite optimal solution and then show that the dual has a solution with the same value. This follows because either problem can be converted to standard form and because the roles of primal and dual are reversible.

Suppose (3) has a finite optimal solution with value z_0. In the space E^{m+1} define the convex set

$$C = \{(r, \mathbf{w}) : r = tz_0 - \mathbf{cx},\ \mathbf{w} = t\mathbf{b} - \mathbf{Ax},\ \mathbf{x} \geqslant \mathbf{0},\ t \geqslant 0\}.$$

It is easily verified that C is in fact a closed convex cone. We show that the point $(1, \mathbf{0})$ is not in C. If $\mathbf{w} = t^0\mathbf{b} - \mathbf{A}\mathbf{x}^0 = \mathbf{0}$ with $t^0 > 0$, $\mathbf{x}^0 \geqslant \mathbf{0}$, then $\mathbf{x} = \mathbf{x}^0/t^0$ is feasible for (3) and hence $r/t_0 = z_0 - \mathbf{c}\mathbf{x} \leqslant 0$; which means $r \leqslant 0$. If $\mathbf{w} = -\mathbf{A}\mathbf{x}^0 = \mathbf{0}$ with $\mathbf{x}^0 \geqslant \mathbf{0}$ and $\mathbf{c}\mathbf{x}^0 = -1$, and if $\mathbf{x}$ is any feasible solution to (3), then $\mathbf{x} + \alpha\mathbf{x}^0$ is feasible for any $\alpha \geqslant 0$ and gives arbitrarily small objective values as α is increased. This contradicts our assumption on the existence of a finite optimum and thus we conclude that no such $\mathbf{x}^0$ exists. Hence $(1, \mathbf{0}) \notin C$.

Now since C is a closed convex set, there is by Theorem 1, Section B.3, a hyperplane separating $(1, \mathbf{0})$ and C. Thus there is a nonzero vector $[s, \boldsymbol{\lambda}] \in E_{m+1}$ and a constant c such that

$$s < c = \inf\{sr + \boldsymbol{\lambda}\mathbf{w} : (r, \mathbf{w}) \in C\}.$$

Now since C is a cone, it follows that $c \geqslant 0$. For if there were $(r, \mathbf{w}) \in C$ such that $sr + \boldsymbol{\lambda}\mathbf{w} < 0$, then $\alpha(r, \mathbf{w})$ for large α would violate the hyperplane inequality. On the other hand, since $(0, \mathbf{0}) \in C$ we must have $c \leqslant 0$. Thus $c = 0$. As a consequence $s < 0$, and without loss of generality we may assume $s = -1$.

We have to this point established the existence of $\boldsymbol{\lambda} \in E_m$ such that

$$-r + \boldsymbol{\lambda}\mathbf{w} \geqslant 0$$

for all $(r, \mathbf{w}) \in C$. Equivalently, using the definition of C,

$$(\mathbf{c} - \boldsymbol{\lambda}\mathbf{A})\mathbf{x} - tz_0 + t\boldsymbol{\lambda}\mathbf{b} \geqslant 0$$

for all $\mathbf{x} \geqslant \mathbf{0}$, $t \geqslant 0$. Setting $t = 0$ yields $\boldsymbol{\lambda}\mathbf{A} \leqslant \mathbf{c}$, which says $\boldsymbol{\lambda}$ is feasible for the dual. Setting $\mathbf{x} = \mathbf{0}$ and $t = 1$ yields $\boldsymbol{\lambda}\mathbf{b} \geqslant z_0$, which in view of Lemma 1 and its corollary, shows that $\boldsymbol{\lambda}$ is optimal for the dual. ▮

It will be noticed that the duality theorem makes no assumption on the rank of the matrix $\mathbf{A}$. This generality is partially responsible for the relative difficulty of the proof. If the existence of a nondegenerate basic feasible solution is assumed, a simpler proof along similar lines can be constructed.

4.3 RELATIONS TO THE SIMPLEX PROCEDURE

In this section the Duality Theorem is proved by making explicit use of the characteristics of the simplex procedure. As a result of this proof it becomes clear that once the primal is solved by the simplex procedure a solution to the dual is readily obtainable.

Suppose that for the linear program

$$\begin{aligned} \text{minimize} \quad & \mathbf{c}\mathbf{x} \\ \text{subject to} \quad & \mathbf{A}\mathbf{x} = \mathbf{b} \\ & \mathbf{x} \geqslant \mathbf{0} \end{aligned} \tag{5}$$

we have the optimal basic feasible solution $\mathbf{x} = (\mathbf{x_B}, \mathbf{0})$ with corresponding basis $\mathbf{B}$. We shall determine a solution of the dual problem

$$\begin{aligned} &\text{maximize} \quad \boldsymbol{\lambda}\mathbf{b} \\ &\text{subject to} \quad \boldsymbol{\lambda}\mathbf{A} \leqslant \mathbf{c} \end{aligned} \tag{6}$$

in terms of $\mathbf{B}$.

We partition $\mathbf{A}$ as $\mathbf{A} = [\mathbf{B}, \mathbf{D}]$. Since the basic feasible solution $\mathbf{x_B} = \mathbf{B}^{-1}\mathbf{b}$ is optimal, the relative cost vector $\mathbf{r} = \mathbf{c} - \mathbf{z}$ must be nonnegative in each component. From Section 3.7 we have

$$\mathbf{r} = \mathbf{c_D} - \mathbf{c_B}\mathbf{B}^{-1}\mathbf{D}$$

and since $\mathbf{r}$ is nonnegative in each component we have $\mathbf{c_B}\mathbf{B}^{-1}\mathbf{D} \leqslant \mathbf{c_D}$.

Now define $\boldsymbol{\lambda} = \mathbf{c_B}\mathbf{B}^{-1}$. We show that this choice of $\boldsymbol{\lambda}$ solves the dual problem. We have

$$\boldsymbol{\lambda}\mathbf{A} = [\boldsymbol{\lambda}\mathbf{B}, \boldsymbol{\lambda}\mathbf{D}] = [\mathbf{c_B}, \mathbf{c_B}\mathbf{B}^{-1}\mathbf{D}] \leqslant [\mathbf{c_B}, \mathbf{c_D}] = \mathbf{c}.$$

Thus since $\boldsymbol{\lambda}\mathbf{A} \leqslant \mathbf{c}$, $\boldsymbol{\lambda}$ is feasible for the dual. On the other hand,

$$\boldsymbol{\lambda}\mathbf{b} = \mathbf{c_B}\mathbf{B}^{-1}\mathbf{b} = \mathbf{c_B}\mathbf{x_B}$$

and thus the value of the dual objective function for this $\boldsymbol{\lambda}$ is equal to the value of the primal problem. This, in view of Lemma 1, Section 4.2, establishes the optimality of $\boldsymbol{\lambda}$ for the dual. We summarize this discussion by a theorem.

Theorem. *Let the linear programming problem* (5) *have an optimal basic feasible solution corresponding to the basis* $\mathbf{B}$. *Then the vector* $\boldsymbol{\lambda} = \mathbf{c_B}\mathbf{B}^{-1}$ *is an optimal solution to the dual problem* (6). *The optimal values of both problems are equal.*

We turn now to a discussion of how the solution of the dual problem can be obtained directly from the final simplex tableau of the primal. Suppose that embedded in the original matrix $\mathbf{A}$ is an $m \times m$ identity matrix. This will be the case if, for example, m slack variables are employed to convert inequalities to equalities. Then in the final tableau the matrix $\mathbf{B}^{-1}$ appears where the identity appeared in the beginning. Furthermore, in the last row the components corresponding to this identity matrix will be $\mathbf{c_I} - \mathbf{c_B}\mathbf{B}^{-1}$, where $\mathbf{c_I}$ is the m-vector representing the cost coefficients of the variables corresponding to the columns of the original identity matrix. Thus by subtracting these cost coefficients from the corresponding elements in the last row, the negative of the solution $\boldsymbol{\lambda} = \mathbf{c_B}\mathbf{B}^{-1}$ to the dual is obtained. In particular, if, as is the case with slack variables, $\mathbf{c_I} = \mathbf{0}$, then the elements in the last row under $\mathbf{B}^{-1}$ are equal to the negative of components of the solution to the dual.

Example. Consider the primal problem

$$\begin{aligned}
\text{minimize} \quad & -x_1 - 4x_2 - 3x_3 \\
\text{subject to} \quad & 2x_1 + 2x_2 + x_3 \leqslant 4 \\
& x_1 + 2x_2 + 2x_3 \leqslant 6 \\
& x_1 \geqslant 0, \quad x_2 \geqslant 0, \quad x_3 \geqslant 0.
\end{aligned}$$

This can be solved by introducing slack variables and using the simplex procedure. The appropriate sequence of tableaus is given below without explanation.

2	(2)	1	1	0	4
1	2	2	0	1	6
−1	−4	−3	0	0	0

1	1	1/2	1/2	0	2
−1	0	(1)	−1	1	2
3	0	−1	2	0	8

3/2	1	0	1	−1/2	1
−1	0	1	−1	1	2
2	0	0	1	1	10

The optimal solution is $x_1 = 0$, $x_2 = 1$, $x_3 = 2$. The corresponding dual problem is

$$\begin{aligned}
\text{maximize} \quad & 4\lambda_1 + 6\lambda_2 \\
\text{subject to} \quad & 2\lambda_1 + \lambda_2 \leqslant -1 \\
& 2\lambda_1 + 2\lambda_2 \leqslant -4 \\
& \lambda_1 + 2\lambda_2 \leqslant -3 \\
& \lambda_1 \leqslant 0, \quad \lambda_2 \leqslant 0.
\end{aligned}$$

The optimal solution to the dual problem is obtained directly from the last row of the simplex tableau under the columns where the identity appeared in the first tableau: $\lambda_1 = -1$, $\lambda_2 = -1$.

Simplex Multipliers

We conclude this section by giving an economic interpretation of the relation between the simplex basis and the vector $\boldsymbol{\lambda}$. At any point in the simplex procedure we may form the vector $\boldsymbol{\lambda} = \mathbf{c_B B}^{-1}$. This vector is not a solution to the dual problem unless $\mathbf{B}$ is an optimal basis for the primal, but nevertheless, it has an economic interpretation. Furthermore, as we have seen in the development of the revised simplex method, this $\boldsymbol{\lambda}$ vector can be used at every step to calculate the relative cost coefficients. For this reason

$\boldsymbol{\lambda} = \mathbf{c_B B}^{-1}$, corresponding to any basis, is often called the vector of *simplex multipliers*.

Let us pursue the economic interpretation of these simplex multipliers. As usual, denote the columns of $\mathbf{A}$ by $\mathbf{a}_1, \mathbf{a}_2, \ldots, \mathbf{a}_n$ and denote by $\mathbf{e}_1, \mathbf{e}_2, \ldots, \mathbf{e}_m$ the m unit vectors in E^m. The components of the $\mathbf{a}_i$'s and $\mathbf{b}$ tell how to construct these vectors from the $\mathbf{e}_i$'s.

Given any basis $\mathbf{B}$, however, consisting of m columns of $\mathbf{A}$, any other vector can be constructed (synthetically) as a linear combination of these basis vectors. If there is a unit cost c_i associated with each basis vector $\mathbf{a}_i$, then the cost of a (synthetic) vector constructed from the basis can be calculated as the corresponding linear combination of the c_i's associated with the basis. In particular, the cost of the jth unit vector, $\mathbf{e}_j$, when constructed from the basis $\mathbf{B}$ is λ_j, the jth component of $\boldsymbol{\lambda} = \mathbf{c_B B}^{-1}$. Thus the λ_j's can be interpreted as synthetic prices of the unit vectors.

Now, any vector can be expressed in terms of the basis $\mathbf{B}$ in two steps: (i) express the unit vectors in terms of the basis, and then (ii) express the desired vector as a linear combination of unit vectors. The corresponding synthetic cost of a vector constructed from the basis $\mathbf{B}$ can correspondingly be computed directly by: (i) finding the synthetic price of the unit vectors, and then (ii) using these prices to evaluate the cost of the linear combination of unit vectors. Thus, the simplex multipliers can be used to quickly evaluate the synthetic cost of any vector that is expressed in terms of the unit vectors. The difference between the true cost of this vector and the synthetic cost is the relative cost. The process of calculating the synthetic cost of a vector, with respect to a given basis, by using the simplex multipliers is sometimes referred to as *pricing out* the vector.

Optimality of the primal corresponds to the situation where every vector $\mathbf{a}_1, \mathbf{a}_2, \ldots, \mathbf{a}_n$ is cheaper when constructed from the basis than when purchased directly at its own price. Thus we have $\boldsymbol{\lambda}\mathbf{a}_i \leqslant c_i$ for $i = 1, 2, \ldots, n$ or equivalently $\boldsymbol{\lambda}\mathbf{A} \leqslant \mathbf{c}$.

4.4 SENSITIVITY AND COMPLEMENTARY SLACKNESS

The optimal values of the dual variables in a linear programming problem can, as we have seen, be interpreted as prices. In this section this interpretation is explored in further detail.

Sensitivity

Suppose in the linear programming problem

$$\begin{aligned} &\text{minimize} && \mathbf{cx} \\ &\text{subject to} && \mathbf{Ax} = \mathbf{b} \\ &&& \mathbf{x} \geqslant \mathbf{0} \end{aligned} \tag{7}$$

the optimal basis is $\mathbf{B}$ with corresponding solution $(\mathbf{x_B}, \mathbf{0})$, where $\mathbf{x_B} = \mathbf{B}^{-1}\mathbf{b}$. A solution to the corresponding dual problem is $\boldsymbol{\lambda} = \mathbf{c_B B}^{-1}$.

Now, assuming nondegeneracy, small changes in the vector $\mathbf{b}$ will not cause the optimal basis to change. Thus for $\mathbf{b} + \Delta\mathbf{b}$ the optimal solution is

$$\mathbf{x} = (\mathbf{x_B} + \Delta\mathbf{x_B}, \mathbf{0})$$

where $\Delta\mathbf{x_B} = \mathbf{B}^{-1}\Delta\mathbf{b}$. Thus the corresponding increment in the cost function is

$$\Delta z = \mathbf{c_B}\Delta\mathbf{x_B} = \boldsymbol{\lambda}\Delta\mathbf{b}. \tag{8}$$

This equation shows that $\boldsymbol{\lambda}$ gives the sensitivity of the optimal cost with respect to small changes in the vector $\mathbf{b}$. In other words, if a new problem were solved with $\mathbf{b}$ changed to $\mathbf{b} + \Delta\mathbf{b}$, the change in the optimal value of the objective function would be $\boldsymbol{\lambda}\Delta\mathbf{b}$.

This interpretation of the dual vector $\boldsymbol{\lambda}$ is intimately related to its interpretation as a vector of simplex multipliers. Since λ_j is the price of the unit vector $\mathbf{e}_j$ when constructed from the basis $\mathbf{B}$, it directly measures the change in cost due to a change in the jth component of the vector $\mathbf{b}$. Thus, λ_j may equivalently be considered as the *marginal price* of the component b_j, since if b_j is changed to $b_j + \Delta b_j$ the value of the optimal solution changes by $\lambda_j \Delta b_j$.

If the linear program is interpreted as a diet problem, for instance, then λ_j is the maximum price per unit that the dietician would be willing to pay for a small amount of the jth nutrient, because decreasing the amount of nutrient that must be supplied by food will reduce his food bill by λ_j dollars per unit. If, as another example, the linear program is interpreted as the problem faced by a manufacturer who must select levels $x_1, x_2, \ldots, x_n$ of n production activities in order to meet certain required levels of output $b_1, b_2, \ldots, b_m$ while minimizing production costs, the λ_i's are the marginal prices of his outputs. They show directly how much his production cost varies if a small change is made in the output levels.

Complementary Slackness

The optimal solutions to primal and dual problems satisfy an additional relation that has an economic interpretation. This relation can be stated for any pair of dual linear programs, but we state it here only for the unsymmetric and the symmetric pairs of problems defined in Section 4.1.

Theorem 1 (Complementary slackness–unsymmetric form). *Let* $\mathbf{x}$ *and* $\boldsymbol{\lambda}$ *be feasible solutions for the primal and dual problems, respectively, in the pair* (2). *A necessary and sufficient condition that they both be optimal solutions is that*† *for all* i

i) $x_i > 0 \Rightarrow \boldsymbol{\lambda}\mathbf{a}_i = c_i$

ii) $x_i = 0 \Leftarrow \boldsymbol{\lambda}\mathbf{a}_i < c_i$.

† The symbol $\Rightarrow$ means "implies" and $\Leftarrow$ means "is implied by."

Proof. If the stated conditions hold, then clearly $(\boldsymbol{\lambda}\mathbf{A} - \mathbf{c})\mathbf{x} = 0$. Thus $\boldsymbol{\lambda}\mathbf{b} = \mathbf{c}\mathbf{x}$, and by the corollary to Lemma 1, Section 4.2, the two solutions are optimal. Conversely, if the two solutions are optimal, it must hold, by the Duality Theorem, that $\boldsymbol{\lambda}\mathbf{b} = \mathbf{c}\mathbf{x}$ and hence that $(\boldsymbol{\lambda}\mathbf{A} - \mathbf{c})\mathbf{x} = 0$. Since each component of $\mathbf{x}$ is nonnegative and each component of $\boldsymbol{\lambda}\mathbf{A} - \mathbf{c}$ is nonpositive, the conditions (i) and (ii) must hold. ▌

Theorem 2 (Complementary slackness—symmetric form). *Let* $\mathbf{x}$ *and* $\boldsymbol{\lambda}$ *be feasible solutions for the primal and dual problems, respectively, in the pair* (1). *A necessary and sufficient condition that they both be optimal solutions is that for all i and j*

i) $x_i > 0 \Rightarrow \boldsymbol{\lambda}\mathbf{a}_i = c_i$

ii) $x_i = 0 \Leftarrow \boldsymbol{\lambda}\mathbf{a}_i < c_i$

iii) $\lambda_j > 0 \Rightarrow \mathbf{a}^j\mathbf{x} = b_j$

iv) $\lambda_j = 0 \Leftarrow \mathbf{a}^j\mathbf{x} > b_j$,

(*where* $\mathbf{a}^j$ *is the j-th row of* $\mathbf{A}$).

Proof. This follows by transforming the previous theorem. ▌

The complementary slackness conditions have a rather obvious economic interpretation. Thinking in terms of the diet problem, for example, which is the primal part of a symmetric pair of dual problems, suppose that the optimal diet supplies more than b_j units of the jth nutrient. This means that the dietician would be unwilling to pay anything for small quantities of that nutrient, since availability of it would not reduce the cost of the optimal diet. This, in view of our previous interpretation of λ_j as a marginal price, implies $\lambda_j = 0$ which is (iv) of Theorem 2. The other conditions have similar interpretations which the reader can work out.

*4.5 THE DUAL SIMPLEX METHOD

Often there is available a basic solution to a linear programming problem which is not feasible but which prices out optimally, that is, the simplex multipliers are feasible for the dual problem. In the simplex tableau this situation corresponds to having no negative elements in the bottom row but an infeasible basic solution. Such a situation may arise, for example, if a solution to a certain linear programming problem is calculated and then a new problem is constructed by changing the vector $\mathbf{b}$. In such situations a basic feasible solution to the dual is available and hence it is desirable to pivot in such a way as to optimize the dual.

Rather than constructing a tableau for the dual problem (which, if the primal is in standard form, involves m free variables and n nonnegative slack variables), it is more efficient to work on the dual from the primal tableau. The complete technique based on this idea is the dual simplex method. In terms of the primal problem, it operates by maintaining the optimality condition of the last row while working toward feasibility. In terms of the dual problem, however, it maintains feasibility while working toward optimality.

Given the linear programming problem

$$\begin{aligned} &\text{minimize} && \mathbf{cx} \\ &\text{subject to} && \mathbf{Ax} = \mathbf{b} \\ &&& \mathbf{x} \geqslant \mathbf{0} \end{aligned} \tag{9}$$

suppose a basis $\mathbf{B}$ is known such that $\boldsymbol{\lambda} = \mathbf{c_B B}^{-1}$ is feasible for the dual. In this case we say that the corresponding basic solution to the primal, $\mathbf{x_B} = \mathbf{B}^{-1}\mathbf{b}$, is *dual feasible.* If $\mathbf{x_B} \geqslant \mathbf{0}$ then this solution is also primal feasible and hence optimal.

The given vector $\boldsymbol{\lambda}$ is feasible for the dual and thus satisfies $\boldsymbol{\lambda}\mathbf{a}_j \leqslant c_j$, for $j = 1, 2, \ldots, n$. Indeed, assuming as usual that the basis is the first m columns of $\mathbf{A}$, there is equality

$$\boldsymbol{\lambda}\mathbf{a}_j = c_j, \quad \text{for} \quad j = 1, 2, \ldots, m \tag{10a}$$

and (barring degeneracy in the dual) there is inequality

$$\boldsymbol{\lambda}\mathbf{a}_j < c_j, \quad \text{for} \quad j = m + 1, \ldots, n. \tag{10b}$$

To develop one cycle of the dual simplex method, we find a new vector $\bar{\boldsymbol{\lambda}}$ such that one of the equalities becomes an inequality and one of the inequalities becomes equality, while at the same time increasing the value of the dual objective function. The m equalities in the new solution then determine a new basis.

Denote the ith row of $\mathbf{B}^{-1}$ by $\mathbf{u}^i$. Then for

$$\bar{\boldsymbol{\lambda}} = \boldsymbol{\lambda} - \varepsilon\mathbf{u}^i \tag{11}$$

we have $\bar{\boldsymbol{\lambda}}\mathbf{a}_j = \boldsymbol{\lambda}\mathbf{a}_j - \varepsilon\mathbf{u}^i\mathbf{a}_j$. Thus, recalling that $z_j = \boldsymbol{\lambda}\mathbf{a}_j$ and noting that $\mathbf{u}^i\mathbf{a}_j = y_{ij}$, the ijth element of the tableau, we have

$$\bar{\boldsymbol{\lambda}}\mathbf{a}_j = c_j, \qquad j = 1, 2, \ldots, m, \quad i \neq j \tag{12a}$$

$$\bar{\boldsymbol{\lambda}}\mathbf{a}_i = c_i - \varepsilon \tag{12b}$$

$$\bar{\boldsymbol{\lambda}}\mathbf{a}_j = z_j - \varepsilon y_{ij}, \qquad j = m + 1, \quad m + 2, \ldots, n. \tag{12c}$$

Also,

$$\bar{\boldsymbol{\lambda}}\mathbf{b} = \boldsymbol{\lambda}\mathbf{b} - \varepsilon x_{\mathbf{B}_i}. \tag{13}$$

These last equations lead directly to the algorithm:

Step 1. Given a dual feasible basic solution $\mathbf{x_B}$, if $\mathbf{x_B} \geqslant \mathbf{0}$ the solution is optimal. If $\mathbf{x_B}$ is not nonnegative, select an index i such that the ith component of $\mathbf{x_B}$, $\mathbf{x}_{\mathbf{B}i} < 0$.

Step 2. If all $y_{ij} \geqslant 0, j = 1, 2, \ldots, n$, then the dual has no maximum (this follows since by (12) $\bar{\boldsymbol{\lambda}}$ is feasible for all $\varepsilon > 0$). If $y_{ij} < 0$ for some j, then let

$$\varepsilon_0 = \frac{z_k - c_k}{y_{ik}} = \min_j \left\{ \frac{z_j - c_j}{y_{ij}} : y_{ij} < 0 \right\}. \tag{14}$$

Step 3. Form a new basis $\mathbf{B}$ by replacing $\mathbf{a}_i$ by $\mathbf{a}_k$. Using this basis determine the corresponding basic dual feasible solution $\mathbf{x_B}$ and return to step 1.

The proof that the algorithm converges to the optimal solution is similar in its details to the proof for the primal simplex procedure. The essential observations are: (a) From the choice of k in (14) and from (12a, b, c) the new solution will again be dual feasible; (b) By (13) and the choice $\mathbf{x}_{\mathbf{B}_i} < 0$, the value of the dual objective will increase; (c) The procedure cannot terminate at a nonoptimum point; and (d) Since there are only a finite number of bases, the optimum must be achieved in a finite number of steps.

Example. A form of problem arising frequently is that of minimizing a positive combination of positive variables subject to a series of "greater than" type inequalities having positive coefficients. Such problems are natural candidates for application of the dual simplex procedure. The classical diet problem is of this type as is the simple example below.

$$\begin{array}{ll} \text{minimize} & 3x_1 + 4x_2 + 5x_3 \\ \text{subject to} & x_1 + 2x_2 + 3x_3 \geqslant 5 \\ & 2x_1 + 2x_2 + x_3 \geqslant 6 \\ & x_1 \geqslant 0, \quad x_2 \geqslant 0, \quad x_3 \geqslant 0. \end{array}$$

By introducing surplus variables and by changing the sign of the inequalities we obtain the initial tableau

−1	−2	−3	1	0	−5
(−2)	−2	−1	0	1	−6
3	4	5	0	0	0

Initial tableau

The basis corresponds to a dual feasible solution since all of the $c_j - z_j$'s are nonnegative. We select any $\mathbf{x}_{\mathbf{B}_i} < 0$, say $x_5 = -6$, to remove from the set of basic variables. To find the appropriate pivot element in the second row we compute the ratios $(z_j - c_j)/y_{2j}$ and select the minimum positive ratio. This yields the pivot indicated. Continuing, the remaining tableaus are:

0	(−1)	−5/2	1	−1/2	−2
1	1	1/2	0	−1/2	3
0	1	7/2	0	3/2	9

Second tableau

0	1	5/2	−1	1/2	2
1	0	−2	1	−1	1
0	0	1	1	1	11

Final tableau

The third tableau yields a feasible solution to the primal which must be optimal. Thus the solution is $x_1 = 1$, $x_2 = 2$, $x_3 = 0$.

*4.6 THE PRIMAL–DUAL ALGORITHM

In this section a procedure is described for solving linear programming problems by working simultaneously on the primal and the dual problems. The procedure begins with a feasible solution to the dual that is improved at each step by optimizing an *associated restricted primal* problem. As the method progresses it can be regarded as striving to achieve the complementary slackness conditions for optimality. Originally, the primal–dual method was developed for solving a special kind of linear program arising in network flow problems, and it continues to be the most efficient procedure for these problems. In this section we describe the generalized version of the algorithm and point out an interesting economic interpretation of it. We consider the problem

$$\begin{aligned} &\text{minimize} && \mathbf{cx} \\ &\text{subject to} && \mathbf{Ax} = \mathbf{b} \\ &&& \mathbf{x} \geqslant \mathbf{0} \end{aligned} \tag{15}$$

and the corresponding dual program

$$\begin{aligned} &\text{maximize} && \boldsymbol{\lambda}\mathbf{b} \\ &\text{subject to} && \boldsymbol{\lambda}\mathbf{A} \leqslant \mathbf{c}. \end{aligned} \tag{16}$$

Given a feasible solution $\boldsymbol{\lambda}$ to the dual problem, define the subset P of $\{1, 2, \ldots, n\}$ by $i \in P$ if $\boldsymbol{\lambda}\mathbf{a}_i = c_i$ where $\mathbf{a}_i$ is the ith column of $\mathbf{A}$. Thus, since $\boldsymbol{\lambda}$ is dual feasible, it follows that $i \notin P$ implies $\boldsymbol{\lambda}\mathbf{a}_i < c_i$. Now corresponding to $\boldsymbol{\lambda}$ and P, we define the *associated restricted primal* problem

$$\begin{aligned} \text{minimize} \quad & \mathbf{1y} \\ \text{subject to} \quad & \mathbf{Ax} + \mathbf{y} = \mathbf{b} \\ & \mathbf{x} \geqslant \mathbf{0}, \qquad x_i = 0 \quad \text{for} \quad i \notin P \\ & \mathbf{y} \geqslant \mathbf{0} \end{aligned} \tag{17}$$

where $\mathbf{1}$ denotes the m-vector $[1, 1, \ldots, 1]$.

The dual of this associated restricted primal is called the *associated restricted dual.* It is

$$\begin{aligned} \text{maximize} \quad & \mathbf{ub} \\ \text{subject to} \quad & \mathbf{ua}_i \leqslant 0, \qquad i \in P \\ & \mathbf{u} \leqslant \mathbf{1}. \end{aligned} \tag{18}$$

The condition for optimality of the primal–dual method is expressed in the following theorem.

Primal–Dual Optimality Theorem. *Suppose that $\boldsymbol{\lambda}$ is feasible for the dual and that $\mathbf{x}$ and $\mathbf{y} = \mathbf{0}$ is feasible (and of course optimal) for the associated restricted primal. Then $\mathbf{x}$ and $\boldsymbol{\lambda}$ are optimal for the original primal and dual problems, respectively.*

Proof. Clearly $\mathbf{x}$ is feasible for the primal. Also we have $\mathbf{cx} = \boldsymbol{\lambda}\mathbf{Ax}$, because $\boldsymbol{\lambda}\mathbf{A}$ is identical to $\mathbf{c}$ on the components corresponding to nonzero elements of $\mathbf{x}$. Thus $\mathbf{cx} = \boldsymbol{\lambda}\mathbf{Ax} = \boldsymbol{\lambda}\mathbf{b}$ and optimality follows from Lemma 1, Section 4.2. ∎

The primal–dual method starts with a feasible solution to the dual and then optimizes the associated restricted primal. If the optimal solution to this associated restricted primal is not feasible for the primal, the feasible solution to the dual is improved and a new associated restricted primal is determined. Here are the details:

Step 1. Given a feasible solution $\boldsymbol{\lambda}^0$ to the dual problem (16), determine the associated restricted primal according to (17).

Step 2. Optimize the associated restricted primal. If the minimal value of this problem is zero, the corresponding solution is optimal for the original primal problem by the Primal–Dual Optimality Theorem.

Step 3. If the minimal value of the associated restricted primal is strictly positive, obtain from the final simplex tableau of the restricted primal, the solution $\mathbf{u}^0$ of the associated restricted dual (18). If there is no j for which

$\mathbf{u}^0\mathbf{a}_j > 0$ conclude the primal has no feasible solutions. If, on the other hand, for at least one j, $\mathbf{u}^0\mathbf{a}_j > 0$, define the new dual feasible vector

$$\boldsymbol{\lambda} = \boldsymbol{\lambda}^0 + \varepsilon^0\mathbf{u}^0$$

where

$$\varepsilon^0 = \frac{c_k - \boldsymbol{\lambda}^0\mathbf{a}_k}{\mathbf{u}^0\mathbf{a}_k} = \min_j \left\{\frac{c_j - \boldsymbol{\lambda}^0\mathbf{a}_j}{\mathbf{u}^0\mathbf{a}_j} : \mathbf{u}^0\mathbf{a}_j > 0\right\}.$$

Now go back to step 1 using this $\boldsymbol{\lambda}$.

To prove convergence of this method a few simple observations and explanations must be made. First we verify the statement made in Step 3 that $\mathbf{u}^0\mathbf{a}_j \leqslant 0$ for all j implies that the primal has no feasible solution. The vector $\boldsymbol{\lambda}^\varepsilon = \boldsymbol{\lambda}^0 + \varepsilon\mathbf{u}^0$ is feasible for the dual problem for all positive ε, since $\mathbf{u}^0\mathbf{A} \leqslant \mathbf{0}$. In addition, $\boldsymbol{\lambda}^\varepsilon\mathbf{b} = \boldsymbol{\lambda}^0\mathbf{b} + \varepsilon\mathbf{u}^0\mathbf{b}$ and, since $\mathbf{u}^0\mathbf{b} = \mathbf{1y} > 0$, we see that as ε is increased we obtain an unbounded solution to the dual. In view of the Duality Theorem, this implies that there is no feasible solution to the primal.

Next suppose that in step 3, for at least one j, $\mathbf{u}^0\mathbf{a}_j > 0$. Again we define the family of vectors $\boldsymbol{\lambda}^\varepsilon = \boldsymbol{\lambda}^0 + \varepsilon\mathbf{u}^0$. Since $\mathbf{u}^0$ is a solution to (18) we have $\mathbf{u}^0\mathbf{a}_i \leqslant 0$ for $i \in P$, and hence for small positive ε the vector $\boldsymbol{\lambda}^\varepsilon$ is feasible for the dual. We increase ε to the first point where one of inequalities $\boldsymbol{\lambda}^\varepsilon\mathbf{a}_j < c_j$, $j \notin P$ becomes an equality. This determines $\varepsilon^0 > 0$ and k. The new $\boldsymbol{\lambda}$ vector corresponds to an increased value of the dual objective $\boldsymbol{\lambda}\mathbf{b} = \boldsymbol{\lambda}^0\mathbf{b} + \varepsilon^0\mathbf{u}^0\mathbf{b}$. In addition, the corresponding new set P now includes the index k. Any other index i that corresponded to a positive value of x_i in the associated restricted primal is in the new set P, because by complementary slackness $\mathbf{u}^0\mathbf{a}_i = 0$ for such an i and thus $\boldsymbol{\lambda}\mathbf{a}_i = \boldsymbol{\lambda}^0\mathbf{a}_i + \varepsilon^0\mathbf{u}^0\mathbf{a}_i = c_i$. This means that the old optimal solution is feasible for the new associated restricted primal and that $\mathbf{a}_k$ can be pivoted into the basis. Since $\mathbf{u}^0\mathbf{a}_k > 0$, pivoting in $\mathbf{a}_k$ will decrease the value of the associated restricted primal.

In summary, it has been shown that at each step either an improvement in the associated primal is made or an unfeasibility condition is detected. Assuming nondegeneracy, this implies that no basis of the associated primal is repeated—and since there are only a finite number of possible bases, the solution is reached in a finite number of steps.

The primal–dual algorithm can be given an interesting interpretation in terms of the manufacturing problem in Example 3, Section 2.2. Suppose we own a facility that is capable of engaging in n different production activities each of which produces various amounts of m commodities. Each activity i can be operated at any level $x_i \geqslant 0$, but when operated at the unity level the ith activity cost c_i dollars and yields the m commodities in the amounts specified by the m-vector $\mathbf{a}_i$. Assuming linearity of the production facility, if we are given a vector $\mathbf{b}$ describing output requirements of the m commodities, and we wish to produce these at minimum cost, ours is the primal problem.

Imagine that an entrepreneur *not knowing* the value of our requirements vector $\mathbf{b}$ decides to sell us these requirements directly. He assigns a price vector $\boldsymbol{\lambda}^0$ to these requirements such that $\boldsymbol{\lambda}^0\mathbf{A} \leqslant \mathbf{c}$. In this way his prices are competitive with our production activities, and he can assure us that purchasing directly from him is no more costly than engaging activities. As owner of the production facilities we are reluctant to abandon our production enterprise but, on the other hand, we deem it not frugal to engage an activity whose output can be duplicated by direct purchase for lower cost. Therefore, we decide to engage only activities that cannot be duplicated cheaper, and at the same time we attempt to minimize the total business volume given the entrepreneur. Ours is the associated restricted primal problem.

Upon receiving our order, the greedy entrepreneur decides to modify his prices in such a manner as to keep them competitive with our activities but increase the cost of our order. As a reasonable and simple approach he seeks new prices of the form

$$\boldsymbol{\lambda} = \boldsymbol{\lambda}^0 + \varepsilon \mathbf{u}^0$$

where he selects $\mathbf{u}^0$ as the solution to

$$\begin{aligned} \text{maximize} \quad & \mathbf{u}\mathbf{y} \\ \text{subject to} \quad & \mathbf{u}\mathbf{a}_i \leqslant 0, \qquad i \in P \\ & \mathbf{u} \leqslant \mathbf{1}. \end{aligned}$$

The first set of constraints is to maintain competitiveness of his new price vector for small ε, while the second set is an arbitrary bound imposed to keep this subproblem bounded. It is easily shown that the solution $\mathbf{u}^0$ to this problem is identical to the solution of the associated dual (18). After determining the maximum ε to maintain feasibility, he announces his new prices.

At this point, rather than concede to the price adjustment, we recalculate the new minimum volume order based on the new prices. As the greedy (and shortsighted) entrepreneur continues to change his prices in an attempt to maximize profit he eventually finds he has reduced his business to zero! At that point we have, with his help, solved the original primal problem.

Example. To illustrate the primal–dual method and to indicate how it can be implemented through use of the tableau format consider the following problem:

$$\begin{aligned} \text{minimize} \quad & 2x_1 + x_2 + 4x_3 \\ \text{subject to} \quad & x_1 + x_2 + 2x_3 = 3 \\ & 2x_1 + x_2 + 3x_3 = 5 \\ & x_1 \geqslant 0, \quad x_2 \geqslant 0, \quad x_3 \geqslant 0. \end{aligned}$$

Because all of the coefficients in the objective function are nonnegative, $\boldsymbol{\lambda} = [0, 0]$ is a feasible vector for the dual. We lay out the simplex tableau shown below

	$\mathbf{a}_1$	$\mathbf{a}_2$	$\mathbf{a}_3$	.	.	$\mathbf{b}$
	1	1	2	1	0	3
	2	1	3	0	1	5
	−3	−2	−5	0	0	−8
$c_i - \boldsymbol{\lambda}\mathbf{a}_i \rightarrow$	2	1	4	.	.	.

First tableau

To form this tableau we have adjoined artificial variables in the usual manner. The third row gives the relative cost coefficients of the associated primal problem—the same as the row that would be used in a phase I procedure. In the fourth row are listed the $c_i - \boldsymbol{\lambda}\mathbf{a}_i$'s for the current $\boldsymbol{\lambda}$. The allowable columns in the associated restricted primal are determined by the zeros in this last row.

Since there are no zeros in the last row, no progress can be made in the associated restricted primal and hence the original solution $x_1 = x_2 = x_3 = 0$, $y_1 = 3$, $y_2 = 5$ is optimal for this $\boldsymbol{\lambda}$. The solution $\mathbf{u}^0$ to the associated restricted dual is $\mathbf{u}^0 = [1, 1]$, and the numbers $-\mathbf{u}^0\mathbf{a}_i$, $i = 1, 2, 3$ are equal to the first three elements in the third row. Thus, we compute the three ratios $\frac{2}{3}$, $\frac{1}{2}$, $\frac{4}{5}$ from which we find $\varepsilon^0 = \frac{1}{2}$. The new values for the fourth row are now found by adding ε^0 times the (first three) elements of the third row to the fourth row.

$\mathbf{a}_1$	$\mathbf{a}_2$	$\mathbf{a}_3$	.	.	$\mathbf{b}$
1	(1)	2	1	0	3
2	1	3	0	1	5
−3	−2	−5	0	0	−8
1/2	0	3/2	.	.	.

Second tableau

Minimizing the new associated restricted primal by pivoting as indicated we obtain

$\mathbf{a}_1$	$\mathbf{a}_2$	$\mathbf{a}_3$	.	.	$\mathbf{b}$
1	1	2	1	0	3
1	0	1	−1	1	2
−1	0	−1	2	0	−2
1/2	0	3/2	.	.	.

Now we again calculate the ratios $\frac{1}{2}$, 3 obtaining $\varepsilon^0 = \frac{1}{2}$, and add this multiple of the third row to the fourth row to obtain the next tableau.

$\mathbf{a}_1$	$\mathbf{a}_2$	$\mathbf{a}_3$	.	.	$\mathbf{b}$
1	1	2	1	0	3
(1)	0	1	−1	1	2
−1	0	−1	2	0	−2
0	0	1	.	.	.

Third tableau

Optimizing the new restricted primal we obtain the tableau:

$\mathbf{a}_1$	$\mathbf{a}_2$	$\mathbf{a}_3$	.	.	$\mathbf{b}$
0	1	1	2	−1	1
1	0	1	−1	1	2
0	0	0	1	1	0
0	0	1	.	.	.

Final tableau

Having obtained feasibility in the primal, we conclude that the solution is also optimal: $x_1 = 2$, $x_2 = 1$, $x_3 = 0$.

4.7 EXERCISES

1. Verify in detail that the dual of the dual of a linear programming problem is the original problem.
2. Show that if a linear inequality in a linear program is changed to equality, the corresponding dual variable becomes free.
3. Find the dual of

$$\begin{aligned} &\text{minimize} && \mathbf{cx} \\ &\text{subject to} && \mathbf{Ax} = \mathbf{b} \\ & && \mathbf{x} \geqslant \mathbf{a} \\ &\text{where} && \mathbf{a} \geqslant \mathbf{0}. \end{aligned}$$

4. Show that in the transportation problem the linear equality constraints are not linearly independent, and that in an optimal solution to the dual problem the dual variables are not unique. Generalize this observation to any linear program having redundant equality constraints.
5. Construct an example of a primal problem that has no feasible solutions and whose corresponding dual also has no feasible solutions.

6. Under the assumption that there is a nondegenerate basic feasible solution to the primal problem, develop a proof of the Duality Theorem that is set in E^m rather than E^{m+1}.

7. There is in general a strong connection between the theories of optimization and free competition, which is illustrated by an idealized model of activity location. Suppose there are n economic activities (various factories, homes, stores, etc.) that are to be individually located on n distinct parcels of land. If activity i is located on parcel j that activity can yield s_{ij} units (dollars) of value.

   If the assignment of activities to land parcels is made by a central authority, it might be made in such a way as to maximize the total value generated. In other words, the assignment would be made so as to maximize $\Sigma_i \Sigma_j s_{ij} x_{ij}$ where

$$x_{ij} = \begin{cases} 1 & \text{if activity } i \text{ is assigned to parcel } j \\ 0 & \text{otherwise.} \end{cases}$$

More explicitly this approach leads to the optimization problem

$$\begin{aligned} \text{maximize} \quad & \sum_i \sum_j s_{ij} x_{ij} \\ \text{subject to} \quad & \sum_j x_{ij} = 1, \qquad i = 1, 2, \ldots, n \\ & \sum_i x_{ij} = 1, \qquad j = 1, 2, \ldots, n \\ & x_{ij} \geqslant 0, \qquad x_{ij} = 0 \text{ or } 1. \end{aligned}$$

Actually, it can be shown that the final requirement ($x_{ij} = 0$ or 1) is automatically satisfied at any extreme point of the set defined by the other constraints, so that in fact the optimal assignment can be found by using the simplex method of linear programming.

   If one views the problem from the viewpoint of free competition, it is assumed that, rather than a central authority determining the assignment, the individual activities bid for the land and thereby establish prices.

a) Show that there exists a set of activity prices p_i, $i = 1, 2, \ldots, n$ and land prices q_j, $j = 1, 2, \ldots, n$ such that

$$p_i + q_j \geqslant s_{ij}, \qquad i = 1, 2, \ldots, n, \qquad j = 1, 2, \ldots, n$$

with equality holding if in an optimal assignment activity i is assigned to parcel j.

b) Show that part (a) implies that if activity i is optimally assigned to parcel j and if j' is any other parcel

$$s_{ij} - q_j \geqslant s_{ij'} - q_{j'}.$$

Give an economic interpretation of this result and explain the relation between free competition and optimality in this context.

c) Assuming that each s_{ij} is positive, show that the prices can all be assumed to be nonnegative.

8. Consider the primal linear programming problem

$$\begin{aligned}\text{minimize} \quad & \mathbf{cx} \\ \text{subject to} \quad & \mathbf{Ax} = \mathbf{b} \\ & \mathbf{x} \geqslant \mathbf{0}.\end{aligned}$$

Suppose that this problem and its dual are feasible. Let $\boldsymbol{\lambda}$ be a known solution to the dual.

a) If the kth equation of the primal is multiplied by $\mu \neq 0$, determine a solution $\mathbf{w}$ to the dual of this new problem.
b) Suppose that, in the original primal, we add μ times the kth equation to the rth equation. What is a solution $\mathbf{w}$ to the corresponding dual problem?
c) Suppose, in the original primal, we add μ times the kth row of $\mathbf{A}$ to $\mathbf{c}$. What is a solution to the corresponding problem?

9. Solve the linear inequalities

$$\begin{aligned}-2x_1 + 2x_2 &\leqslant -1 \\ 2x_1 - x_2 &\leqslant 2 \\ -4x_2 &\leqslant 3 \\ -15x_1 - 12x_2 &\leqslant -2 \\ 12x_1 + 20x_2 &\leqslant -1\end{aligned}$$

Note that x_1 and x_2 are *not* restricted to be positive. Solve this problem by considering the problem of maximizing $0 \cdot x_1 + 0 \cdot x_2$ subject to these constraints, taking the dual and using the simplex method.

10. a) Using the simplex method solve

$$\begin{aligned}\text{minimize} \quad & 2x_1 - x_2 \\ \text{subject to} \quad & 2x_1 - x_2 - x_3 \geqslant 3 \\ & x_1 - x_2 + x_3 \geqslant 2 \\ & x_i \geqslant 0, \qquad i = 1, 2, 3.\end{aligned}$$

Hint: Note that $x_1 = 2$ gives a feasible solution.

b) What is the dual problem and its optimal solution?

11. a) Using the simplex method solve

$$\begin{aligned}\text{minimize} \quad & 2x_1 + 3x_2 + 2x_3 + 2x_4 \\ \text{subject to} \quad & x_1 + 2x_2 + x_3 + 2x_4 = 3 \\ & x_1 + x_2 + 2x_3 + 4x_4 = 5 \\ & x_i \geqslant 0, \qquad i = 1, 2, 3, 4.\end{aligned}$$

b) Using the work done in part (a) and the dual simplex method, solve the same problem but with the right hand sides of the equations changed to 8 and 7 respectively.

12. For the problem

$$\begin{aligned} \text{minimize} \quad & 5x_1 - 3x_2 \\ \text{subject to} \quad & 2x_1 - x_2 + 4x_3 \leqslant 4 \\ & x_1 + x_2 + 2x_2 \leqslant 5 \\ & 2x_1 - x_2 + x_3 \geqslant 1 \\ & x_1 \geqslant 0, \quad x_2 \geqslant 0, \quad x_3 \geqslant 0: \end{aligned}$$

a) Using a single pivot operation with pivot element 1, find a feasible solution.

b) Using the simplex method, solve the problem.

c) What is the dual problem?

d) What is the solution to the dual?

13. Prove the validity of the dual simplex method.

14. Consider the problem

$$\begin{aligned} \text{minimize} \quad & 2x_1 + x_2 + 4x_3 \\ \text{subject to} \quad & x_1 + x_2 + 2x_3 = 3 \\ & 2x_1 + x_2 + 3x_3 = 5 \\ & x_1 \geqslant 0, \quad x_2 \geqslant 0, \quad x_3 \geqslant 0 \end{aligned}$$

a) What is the dual problem?

b) Note that $\boldsymbol{\lambda} = [1, 0]$ is feasible for the dual. Starting with this $\boldsymbol{\lambda}$, solve the primal using the primal–dual algorithm.

15. Show that in the associated restricted dual of the primal–dual method the objective $\boldsymbol{\lambda}\mathbf{b}$ can be replaced by $\boldsymbol{\lambda}\mathbf{y}$.

REFERENCES

4.1.–4.4. Again most of the material in this chapter is now quite standard. See the references of Chapter 2. A particularly careful discussion of duality can be found in Simonnard [S3]. The direct proof in Section 4.3 is new, although its spirit is similar to classical proofs.

4.5. The dual simplex method is due to Lemke [L2].

4.6. The general primal–dual algorithm is due to Dantzig, Ford and Fulkerson [D6]. See also Ford and Fulkerson [F7]. The economic interpretation given in this section is apparently new.

Chapter 5 REDUCTION OF LINEAR INEQUALITIES

5.1 INTRODUCTION

In the decision sciences, and especially in applied optimization theory, systems of linear inequalities must be regarded as a fundamental mathematical structure. In a majority of applied problem areas they provide the natural descriptions of economic, physical, or social interactions and constraints. This is typified, of course, by many of the examples discussed throughout the book.

The classical background of linear programming, discussed in the previous three chapters, provides an important analysis of these inequalities. This analysis is based primarily on the concept of a basic feasible solution or, from the geometric viewpoint, on the concept of an extreme point; and led, as we have witnessed, to the development of the simplex method—the most efficient technique known for finding a single solution to a system of inequalities or for minimizing a linear form subject to linear inequalities. With linear programming, duality theory, and the simplex method in hand, it is worthwhile to redirect our attention to systems of linear inequalities with an aim toward furthering our understanding of this fundamental structure.

Consider a system of linear inequalities in standard form

$$\begin{aligned} \mathbf{Ax} &= \mathbf{b}, \\ \mathbf{x} &\geqslant \mathbf{0} \end{aligned} \tag{1}$$

where $\mathbf{A}$ is an $m \times n$ matrix, $\mathbf{b}$ is a constant nonzero m-vector, and $\mathbf{x}$ is a variable n-vector. Any point $\mathbf{x}$ satisfying these conditions is called a *solution*. The set of solutions is denoted by S.

It is the set S that is of primary interest in most problems involving systems of inequalities—the inequalities themselves acting merely to provide a description of S. Alternative systems having the same solution set S are, from this viewpoint, equivalent. In many cases, therefore, the system of linear inequalities originally used to define S may not be the simplest, and it may be possible to find another system having fewer inequalities or fewer variables while defining the same solution set S. It is this general problem that is explored in this chapter. It is shown that corresponding to any system of linear inequalities there is, in a sense to be precisely defined later, a minimal representation defining the solution set S.

5.2 REDUNDANT EQUATIONS

One way that a system of linear inequalities can sometimes be simplified is by the elimination of redundant equations. This leads to a new equivalent system having the same number of variables but fewer equations.

Definition. Corresponding to the system of linear inequalities

$$\begin{aligned} \mathbf{Ax} &= \mathbf{b}, \\ \mathbf{x} &\geqslant \mathbf{0} \end{aligned} \tag{1}$$

we say the system has *redundant equations* if there is a nonzero $\boldsymbol{\lambda} \in E_m$ satisfying

$$\begin{aligned} \boldsymbol{\lambda}\mathbf{A} &= \mathbf{0}, \\ \boldsymbol{\lambda}\mathbf{b} &= 0. \end{aligned} \tag{2}$$

This definition is equivalent, as the reader is aware, to the statement that a system of equations is redundant if one of the equations can be expressed as a linear combination of the others. In most of our previous analysis we have assumed, for simplicity, that such redundant equations were not present in our given system or that they were eliminated prior to further computation. Indeed, such redundancy presents no real computational difficulty, since redundant equations are detected and can be eliminated during application of the phase I procedure for determining a basic feasible solution.

5.3 NULL VARIABLES

Definition. Corresponding to the system of linear inequalities

$$\begin{aligned} \mathbf{Ax} &= \mathbf{b}, \\ \mathbf{x} &\geqslant \mathbf{0}, \end{aligned} \tag{1}$$

a variable x_i is said to be a *null variable* if $x_i = 0$ in every solution.

It is clear that if it were known that a variable x_i were a null variable, then the solution set S could be equivalently described by the system of linear inequalities obtained from (1) by deleting the ith column of $\mathbf{A}$, deleting the inequality $x_i \geqslant 0$, and adjoining the equality $x_i = 0$. This yields an obvious simplification in the description of the solution set S. It is perhaps not so obvious how null variables can be identified.

Example. As a simple example of how null variables may appear consider the system

$$\begin{aligned} 2x_1 + 3x_2 + 4x_3 + 4x_4 &= 6 \\ x_1 + x_2 + 2x_3 + x_4 &= 3 \end{aligned}$$

$$x_1 \geqslant 0, \quad x_2 \geqslant 0, \quad x_3 \geqslant 0, \quad x_4 \geqslant 0.$$

By subtracting twice the second equation from the first we obtain

$$x_2 + 2x_4 = 0.$$

Since the x_i's must all be nonnegative, it follows immediately that x_2 and x_4 are zero in any solution. Thus x_2 and x_4 are null variables.

Generalizing from the above example it is clear that if a linear combination of the equations can be found such that the right hand side is zero while the coefficients on the left side are all either zero or positive; then the variables corresponding to the positive coefficients in this equation are null variables. In other words, if from the original system it is possible to combine equations so as to yield

$$\xi_1 x_1 + \xi_2 x_2 + \cdots + \xi_n x_n = 0$$

with $\xi_i \geqslant 0$, $i = 1, 2, \ldots, n$. Then $\xi_i > 0$ implies that x_i is a null variable.

The above elementary observations clearly can be used to identify null variables in some cases. A more surprising result is that the technique described above can be used to identify all null variables. The proof of this fact is based on the Duality Theorem.

Null Variable Theorem. *If S is not empty, the variable x_i is a null variable in the system* (1) *if and only if there is a nonzero vector $\boldsymbol{\lambda} \in E_m$ such that*

$$\begin{aligned} \boldsymbol{\lambda}\mathbf{A} &\geqslant \mathbf{0} \\ \boldsymbol{\lambda}\mathbf{b} &= 0 \end{aligned} \tag{3}$$

and the i-th component of $\boldsymbol{\lambda}\mathbf{A}$ is strictly positive.

Proof. The "if" part follows immediately from the discussion above. To prove the "only if" part, suppose that x_i is a null variable, and suppose that S is not empty. Consider the program

$$\begin{aligned} \text{minimize} \quad & -\mathbf{e}^i\mathbf{x} \\ \text{subject to} \quad & \mathbf{A}\mathbf{x} = \mathbf{b}, \\ & \mathbf{x} \geqslant \mathbf{0} \end{aligned}$$

where $\mathbf{e}^i$ is the ith unit row vector. By our hypotheses, there is a feasible solution and the optimal value is zero. By the Duality Theorem the dual program

$$\begin{aligned}&\text{maximize } \boldsymbol{\lambda}\mathbf{b}\\&\text{subject to } \boldsymbol{\lambda}\mathbf{A} \leqslant -\mathbf{e}^i\end{aligned}$$

is also feasible and has optimal value zero. Thus there is a $\boldsymbol{\lambda}$ with

$$\begin{aligned}\boldsymbol{\lambda}\mathbf{A} &\leqslant -\mathbf{e}^i,\\ \boldsymbol{\lambda}\mathbf{b} &= 0.\end{aligned}$$

Changing the sign of $\boldsymbol{\lambda}$ proves the theorem. ▌

Next we indicate briefly how the above results can be used in conjunction with the simplex method to determine the existence of null variables. We shall find that the existence of null variables is strongly related to the notion of degenerate basic feasible solutions.

Assume, without loss of generality, that a basic feasible solution is known and that, correspondingly, the system has been put in canonical form so that $\mathbf{A}$ can be partitioned as

$$\mathbf{A} = [\mathbf{I}, \mathbf{G}]$$

where $\mathbf{I}$ is an $m \times m$ identity, and $\mathbf{G}$ is an $m \times (n-m)$ matrix; and $\mathbf{b} \geqslant \mathbf{0}$.

According to the Null Variable Theorem we seek a nonzero vector $\boldsymbol{\lambda}$ such that $\boldsymbol{\lambda}\mathbf{A} \geqslant \mathbf{0}$, $\boldsymbol{\lambda}\mathbf{b} = 0$. In canonical form, however, $\boldsymbol{\lambda}\mathbf{A} \geqslant \mathbf{0}$ implies

$$[\boldsymbol{\lambda}, \boldsymbol{\lambda}\mathbf{G}] \geqslant \mathbf{0}$$

thus, in particular, $\boldsymbol{\lambda} \geqslant \mathbf{0}$. Furthermore, since $\mathbf{b} \geqslant \mathbf{0}$, $\boldsymbol{\lambda} \geqslant \mathbf{0}$ implies $\lambda_i = 0$ if $b_i > 0$. Thus if $\mathbf{b} > \mathbf{0}$, we conclude that there are no null variables. From this observation we deduce an important corollary of the Null Variable Theorem that can often be used to quickly terminate the search to null variables.

Corollary. *If any basic feasible solution to* (1) *is nondegenerate, the system contains no null variables.*

If there are zero components in $\mathbf{b}$, we partition $\mathbf{b}$ into its zero and positive components and then, by a reordering of equations and variables, we have the induced partitioning

$$\mathbf{A} = \left[\begin{array}{c|c|c}\mathbf{I}_0 & \mathbf{0} & \mathbf{G}_0\\ \hline \mathbf{0} & \mathbf{I}_1 & \mathbf{G}_1\end{array}\right], \qquad \mathbf{b} = \left[\begin{array}{c}\mathbf{0}\\ \hline \mathbf{b}_1\end{array}\right]$$

where $\mathbf{b}_1 > \mathbf{0}$ and where $\mathbf{I}_0$, $\mathbf{G}_0$ correspond to the zero components of $\mathbf{b}$. Let q be the number of zero components in $\mathbf{b}$. Then from the partitioning it is clear that we seek a nonzero q-dimensional row vector $\mathbf{y}$ such that $\mathbf{y} \geqslant \mathbf{0}, \mathbf{y}\mathbf{G}_0 \geqslant \mathbf{0}$. The existence of such a $\mathbf{y}$ can be determined by seeking a solution to the linear inequalities

$$\mathbf{y}\mathbf{G}_0 \geqslant \mathbf{0}, \qquad \mathbf{y}\mathbf{1} = 1, \qquad \mathbf{y} \geqslant \mathbf{0} \tag{4}$$

where $\mathbf{1}$ denotes the q-vector $\mathbf{1} = (1, 1, \ldots, 1)$. If there is no solution to these inequalities, there are no null variables. If there is a solution $\mathbf{y}$, then the components of $\mathbf{x}$ corresponding to positive elements of $[\mathbf{y}, \mathbf{0}, \mathbf{y}\mathbf{G}_0]$ are null variables, and can be eliminated.

5.4 NONEXTREMAL VARIABLES

Example 1. Consider the system of linear inequalities

$$\begin{aligned} x_1 + 3x_2 + 4x_3 &= 4 \\ 2x_1 + x_2 + 3x_3 &= 6 \\ x_1 \geqslant 0, \quad x_2 \geqslant 0, \quad x_3 &\geqslant 0. \end{aligned} \tag{5}$$

By subtracting the second equation from the first and rearranging, we obtain

$$x_1 = 2 + 2x_2 + x_3 \tag{6}$$

From this we observe that since x_2 and x_3 are nonnegative, the value of x_1 is greater than or equal to 2 in any solution to the equalities. This means that the inequality $x_1 \geqslant 0$ can be dropped from the original set, and x_1 can be treated as a free variable even though the remaining inequalities actually do not allow complete freedom. Hence x_1 can be replaced everywhere by (6) in the original system (5) leading to

$$\begin{aligned} 5x_2 + 5x_3 &= 2, \\ x_2 \geqslant 0, \qquad x_3 &\geqslant 0 \\ x_1 &= 2 + 2x_2 + x_3. \end{aligned} \tag{7}$$

The first two lines of (7) represent a system of linear inequalities in standard form with one less variable and one less equation than the original system. The last equation is a simple linear equation from which x_1 is determined by a solution to the smaller system of inequalities.

This example illustrates and motivates the concept of a nonextremal variable. As illustrated, the identification of such nonextremal variables results in a significant simplification of a system of linear inequalities.

Definition. A variable x_i in the system of linear inequalities

$$\begin{aligned} \mathbf{A}\mathbf{x} &= \mathbf{b}, \\ \mathbf{x} &\geqslant \mathbf{0} \end{aligned} \tag{8}$$

is *nonextremal* if the inequality $x_i \geqslant 0$ in (8) is redundant.

A nonextremal variable can be treated as a free variable, and thus can be eliminated from the system by using one equation to define that variable in terms of the other variables. The result is a new system having one less variable and one less equation. Solutions to the original system can be obtained from solutions to the new system by substituting into the expression for the value of the free variable.

It is clear that if, as in the example, a linear combination of the equations in the system can be found that imply that x_i is nonnegative if all other variables are nonnegative, then x_i is nonextremal. That the converse of this statement is also true is perhaps not so obvious. Again the proof of this is based on the Duality Theorem.

Nonextremal Variable Theorem. *If S is not empty, the variable x_j is a nonextremal variable for the system* (8) *if and only if there is $\boldsymbol{\lambda} \in E_m$ and $\mathbf{d} \in E_n$ such that*

$$\boldsymbol{\lambda}\mathbf{A} = \mathbf{d} \tag{9}$$

where

$$d_j = -1, \qquad d_i \geqslant 0 \quad \text{for} \quad i \neq j;$$

and such that

$$\boldsymbol{\lambda}\mathbf{b} = -\beta, \tag{10}$$

for some $\beta \geqslant 0$.

Proof. The "if" part of the result is trivial, since forming the corresponding linear combination of the equations in (8) yields

$$x_j = \beta + d_1x_1 + \cdots + d_{j-1}x_{j-1} + d_{j+1}x_{j+1} + \cdots + d_nx_n$$

which implies that x_j is nonextremal.

To prove the "only if" part, let $\mathbf{a}_i, i = 1, 2, \ldots, n$ denote the ith column of $\mathbf{A}$. Let us assume that the solution set S is nonempty and that x_j is nonextremal. Consider the linear program

$$\begin{aligned} &\text{minimize} \quad x_j \\ &\text{subject to} \quad \mathbf{Ax} = \mathbf{b}, \\ &\qquad\qquad x_i \geqslant 0, \quad i \neq j. \end{aligned} \tag{11}$$

By hypothesis the minimum value is nonnegative, say it is $\beta \geqslant 0$. Then by the Duality Theorem the value of the dual program

$$\begin{aligned} &\text{maximize} \quad \boldsymbol{\lambda}\mathbf{b} \\ &\text{subject to} \quad \boldsymbol{\lambda}\mathbf{a}_i \leqslant 0, \quad i \neq j \\ &\qquad\qquad \boldsymbol{\lambda}\mathbf{a}_j = 1 \end{aligned}$$

is also β. Taking the negative of the optimal solution to the dual yields the desired result. ∎

Unlike null variables which occur only when there is degeneracy, nonextremal variables occur frequently in systems of linear inequalities. It can be shown, for instance, that every system having three nonnegative variables and two (independent) equations can be reduced to two nonnegative variables and one equation.

Example 2. Consider the system

$$\begin{aligned} x_1 + 2x_2 + 3x_3 &= 4 \\ x_1 + x_2 + x_3 &= 3 \\ x_1 \geqslant 0, \quad x_2 \geqslant 0, &\quad x_3 \geqslant 0. \end{aligned}$$

The easiest way to reduce a system like this is to look for a linear combination of the equations to yield a zero right hand side. Thus, subtracting $\frac{3}{4}$ of the first equation from the second we find

$$\tfrac{1}{4}x_1 - \tfrac{1}{2}x_2 - \tfrac{5}{4}x_3 = 0$$

or

$$x_1 = 2x_2 + 5x_3. \tag{12}$$

Substituting this last relation in the original system and dropping one equation to account for (12), we obtain the equivalent system

$$\begin{aligned} x_2 + 2x_3 &= 1, \\ x_2 \geqslant 0, \quad x_3 &\geqslant 0 \end{aligned}$$

where x_1 is determined through (12).

5.5 REDUCED PROBLEMS

In this section we show that the elimination of redundant equations, null variables, and nonextremal variables, as discussed in the previous sections, is in some sense a complete reduction of the system of inequalities to minimal order. At the same time we investigate the geometric interpretation of null and nonextremal variables.

Similar Systems

As in previous sections we denote by S the set of all solutions to the system

$$\begin{aligned} \mathbf{Ax} &= \mathbf{b} \\ \mathbf{x} &\geqslant \mathbf{0}. \end{aligned} \tag{13}$$

In addition, we define the subspace Σ of $X = E^n$ as the subspace generated by S; that is, Σ is the smallest subspace of X that contains the solution set S. We refer to Σ as the *solution subspace*.

We have seen that different systems of linear inequalities may, in some sense, define the same solution sets. We can now define this notion precisely.

Definition. Two systems of linear inequalities on the spaces X and $\bar{X}$ having solutions sets S and $\bar{S}$ and solution spaces Σ and $\bar{\Sigma}$, respectively, are *similar* if there is a linear invertible transformation $\mathbf{T}$ from Σ to $\bar{\Sigma}$ that maps S onto $\bar{S}$. A transformation $\mathbf{T}$ with this property is called a *similarity transformation.*

This is a very general encapsulation of the idea that two systems of linear inequalities have the same essential structure. It allows any arbitrary linear redefinition of the underlying variables.

Next we define what it means for one system of linear inequalities to be a reduction of another.

Definition. Suppose two systems of linear inequalities on the spaces X and $\bar{X}$ are similar. The second system is said to be a *reduction* of the first system if the dimension of the space $\bar{X}$ is less than the dimension of X; or if the dimensions are equal but the second system has fewer equations than the first.

Reduction Theorems

In the remainder of this section it is shown that any system of linear inequalities has a minimal similar representation of lowest dimension and lowest number of equations. It is further shown that reducibility of a system can be determined by examining whether it contains redundant equations, null variables, or nonextremal variables. This theory of reduction is fairly complex, so we begin by stating the main results and then devote the remainder of the section to their proofs.

It is convenient, in order to somewhat simplify our analysis and at the same time develop as symmetric a theory as possible, to focus primarily on *homogeneous* systems of linear inequalities in the standard form

$$\begin{aligned} \mathbf{Ax} &= \mathbf{0}, \\ \mathbf{x} &\geqslant \mathbf{0}. \end{aligned} \tag{14}$$

Inhomogeneous systems such as

$$\begin{aligned} \mathbf{Ax} &= \mathbf{b} \\ \mathbf{x} &\geqslant \mathbf{0} \end{aligned} \tag{15}$$

are treated by consideration of the related homogeneous system

$$\begin{aligned} \mathbf{Ax} - \mathbf{b}x_{n+1} &= \mathbf{0}, \\ \mathbf{x} \geqslant \mathbf{0}, \qquad x_{n+1} &\geqslant 0. \end{aligned} \tag{16}$$

The solution set of the inhomogeneous system consists of the subset of solutions to (16) having $x_{n+1} = 1$.

Reduction Theorem 1. *A homogeneous system of linear inequalities* (14) *is reducible if and only if it contains redundant equations, null variables, or nonextremal variables.*

The corresponding result for inhomogeneous systems requires the introduction of a boundedness assumption.

Reduction Theorem 2. *Suppose $n > 1$ and the solution set of the system of linear inequalities* (15) *is nonempty and bounded. Then* (15) *is reducible if and only if it contains redundant equations, null variables, or nonextremal variables.*

Proof of "if" portion of Reduction Theorem

Theorem 2 is really just a corollary of Theorem 1, so we shall devote most effort to the proof of Theorem 1. The proof of the "if" part of this theorem (that existence of redundant equations, null variables, or nonextremal variables, implies that reduction is possible) is the easy part of the proof and follows directly from our earlier discussions.

First, if there are redundant equations the given system can be reduced by eliminating one equation. In this case the new (reduced) system has $\bar{X} = X, \bar{S} = S, \bar{\Sigma} = \Sigma$, and hence the required transformation is the identity mapping.

Suppose that x_j is a null variable or nonextremal variable. Then in view of the Null Variable Theorem in Section 5.3 and the Nonextremal Variable Theorem in Section 5.4, it can be assumed that the jth column of $\mathbf{A}$ is

$$\mathbf{a}_j = (0, 0, \ldots, 0, d_j)$$

where $|d_j| = 1$, the last row of $\mathbf{A}$ is

$$\mathbf{a}^m = [d_1, d_2, \ldots, d_n]$$

with $d_i \geqslant 0, i \neq j$, and $b_m = 0$. The zeros can be put into $\mathbf{a}_j$ by one step of a Gaussian elimination scheme and the required row operations on $\mathbf{A}$ do not change S.

Consider the corresponding new system with the jth column of $\mathbf{A}$ removed and the last equation eliminated. This system on E^{n-1} has a solution set $\bar{S}$. Define $\mathbf{T}$ from Σ to $\bar{\Sigma}$ as the transformation that eliminates the jth component. For any $\mathbf{x} \in S$, clearly $\mathbf{Tx} \in \bar{S}$.

If $d_j = 1$, then x_j is a null variable and $x_j = 0$ on Σ, and hence $\mathbf{T}$ is one-to-one and onto. If $d_j = -1$ then for any $\bar{\mathbf{x}} \in \bar{S}$, the $\mathbf{x}$ obtained by reconstructing x_j through $x_j = \sum_{i \neq j} d_i \bar{x}_i$ is a member of S; thus $\mathbf{T}$ is again one-to-one and onto. ▌

Proof of "only if" portion of Reduction Theorem

We now turn to the difficult task of proving the "only if" part of the Reduction Theorem. The arguments used in the remainder of this section, although geometrical in character, are of a more advanced nature than other parts of the chapter and hence the reader may wish to proceed directly to the next section.

We consider the homogeneous system

$$\mathbf{Ax} = \mathbf{0}, \qquad (14)$$
$$\mathbf{x} \geqslant \mathbf{0}$$

with corresponding solution set S and solution subspace Σ. Since the system is homogeneous, the solution set S is a convex cone in $X = E^n$.

Corresponding to each unit coordinate vector $\mathbf{e}_i \in E^n$, let the vector $\mathbf{x}_i^*$ be defined as the orthogonal projection of $\mathbf{e}_i$ onto Σ. The resulting $\mathbf{x}_i^*$'s will not be linearly independent (unless $\Sigma = E^n$) but they are all that is needed to uniquely determine components of vectors in Σ as follows: If $\mathbf{x} \in \Sigma$ then

$$x_i = \mathbf{x}'\mathbf{e}_i = \mathbf{x}'\mathbf{x}_i^*. \qquad (17)$$

In other words, for determining components of a solution, the $\mathbf{x}_i^*$'s serve as well as the $\mathbf{e}_i$'s. The vector $\mathbf{x}_i^*$ is called the *adjoint vector* of x_i. See Fig. 5.1.

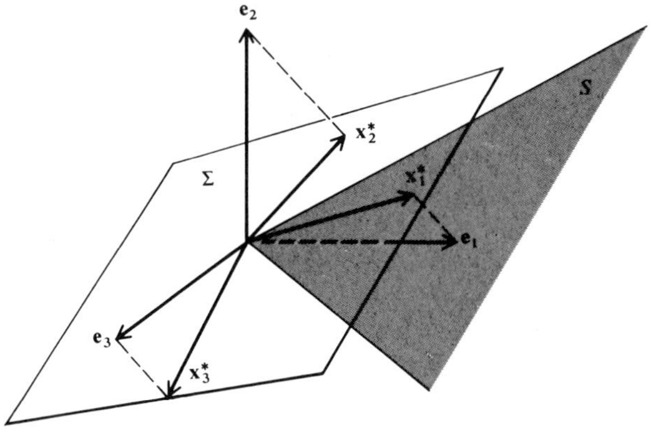

Fig. 5.1 Adjoint vectors

We define the *adjoint solution set* S^* to be the convex cone generated by the adjoint vectors. Thus S^* consists of all vectors of the form

$$\mathbf{x}^* = \alpha_1\mathbf{x}_1^* + \alpha_2\mathbf{x}_2^* + \cdots + \alpha_n\mathbf{x}_n^*, \qquad \alpha_i \geqslant 0, \qquad i = 1, 2, \ldots, n.$$

In terms of the adjoint vectors and the adjoint solution set we have alternative, more geometric, definitions of null and nonextremal variables.

Proposition 1. *The variable x_i is a null variable for* (14) *if and only if* $\mathbf{x}_i^*$ *is zero.*

Proof. Left to the reader.

Proposition 2. *The variable x_j is a nonextremal variable for* (14) *if and only if* $\mathbf{x}_j^*$ *lies in the convex cone generated by the other adjoint variables, i.e., if the convex cone generated by* $\{\mathbf{x}_1^*, \mathbf{x}_2^*, \ldots, \mathbf{x}_{j-1}^*, \mathbf{x}_{j+1}^*, \ldots, \mathbf{x}_n^*\}$ *is* S^*.

Proof. The variable x_j is nonextremal only if there is a $\boldsymbol{\lambda}$ satisfying the conditions of the Nonextremal Variable Theorem, Section 5.4. Since $\mathbf{b} = \mathbf{0}$, we have $\beta = 0$. Thus for all solutions

$$x_j = d_1 x_1 + \cdots + d_{j-1} x_{j-1} + d_{j+1} x_{j+1} + \cdots + d_n x_n$$

where

$$d_i \geqslant 0, \qquad i \neq j$$

and by linearity this relation must hold in Σ as well as S. Thus

$$\mathbf{x}'\mathbf{x}_j^* = \mathbf{x}'(d_1\mathbf{x}_1^* + \cdots + d_{j-1}\mathbf{x}_{j-1}^* + d_{j+1}\mathbf{x}_{j+1}^* + \cdots + d_n\mathbf{x}_n^*)$$

for all $\mathbf{x} \in \Sigma$, which implies that

$$\mathbf{x}_j^* = d_1\mathbf{x}_1^* + \cdots + d_{j-1}\mathbf{x}_{j-1}^* + d_{j+1}\mathbf{x}_{j+1}^* + \cdots + d_n\mathbf{x}_n^*$$

and hence $\mathbf{x}_j^*$ is in the convex cone generated by the remaining adjoint variables.

The converse statement is obvious. ∎

Extremal variables essentially correspond to extreme vectors (or edges) of the adjoint solution set S^*. There are, however, two exceptions to this characterization. The first is where two adjoint variables are proportional and each define the same edge of S^*; and consequently both vectors correspond to nonextremal variables. The second is where a variable is both extreme and null; corresponding to the zero vector.

Associated with the homogeneous system of linear inequalities (14) we have defined two convex cones S and S^* corresponding to the solution set and adjoint solution set. Earlier the concept of similarity between two systems was defined in terms of the existence of a linear invertible mapping between the corresponding solution sets. Our aim now is to show that for similar systems there is also a linear invertible map between the corresponding adjoint solution sets. Since, in view of Propositions 1 and 2 above, the structure of the adjoint solution set is directly related to null and nonextremal variables, the relation between these concepts and similarity becomes apparent. Thus, our immediate aim, encompassed by the next four propositions, is to more fully characterize the relationship between S and S^* and show that similarity between the solution sets of two systems implies similarity between their adjoint solution sets as well.

Proposition 3. *The subspace generated by S^* is Σ.*

Proof. By definition $S^* \subset \Sigma$, so the subspace M generated by S^* is contained in Σ. If M is not equal to Σ, then there is an $\mathbf{x} \in \Sigma$, $\mathbf{x} \neq \mathbf{0}$ that is perpendicular to M. Thus $\mathbf{x}'\mathbf{x}_i^* = x_i = 0$ for $i = 1, 2, \ldots, n$. Thus $\mathbf{x} = \mathbf{0}$ which is a contradiction. ▮

Definition. Corresponding to the set S in Σ we define $S^\oplus = \{\mathbf{y} : \mathbf{y} \in \Sigma, \mathbf{x}'\mathbf{y} \geqslant 0 \text{ for all } \mathbf{x} \in S\}$.

Proposition 4. $S^* = S^\oplus$.

Proof. It is clear that $S^* \subset S^\oplus$, because for $\mathbf{x} \in S$ we have $\mathbf{x}'\mathbf{x}_i^* \geqslant 0$, $i = 1, 2, \ldots, n$ and hence any $\mathbf{y}$ that is a positive combination of $\mathbf{x}_i^*$'s also has $\mathbf{x}'\mathbf{y} \geqslant 0$.

Let $\mathbf{y} \in \Sigma, \mathbf{y} \notin S^*$, then by the Separating Hyperplane Theorem applied to Σ there is a hyperplane separating $\mathbf{y}$ and S^*, that is, there is an $\mathbf{x} \in \Sigma$ such that $\mathbf{x}'\mathbf{x}_i^* \geqslant 0$ for all $i = 1, 2, \ldots, n$ and $\mathbf{x}'\mathbf{y} < 0$. From $\mathbf{x} \in \Sigma$ it follows that $\mathbf{x}'\mathbf{x}_i^* = \mathbf{x}'\mathbf{e}_i = x_i \geqslant 0$ and hence $\mathbf{x} \geqslant \mathbf{0}$. This implies $\mathbf{x} \in S$, and hence $\mathbf{x}'\mathbf{y} < 0$ implies $\mathbf{y} \notin S^\oplus$. Therefore $S^* \supset S^\oplus$, which together with $S^* \subset S^\oplus$ completes the proof. ▮

Proposition 5. $S = (S^\oplus)^\oplus$.

Proof. The proof of this proposition is quite similar to that of Proposition 4 and is left as an exercise.

Proposition 6. *Corresponding to the homogeneous system* (14),

$$S^\oplus = \{\mathbf{y} : \mathbf{y}' \geqslant \boldsymbol{\lambda}\mathbf{A} \text{ for some } \boldsymbol{\lambda} \in E_m\}.$$

Proof. If $\mathbf{y}' \geqslant \boldsymbol{\lambda}\mathbf{A}$ for some $\boldsymbol{\lambda}$, then for any $\mathbf{x} \in S$ there holds

$$\mathbf{y}'\mathbf{x} \geqslant \boldsymbol{\lambda}\mathbf{A}\mathbf{x} = 0,$$

so we conclude $S^\oplus \supset \{\mathbf{y} : \mathbf{y}' \geqslant \boldsymbol{\lambda}\mathbf{A}\}$.

To prove the opposite inclusion, let $\mathbf{z} \in S^\oplus$. Then the linear program

$$\begin{aligned} \text{minimize} \quad & \mathbf{z}'\mathbf{x} \\ \text{subject to} \quad & \mathbf{A}\mathbf{x} = \mathbf{0}, \\ & \mathbf{x} \geqslant \mathbf{0} \end{aligned}$$

has optimal value equal to zero. Thus the dual problem has feasible solutions, which implies $\boldsymbol{\lambda}\mathbf{A} \leqslant \mathbf{z}'$ for some $\boldsymbol{\lambda}$. ▮

With the above array of propositions we have characterized the adjoint solution set quite profoundly. We now can establish the last required proposition on similarity of adjoint solution sets and then, finally, the Reduction Theorems.

Proposition 7. *If* $\mathbf{T}$ *is a similarity transformation between the solution sets* S *and* $\bar{S}$ *of two homogeneous systems, then* $\mathbf{T}'$, *the transpose* (*or adjoint*) *of* $\mathbf{T}$, *is a similarity transformation between* $\bar{S}^*$ *and* S^*.

Proof. The transformation $\mathbf{T}$ is a linear invertible transformation from Σ to $\bar{\Sigma}$. Thus $\mathbf{T}'$, the transformation satisfying $\mathbf{x}'\mathbf{T}'\bar{\mathbf{x}} = \bar{\mathbf{x}}'\mathbf{T}\mathbf{x}$ for all $\mathbf{x} \in \Sigma$, $\bar{\mathbf{x}} \in \bar{\Sigma}$, is a linear invertible transformation from $\bar{\Sigma}$ to Σ. By Proposition 3, Σ and $\bar{\Sigma}$ serve as solution subspaces for S^* and $\bar{S}^*$. In view of Proposition 4, it is only necessary to show that $\mathbf{T}'(\bar{S}^{\oplus}) = S^{\oplus}$, i.e. that $\mathbf{T}'$ maps $\bar{S}^{\oplus}$ onto $S^{\oplus}$.

Suppose $\bar{\mathbf{x}} \in \bar{S}^{\oplus}$. Then for any $\mathbf{x} \in S$ we have $\mathbf{x}'\mathbf{T}'\bar{\mathbf{x}} = \bar{\mathbf{x}}'\mathbf{T}\mathbf{x} \geqslant 0$, since $\mathbf{T}\mathbf{x} \in \bar{S}$. Thus $\mathbf{T}'\bar{\mathbf{x}} \in S^{\oplus}$. We conclude that $\mathbf{T}'(\bar{S}^{\oplus}) \subset S^{\oplus}$.

Suppose $\mathbf{y} \in \Sigma$, $\mathbf{y} \notin S^{\oplus}$. Then by the Separating Hyperplane Theorem applied in Σ, there is an $\mathbf{x} \in \Sigma$ such that $\mathbf{y}'\mathbf{x} < 0$ and $\mathbf{z}'\mathbf{x} \geqslant 0$ for all $\mathbf{z} \in S^{\oplus}$. In view of Proposition 5, $\mathbf{x} \in S$. Then $\mathbf{T}\mathbf{x} \in S$ and hence for any $\bar{\mathbf{x}} \in S^{\oplus}$, $\bar{\mathbf{x}}'\mathbf{T}\mathbf{x} \geqslant 0$. Thus $\mathbf{x}'\mathbf{T}\bar{\mathbf{x}} \geqslant 0$ for all $\bar{\mathbf{x}} \in S^{\oplus}$, which means $\mathbf{y} \neq \mathbf{T}'\mathbf{x}$ for any $\bar{\mathbf{x}} \in S^{\oplus}$. Thus $\mathbf{y} \notin \mathbf{T}'(S^{\oplus})$. We conclude that $\mathbf{T}'(\bar{S}^{\oplus}) \supset S^{\oplus}$. ▮

Reduction Theorem 1. *A homogeneous system of linear inequalities* (14) *is reducible if and only if it contains redundant equations, null variables, or extremal variables.*

Proof. The "if" part was proved earlier.

Suppose we have two systems on the spaces X and $\bar{X}$ with solution sets S and $\bar{S}$, and solution subspaces Σ and $\bar{\Sigma}$; and suppose $\mathbf{T}$ is a similarity transformation from Σ to $\bar{\Sigma}$. Since, by Proposition 7, $\mathbf{T}'$ maps $\bar{S}^*$ onto S^* in a one-to-one manner, it establishes a one-to-one correspondence between edges of S^* and edges of $\bar{S}^*$.

Let all variables of the first system be extremal and nonnull. Each extremal variable corresponds to a different edge variable of the adjoint solution set, thus S^* and $\bar{S}^*$ have at least as many edges as the dimension of X. On the other hand, there is at least one adjoint vector on each edge of the solution set, and hence the dimension of $\bar{X}$ is at least as great as the number of edges in S^* and $\bar{S}^*$. Thus the dimension of $\bar{X}$ is at least as great as the dimension of X. This proves that if there are no null variables or nonextremal variables, there cannot exist a similar system of lower dimension.

Suppose now that the second system has the same dimension as the first system and the first system contains no redundant equations and no null or nonextremal variables, but that the second system has fewer equations than the first system. Since X and $\bar{X}$ have the same dimension, the transformation $\mathbf{T}$ can be extended (in many ways) so that it is a linear invertible transformation between X and $\bar{X}$. Thus any equation $\mathbf{a}'\mathbf{x} = 0$ in X that is satisfied by all $\mathbf{x} \in S$ can be transformed into an equivalent equation $\bar{\mathbf{a}}'\bar{\mathbf{x}} = 0$ satisfied by all $\bar{\mathbf{x}} \in \bar{X}$ by setting $\bar{\mathbf{a}}' = \mathbf{a}'\mathbf{T}^{-1}$. This follows since $\mathbf{a}'\mathbf{x} = \mathbf{a}'\mathbf{T}^{-1}\mathbf{T}\mathbf{x}$. Transforming all the equations of the first system in this manner, there will be at least one which is linearly independent of the equations comprising the second system. Thus if $\bar{\mathbf{A}}$ denotes the matrix of the second system, there is an $\bar{\mathbf{a}} \neq \mathbf{0}$ such that $\bar{\mathbf{a}}'$ is not a linear combination of the rows of $\bar{\mathbf{A}}$ and such that $\bar{\mathbf{a}}'\bar{\mathbf{x}} = 0$ for all $\bar{\mathbf{x}} \in \bar{S}$. Thus $\bar{\mathbf{a}} \in \bar{S}^{\oplus}$, and by Proposition 6 there is a $\bar{\boldsymbol{\lambda}}$ such that $\mathbf{0} \neq \bar{\mathbf{a}}' - \bar{\boldsymbol{\lambda}}\bar{\mathbf{A}} \geqslant \mathbf{0}$. This implies that the second system contains a null variable (why?). Thus the dimension of the second system can be reduced, contradicting the first part of the proof. Hence if the dimensions of X and $\bar{X}$ are equal, the second system must have at least as many equations as the first. ▮

Inhomogeneous Systems

It is not necessary to derive a special theory of reduction for inhomogeneous systems, since as pointed out before they can be regarded as a special case of a homogeneous system of the form

$$\begin{aligned} \mathbf{A}\mathbf{x} - x_{n+1}\mathbf{b} &= \mathbf{0}, \\ \mathbf{x} \geqslant \mathbf{0}, \qquad x_{n+1} &\geqslant 0. \end{aligned} \tag{16}$$

Nevertheless, it is possible to specialize the results of Reduction Theorem 1 so as to obtain a special reduction theorem for inhomogeneous systems.

Lemma. *If the solution set corresponding to*

$$\begin{aligned} \mathbf{A}\mathbf{x} &= \mathbf{b}, \\ \mathbf{x} &\geqslant \mathbf{0} \end{aligned} \tag{15}$$

is nonempty, then it is bounded only if x_{n+1} is nonextremal for (16).

Proof. If x_{n+1} is extremal, then the constraint $x_{n+1} \geqslant 0$ is not redundant and hence there are positive solutions to $\mathbf{A}\mathbf{x} = \mathbf{b}$. This together with a nonempty solution set implies the solution set is unbounded. ▮

Reduction Theorem 2. *Suppose $n > 1$ and the solution set corresponding to the system of inequalities* (15) *is nonempty and bounded. Then* (15) *is reducible if and only if it contains redundant equations, null variables or nonextremal variables.*

Proof. In view of the above lemma, x_{n+1} is nonextremal for (16) and hence we may, by application of the Nonextremal Variable Theorem and

Gaussian elimination, assume that (15) has the form

$$\mathbf{d}'\mathbf{x} = 1 \tag{18}$$

$$\mathbf{B}\mathbf{x} = \mathbf{0} \tag{19a}$$

$$\mathbf{x} \geqslant \mathbf{0} \tag{19b}$$

where $\mathbf{d} \geqslant \mathbf{0}$ and $\mathbf{B}$ is $(m-1) \times n$.

The "if" part of the theorem now follows by noting that for $n > 1$, (19a) will be nonvacuous and (19) can be reduced if there are redundant equations, null variables, or nonextremal variables.

To prove the "only if" part we note that if (15) can be reduced, the reduced form can, by the same reasoning, be expressed as

$$\bar{\mathbf{d}}'\bar{\mathbf{x}} = 1 \tag{20}$$

$$\bar{\mathbf{B}}\bar{\mathbf{x}} = \mathbf{0} \tag{21a}$$

$$\bar{\mathbf{x}} \geqslant \mathbf{0}. \tag{21b}$$

It follows that (21) is a reduction of (19), and the conclusion follows from Reduction Theorem 1. ▮

5.6 APPLICATIONS OF REDUCTION

A limited number of computational experiments directed toward the ideas of this chapter indicate that many systems of linear inequalities arising from applications can be substantially reduced. There are particular areas, however, possessing perhaps a good deal of inherent redundancy that are more reducible than others.

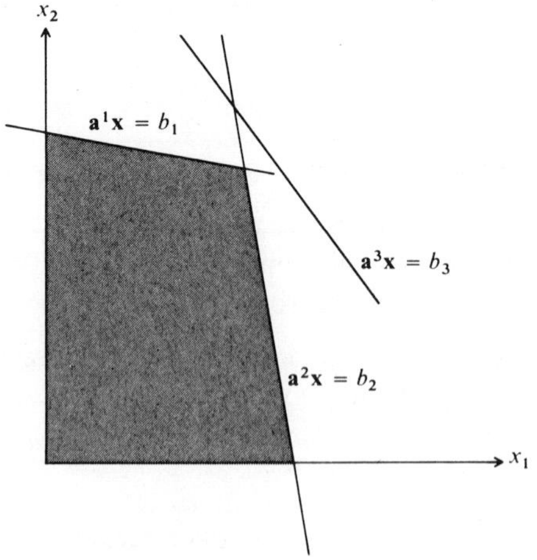

Fig. 5.2 Redundant inequality

One interesting area of application is the elimination of redundant inequality constraints. Consider the region shown in Fig. 5.2 defined by the nonnegativity constraint and three other linear inequalities. The system can be expressed as

$$\mathbf{a}^1\mathbf{x} \leqslant b_1, \qquad \mathbf{a}^2\mathbf{x} \leqslant b_2, \qquad \mathbf{a}^3\mathbf{x} \leqslant b_3, \qquad \mathbf{x} \geqslant \mathbf{0} \tag{22}$$

which in standard form is

$$\mathbf{a}^1\mathbf{x} + y_1 = b_1, \qquad \mathbf{a}^2\mathbf{x} + y_2 = b_2, \qquad \mathbf{a}^3\mathbf{x} + y_3 = b_3, \qquad \mathbf{x} \geqslant \mathbf{0}, \qquad \mathbf{y} \geqslant \mathbf{0}, \tag{23}$$

The third constraint is, as seen from the figure, redundant and can be eliminated without changing the solution set. In the standard form (23) this is reflected in the fact that y_3 is nonextremal and hence it, together with the third constraint, can be eliminated. This special example generalizes, of course, to higher dimensional problems involving many inequalities where, in general, redundant inequalities show up as having nonextremal slack variables. The detection and elimination of such redundant inequalities can be helpful in the cutting-plane methods (discussed in Chapter 13) where inequalities are continually appended to a system as the method progresses.

Whenever an operation more difficult than that of solving a single linear program is to be performed with respect to the solution set S defined by a system of linear inequalities, it will be advantageous to reduce the system to a minimal representation. Thus, when seeking all extreme points of S, when minimizing a nonlinear objective function over S, when solving an integer programming problem over S, or when faced with a series of linear programming problems with different objective functions over S, preliminary computation should be devoted to reduction.

5.7 EXERCISES

1. Show how the Phase I procedure can be used to detect and eliminate redundant equations.

2. Given the system of linear inequalities (1), what is implied by the existence of a $\boldsymbol{\lambda}$ satisfying $\boldsymbol{\lambda}\mathbf{A} = \mathbf{0}, \boldsymbol{\lambda}\mathbf{b} \neq 0$?

3. Suppose a system of linear inequalities possesses null variables. Show that when the null variables are eliminated, by setting them identically to zero, the resulting system will have redundant equations. Verify this for the example in Section 5.3.

4. Show directly, without going to the dual, that if a system of linear inequalities in standard form has a nondegenerate basic feasible solution, it has no null variables.

5. Prove that any system of linear inequalities in standard form having two equations and three variables can be reduced.

6. Show that if a system of linear inequalities in standard form has a nondegenerate basic feasible solution, the corresponding nonbasic variables are extremal.

7. Eliminate the null variables in the system

$$\begin{aligned} 2x_1 + x_2 - x_3 + x_4 + x_5 &= 2 \\ -x_1 + 2x_2 + x_3 + 2x_4 + x_5 &= -1 \\ -x_1 - x_2 \qquad - 3x_4 + 2x_5 &= -1 \end{aligned}$$
$$x_1 \geqslant 0, \quad x_2 \geqslant 0, \quad x_3 \geqslant 0, \quad x_4 \geqslant 0, \quad x_5 \geqslant 0.$$

8. Develop an algorithm for detecting nonextremal variables.

9. Reduce to minimal size

$$\begin{aligned} x_1 + x_2 + 2x_3 + x_4 + x_5 &= 6 \\ 3x_2 + x_3 + 5x_4 + 4x_5 &= 4 \\ x_1 + x_2 - x_3 + 2x_4 + 2x_5 &= 3 \end{aligned}$$
$$x_1 \geqslant 0, \quad x_2 \geqslant 0, \quad x_3 \geqslant 0, \quad x_4 \geqslant 0, \quad x_5 \geqslant 0.$$

10. Prove Proposition 5, Section 5.5.

11. Suppose that the definition of similar systems of linear inequalities is relaxed so as to allow the transformation $\mathbf{T}$ to be affine (that is, of the form $\mathbf{Tx} = \mathbf{Fx} + \mathbf{c}$ where $\mathbf{F}$ is linear and $\mathbf{c}$ is a fixed vector). Develop a reduction theorem for inhomogeneous systems using this relaxed definition.

12. Show that if an inhomogeneous system is similar by an affine transformation (see Exercise 11) to a homogeneous system, then the inhomogeneous system has nonextremal variables.

13. Suppose $\mathbf{A}$ is an $m \times n$ nonsingular matrix and $\mathbf{b}$ is an n-vector with all of its components $b_i > 0$. Suppose also the vector $\mathbf{A}^{-1}\mathbf{b}$ has all of its components strictly positive. Show that the system of linear inequalities

$$\mathbf{Ax} \leqslant \mathbf{b}, \qquad \mathbf{x} \geqslant \mathbf{0}$$

(or more precisely the corresponding standard form) is irreducible.

REFERENCES

The concept of reduction and about one-half of the results in this chapter are due to Shefi [S2]. The presentation given in this chapter follows his original ideas closely, although in substantially modified form.

PART II
UNCONSTRAINED PROBLEMS

Chapter 6 BASIC PROPERTIES OF SOLUTIONS AND ALGORITHMS

In this chapter we consider optimization problems of the form

$$\begin{aligned} &\text{minimize} \quad f(\mathbf{x}) \\ &\text{subject to} \quad \mathbf{x} \in \Omega \end{aligned} \tag{1}$$

where f is a real-valued function and Ω, the feasible set, is a subset of E^n. Throughout most of the chapter attention is restricted to the case where $\Omega = E^n$, corresponding to the completely unconstrained case, but occasionally we consider cases where Ω is some particularly simple subset of E^n.

The first two sections of the chapter characterize the conditions that must hold at a solution point of (1). These conditions are simply extensions to E^n of the well-known derivative conditions for a function of a single variable that hold at a maximum or a minimum point. The third and fourth sections of the chapter introduce the important classes of convex and concave functions that provide a natural formulation for a global theory of optimization and provide geometric interpretations of the derivative conditions derived in the first two sections.

The final sections of the chapter are devoted to basic convergence characteristics of algorithms. Although this material is not exclusively applicable to optimization problems but applies to general iterative algorithms for solving other problems as well, it can be regarded as a fundamental prerequisite for a modern treatment of optimization techniques. Two essential questions are addressed concerning iterative algorithms. The first question, which is qualitative in nature, is whether a given algorithm in some sense yields, at least in the limit, a solution to the original problem. This question is treated in Section 6.5, and conditions sufficient to guarantee appropriate convergence are established. The second question, the more quantitative one, is related to how fast the algorithm converges to a solution. This question is defined more precisely in Section 6.6. Several special types of convergence, which arise frequently in the development of algorithms for optimization, are explored.

6.1 NECESSARY CONDITIONS FOR LOCAL MINIMA

In an investigation of the general problem (1) we distinguish two kinds of solution points: *local minimum points*, and *global minimum points.*

Definition. A point $\mathbf{x}^* \in \Omega$ is said to be a *relative minimum point* or a *local minimum point* of f over Ω if there is an $\varepsilon > 0$ such that $f(\mathbf{x}) \geqslant f(\mathbf{x}^*)$ for all $\mathbf{x} \in \Omega$ within a distance ε of $\mathbf{x}^*$, (that is, $\mathbf{x} \in \Omega$ and $|\mathbf{x} - \mathbf{x}^*| < \varepsilon$). If $f(\mathbf{x}) > f(\mathbf{x}^*)$ for all $\mathbf{x} \in \Omega$, $\mathbf{x} \neq \mathbf{x}^*$, within a distance ε of $\mathbf{x}^*$, then $\mathbf{x}^*$ is said to be a *strict relative minimum point* of f over Ω.

Definition. A point $\mathbf{x}^* \in \Omega$ is said to be a *global minimum point* of f over Ω if $f(\mathbf{x}) \geqslant f(\mathbf{x}^*)$ for all $\mathbf{x} \in \Omega$. If $f(\mathbf{x}) > f(\mathbf{x}^*)$ for all $\mathbf{x} \in \Omega$, $\mathbf{x} \neq \mathbf{x}^*$, then $\mathbf{x}^*$ is said to be a *strict global minimum point* of f over Ω.

In formulating and attacking problem (1) we are, by definition, explicitly asking for a global minimum point of f over the set Ω. Practical reality, however, both from the theoretical and computational viewpoint, dictates that we must in many circumstances be content with a relative minimum point. In deriving necessary conditions based on the differential calculus, for instance, or when searching for the minimum point by a convergent stepwise procedure, comparisons of the values of nearby points is all that is possible and attention focuses on relative minimum points. Global conditions and global solutions can, as a rule, only be found if the problem possesses certain convexity properties that essentially guarantee that any relative minimum is a global minimum. Thus, in formulating and attacking problem (1) we shall, by the dictates of practicality, usually consider, implicitly, that we are asking for a relative minimum point.

Feasible Directions

To derive necessary conditions satisfied by a relative minimum point $\mathbf{x}^*$, the basic idea is to consider movement away from the point in some given direction. Along any given direction the objective function can be regarded as a function of a single variable, the parameter defining movement in this direction, and hence the ordinary calculus of a single variable is applicable. Thus given $\mathbf{x} \in \Omega$ we are motivated to say that a vector $\mathbf{d}$ is a *feasible direction at* $\mathbf{x}$ if there is an $\bar{\alpha} > 0$ such that $\mathbf{x} + \alpha\mathbf{d} \in \Omega$ for all α, $0 \leqslant \alpha \leqslant \bar{\alpha}$. With this simple concept we can state some simple conditions satisfied by relative minimum points.

Proposition 1 (First-order necessary conditions). *Let Ω be a subset of E^n and let $f \in C^1$ be a function on Ω. If $\mathbf{x}^*$ is a relative minimum point of f over Ω, then for any $\mathbf{d} \in E^n$ that is a feasible direction at $\mathbf{x}^*$, we have $\nabla f(\mathbf{x}^*)\,\mathbf{d} \geqslant 0$.*

Proof. For any α, $0 \leqslant \alpha \leqslant \bar{\alpha}$, the point $\mathbf{x}(\alpha) = \mathbf{x}^* + \alpha\mathbf{d} \in \Omega$. For $0 \leqslant \alpha \leqslant \bar{\alpha}$ define the function $g(\alpha) = f(\mathbf{x}(\alpha))$. Then g has a relative minimum at $\alpha = 0$. A typical g is shown in Fig. 6.1. By the ordinary calculus we have

$$g(\alpha) - g(0) = g'(0)\,\alpha + o(\alpha). \tag{2}$$

If $g'(0) < 0$ then, for sufficiently small values of $\alpha > 0$, the right side of (2) will be negative, and hence $g(\alpha) - g(0) < 0$, which contradicts the minimal nature of $g(0)$. Thus $g'(0) = \nabla f(\mathbf{x}^*)\mathbf{d} \geqslant 0$. ∎

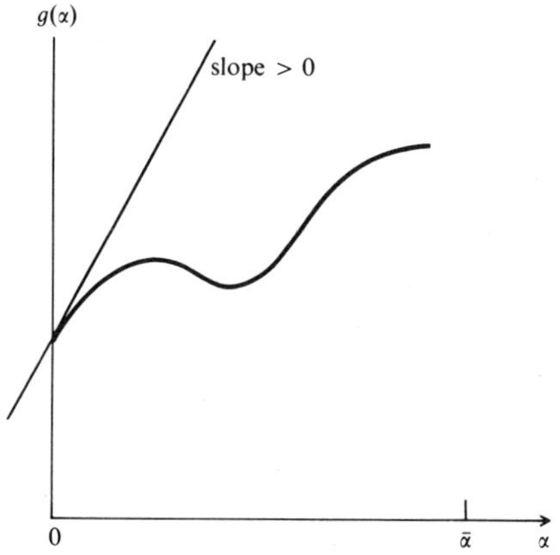

Fig. 6.1 Construction for proof

Example 1. Consider the problem

$$\begin{aligned} &\text{minimize} \quad f(x_1, x_2) = x_1^2 - x_1 + x_2 + x_1 x_2 \\ &\text{subject to} \quad x_1 \geqslant 0, \quad x_2 \geqslant 0. \end{aligned}$$

This problem has a global minimum at $x_1 = \frac{1}{2}$, $x_2 = 0$. At this point

$$\frac{\partial f}{\partial x_1} = 2x_1 - 1 + x_2 = 0$$

$$\frac{\partial f}{\partial x_2} = 1 + x_1 = \tfrac{3}{2}.$$

Thus the partial derivatives do not both vanish at the solution, but since any feasible direction must have an x_2 component greater than or equal to zero, we have $\nabla f(\mathbf{x}^*)\mathbf{d} \geqslant 0$ for all $\mathbf{d} \in E^2$ such that $\mathbf{d}$ is a feasible direction at the point $(\frac{1}{2}, 0)$.

Of special interest is the case where the point $\mathbf{x}^*$ is in the interior of Ω. In this case there are feasible vectors emanating in every direction from $\mathbf{x}^*$, and hence $\nabla f(\mathbf{x}^*)\mathbf{d} \geqslant 0$ for all $\mathbf{d} \in E^n$. This implies that $\nabla f(\mathbf{x}^*) = \mathbf{0}$.

The proof of Proposition 1 is based on making a first-order approximation to the function f in the neighborhood of the relative minimum point. Additional conditions can be obtained by considering higher-order approximations. The second-order conditions, which are defined in terms of the Hessian matrix $\nabla^2 f$ of second partial derivatives of f (see Appendix A), are of extreme theoretical importance and dominate much of the analysis presented in later chapters.

Proposition 2 (Second-order necessary conditions). *Let Ω be a subset of E^n and let $f \in C^2$ be a function on Ω. If $\mathbf{x}^*$ is a relative minimum point of f over Ω, then for any $\mathbf{d} \in E^n$ that is a feasible direction at $\mathbf{x}^*$ we have*

i) $\nabla f(\mathbf{x}^*)\mathbf{d} \geqslant 0$ (3)

ii) *if* $\nabla f(\mathbf{x}^*)\mathbf{d} = 0$, *then* $\mathbf{d}'\nabla^2 f(\mathbf{x}^*)\mathbf{d} \geqslant 0$. (4)

Proof. The first condition is just Proposition 1, and the second applies only if $\nabla f(\mathbf{x}^*)\mathbf{d} = 0$. In this case, introducing $\mathbf{x}(\alpha) = \mathbf{x}^* + \alpha\mathbf{d}$ and $g(\alpha) = f(\mathbf{x}(\alpha))$ as before, we have, in view of $g'(0) = 0$,

$$g(\alpha) - g(0) = \tfrac{1}{2}g''(0)\alpha^2 + o(\alpha^2).$$

If $g''(0) < 0$ the right side of the above equation is negative for sufficiently small α which contradicts the relative minimum nature of $g(0)$. Thus

$$g''(0) = \mathbf{d}'\nabla^2 f(\mathbf{x}^*)\mathbf{d} \geqslant 0. \blacksquare$$

Example 2. For the same problem as Example 1, we have for $\mathbf{d} = (d_1, d_2)$

$$\nabla f(\mathbf{x}^*)\mathbf{d} = \tfrac{3}{2}d_2.$$

Thus condition (ii) of Proposition 2 applies only if $d_2 = 0$. In that case we have $\mathbf{d}'\nabla^2 f(\mathbf{x}^*)\mathbf{d} = 2d_1{}^2 \geqslant 0$, so condition (ii) is satisfied.

Again of special interest is the case where the minimizing point is an interior point of Ω, as, for example, in the case of completely unconstrained problems. We then obtain the following classical result.

Proposition 3 (Second-order necessary conditions—unconstrained case). *Let $\mathbf{x}^*$ be an interior point of the set Ω, and suppose $\mathbf{x}^*$ is a relative minimum point over Ω of the function $f \in C^2$. Then*

i) $\nabla f(\mathbf{x}^*) = \mathbf{0}$ (5)

ii) *For all* $\mathbf{d}$, $\mathbf{d}'\nabla^2 f(\mathbf{x}^*)\mathbf{d} \geqslant 0$. (6)

For notational simplicity we often denote $\nabla^2 f(\mathbf{x})$, the $n \times n$ matrix of the second partial derivatives of f, the Hessian of f, by the alternative notation $\mathbf{F}(\mathbf{x})$. Condition (ii) is equivalent to stating that the matrix $\mathbf{F}(\mathbf{x}^*)$

is positive semidefinite. As we shall see, the matrix $\mathbf{F}(\mathbf{x}^*)$, which arises here quite naturally in a discussion of necessary conditions, plays a fundamental role in the analysis of iterative methods for solving unconstrained optimization problems. The structure of this matrix is the primary determinant of the rate of convergence of algorithms designed to minimize the function f.

Example 3. Consider the problem

$$\begin{aligned}&\text{minimize} \quad f(x_1, x_2) = x_1{}^3 - x_1{}^2 x_2 + 2x_2{}^2\\&\text{subject to} \quad x_1 \geqslant 0, \quad x_2 \geqslant 0.\end{aligned}$$

If we assume that the solution is in the interior of the feasible set, that is, if $x_1 > 0$, $x_2 > 0$, then the first-order necessary conditions are

$$3x_1{}^2 - 2x_1 x_2 = 0, \qquad -x_1{}^2 + 4x_2 = 0.$$

There is a solution to these at $x_1 = x_2 = 0$ which is a boundary point, but there is also a solution at $x_1 = 6$, $x_2 = 9$. We note that for x_1 fixed at $x_1 = 6$, the objective attains a relative minimum with respect to x_2 at $x_2 = 9$. Conversely, with x_2 fixed at $x_2 = 9$, the objective attains a relative minimum with respect to x_1 at $x_1 = 6$. Despite this fact, the point $x_1 = 6, x_2 = 9$ is not a relative minimum point, because the Hessian matrix is

$$\mathbf{F} = \begin{bmatrix} 6x_1 - 2x_2 & -2x_1 \\ -2x_1 & 4 \end{bmatrix}$$

which, evaluated at the proposed solution $x_1 = 6$, $x_2 = 9$, is

$$\mathbf{F} = \begin{bmatrix} 18 & -12 \\ -12 & 4 \end{bmatrix}.$$

This matrix is not positive semidefinite, since its determinant is negative. Thus the proposed solution is not a relative minimum point.

6.2 SUFFICIENT CONDITIONS FOR A RELATIVE MINIMUM

By slightly strengthening the second condition of Proposition 3 in the last section, we obtain a set of conditions that imply that the point $\mathbf{x}^*$ is a relative minimum. We give here the conditions that apply only to unconstrained problems, or to problems where the minimum point is interior to the feasible region, since the corresponding conditions for problems where the minimum is achieved on a boundary point of the feasible set are a good deal more difficult and of marginal practical or theoretical value. A more general result, applicable to problems with functional constraints, is given in Chapter 10.

Proposition (Second-order sufficient conditions—unconstrained case). *Let $f \in C^2$ be a function defined on a region in which the point $\mathbf{x}^*$ is an interior point. Suppose in addition that*

i) $\nabla f(\mathbf{x}^*) = \mathbf{0}$ (7)

ii) $\mathbf{F}(\mathbf{x}^*)$ *is positive definite.* (8)

Then $\mathbf{x}^*$ *is a strict relative minimum point of* f.

Proof. Since $\mathbf{F}(\mathbf{x}^*)$ is positive definite, there is an $a > 0$ such that for all $\mathbf{d}$, $\mathbf{d}'\mathbf{F}(\mathbf{x}^*)\mathbf{d} \geqslant a|\mathbf{d}|^2$. Thus by the Mean Value Theorem (with remainder)

$$\begin{aligned} f(\mathbf{x}^* + \mathbf{d}) - f(\mathbf{x}^*) &= \tfrac{1}{2}\mathbf{d}'\mathbf{F}(\mathbf{x}^*)\mathbf{d} + o(|\mathbf{d}|^2) \\ &\geqslant (a/2)\,|\mathbf{d}|^2 + o(|\mathbf{d}|^2). \end{aligned}$$

For small $|\mathbf{d}|$ the first term on the right dominates the second, implying that both sides are positive for small $\mathbf{d}$. ∎

6.3 CONVEX AND CONCAVE FUNCTIONS

In order to develop a theory directed toward characterizing global, rather than local, minimum points, it is necessary to introduce some sort of convexity assumptions. This results not only in a more potent, although more restrictive, theory but also provides an interesting geometric interpretation of the second-order sufficiency result derived above.

Definition. A function f defined on a convex set Ω is said to be *convex* if, for every $\mathbf{x}_1, \mathbf{x}_2 \in \Omega$ and every α, $0 \leqslant \alpha \leqslant 1$, there holds

$$f(\alpha\mathbf{x}_1 + (1-\alpha)\mathbf{x}_2) \leqslant \alpha f(\mathbf{x}_1) + (1-\alpha) f(\mathbf{x}_2).$$

If, for every $0 < \alpha < 1$ and $\mathbf{x}_1 \neq \mathbf{x}_2$, there holds

$$f(\alpha\mathbf{x}_1 + (1-\alpha)\mathbf{x}_2) < \alpha f(\mathbf{x}_1) + (1-\alpha) f(\mathbf{x}_2),$$

then f is said to be *strictly convex*.

Several examples of convex or nonconvex functions are shown in Fig. 6.2. Geometrically, a function is convex if the line joining two points on its graph lies nowhere below the graph, as shown in Fig. 6.2a, or, thinking of a function in two dimensions, it is convex if its graph is bowl shaped.

Next we turn to the definition of a concave function.

Definition. A function g defined on a convex set Ω is said to be *concave* if the function $f = -g$ is convex. The function g is *strictly concave* if $-g$ is strictly convex.

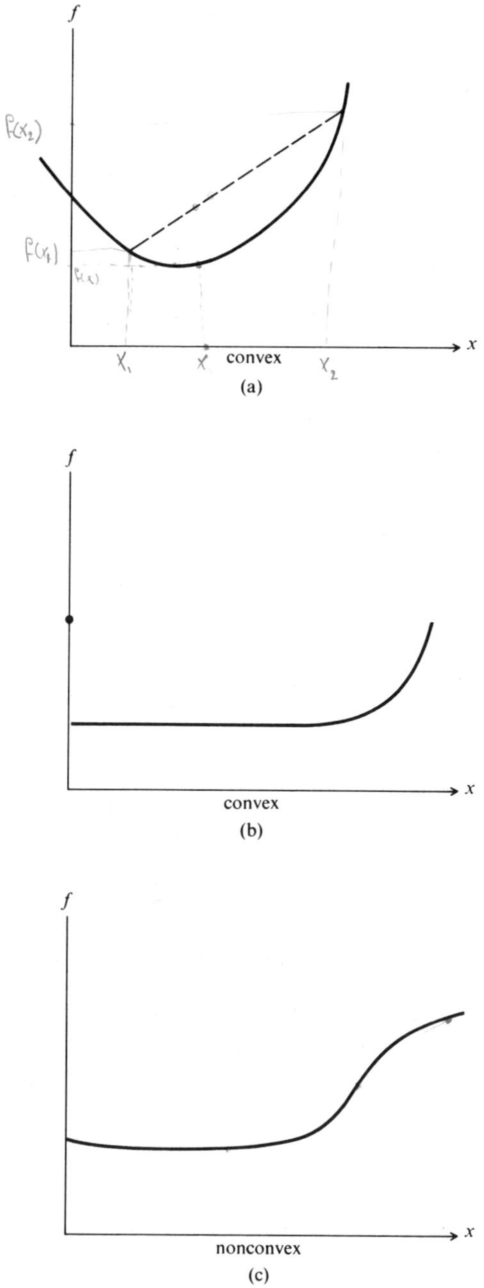

Fig. 6.2 Convex and nonconvex functions

Combinations of Convex Functions

We show that convex functions can be combined to yield new convex functions and that convex functions when used as constraints yield convex constraint sets.

Proposition 1. *Let f_1 and f_2 be convex functions on the convex set Ω. Then the function $f_1 + f_2$ is convex on Ω.*

Proof. Let $\mathbf{x}_1, \mathbf{x}_2 \in \Omega$, and $0 < \alpha < 1$. Then

$$f_1(\alpha\mathbf{x}_1 + (1-\alpha)\mathbf{x}_2) + f_2(\alpha\mathbf{x}_1 + (1-\alpha)\mathbf{x}_2) \leqslant \alpha[f_1(\mathbf{x}_1) + f_2(\mathbf{x}_1)] + (1-\alpha)[f_1(\mathbf{x}_2) + f_2(\mathbf{x}_2)]. \blacksquare$$

Proposition 2. *Let f be a convex function over the convex set Ω. Then the function af is convex for any $a \geqslant 0$.*

Proof. Immediate.

Note that through repeated application of the above two propositions it follows that a positive combination $a_1 f_1 + a_2 f_2 + \cdots + a_m f_m$ of convex functions is again convex.

Finally, we consider sets defined by convex inequality constraints.

Proposition 3. *Let f be a convex function on a convex set Ω. The set $\Gamma_c = \{\mathbf{x} : \mathbf{x} \in \Omega, f(\mathbf{x}) \leqslant c\}$ is convex for every real number c.*

Proof. Let $\mathbf{x}_1, \mathbf{x}_2 \in \Gamma_c$. Then $f(\mathbf{x}_1) \leqslant c, f(\mathbf{x}_2) \leqslant c$ and for $0 < \alpha < 1$,

$$f(\alpha\mathbf{x}_1 + (1-\alpha)\mathbf{x}_2) \leqslant \alpha f(\mathbf{x}_1) + (1-\alpha)f(\mathbf{x}_2) \leqslant c.$$

Thus $\alpha\mathbf{x}_1 + (1-\alpha)\mathbf{x}_2 \in \Gamma_c$. $\blacksquare$

We note that, since the intersection of convex sets is also convex, the set of points simultaneously satisfying

$$f_1(\mathbf{x}) \leqslant c_1, \quad f_2(\mathbf{x}) \leqslant c_2, \ldots, f_m(\mathbf{x}) \leqslant c_m,$$

where each f_i is a convex function, defines a convex set.

Properties of Differentiable Convex Functions

If a function f is differentiable, then there are alternative characterizations of convexity.

Proposition 4. *Let $f \in C^1$. Then f is convex over a convex set Ω if and only if*

$$f(\mathbf{y}) \geqslant f(\mathbf{x}) + \nabla f(\mathbf{x})(\mathbf{y} - \mathbf{x}) \tag{9}$$

for all $\mathbf{x}, \mathbf{y} \in \Omega$.

Proof. First suppose f is convex. Then for all α, $0 \leqslant \alpha \leqslant 1$,

$$f(\alpha\mathbf{y} + (1-\alpha)\mathbf{x}) \leqslant \alpha f(\mathbf{y}) + (1-\alpha)f(\mathbf{x}).$$

Thus for $0 < \alpha \leqslant 1$

$$\frac{f(\mathbf{x} + \alpha(\mathbf{y} - \mathbf{x})) - f(\mathbf{x})}{\alpha} \leqslant f(\mathbf{y}) - f(\mathbf{x}).$$

Letting $\alpha \to 0$ we obtain

$$\nabla f(\mathbf{x})(\mathbf{y} - \mathbf{x}) \leqslant f(\mathbf{y}) - f(\mathbf{x}).$$

This proves the "only if" part.

Now assume

$$f(\mathbf{y}) \geqslant f(\mathbf{x}) + \nabla f(\mathbf{x})(\mathbf{y} - \mathbf{x})$$

for all $\mathbf{x}, \mathbf{y} \in \Omega$. Fix $\mathbf{x}_1, \mathbf{x}_2 \in \Omega$ and α, $0 \leqslant \alpha \leqslant 1$. Setting $\mathbf{x} = \alpha\mathbf{x}_1 + (1 - \alpha)\mathbf{x}_2$ and alternatively $\mathbf{y} = \mathbf{x}_1$ or $\mathbf{y} = \mathbf{x}_2$, we have

$$f(\mathbf{x}_1) \geqslant f(\mathbf{x}) + \nabla f(\mathbf{x})(\mathbf{x}_1 - \mathbf{x}) \tag{10}$$

$$f(\mathbf{x}_2) \geqslant f(\mathbf{x}) + \nabla f(\mathbf{x})(\mathbf{x}_2 - \mathbf{x}). \tag{11}$$

Multiplying (10) by α and (11) by $(1 - \alpha)$ and adding, we obtain

$$\alpha f(\mathbf{x}_1) + (1 - \alpha) f(\mathbf{x}_2) \geqslant f(\mathbf{x}) + \nabla f(\mathbf{x})[\alpha\mathbf{x}_1 + (1 - \alpha)\mathbf{x}_2 - \mathbf{x}].$$

But substituting $\mathbf{x} = \alpha\mathbf{x}_1 + (1 - \alpha)\mathbf{x}_2$, we obtain

$$\alpha f(\mathbf{x}_1) + (1 - \alpha) f(\mathbf{x}_2) \geqslant f(\alpha\mathbf{x}_1 + (1 - \alpha)\mathbf{x}_2). \quad \blacksquare$$

The statement of the above proposition is illustrated in Fig. 6.3. It can be regarded as a sort of dual characterization of the original definition illustrated in Fig. 6.2. The original definition essentially states that linear interpolation between two points overestimates the function, while the above proposition states that linear approximation based on the local derivative underestimates the function.

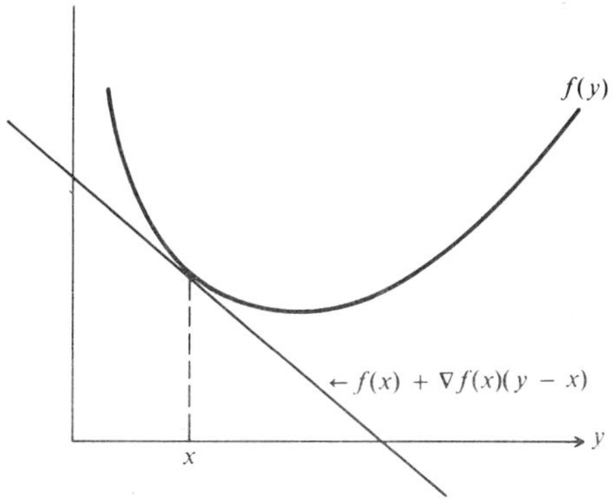

Fig. 6.3 Illustration of Proposition 4

For twice continuously differentiable functions, there is another characterization of convexity.

Proposition 5. *Let $f \in C^2$. Then f is convex over a convex set Ω containing an interior point if and only if the Hessian matrix $\mathbf{F}$ of f is positive semidefinite throughout Ω.*

Proof. By Taylor's theorem we have

$$f(\mathbf{y}) = f(\mathbf{x}) + \nabla f(\mathbf{x})(\mathbf{y} - \mathbf{x}) + \tfrac{1}{2}(\mathbf{y} - \mathbf{x})'\mathbf{F}(\mathbf{x} + \alpha(\mathbf{y} - \mathbf{x}))(\mathbf{y} - \mathbf{x}) \qquad (12)$$

for some α, $0 \leqslant \alpha \leqslant 1$. Clearly, if the Hessian is everywhere positive semidefinite, we have

$$f(\mathbf{y}) \geqslant f(\mathbf{x}) + \nabla f(\mathbf{x})(\mathbf{y} - \mathbf{x}) \qquad (13)$$

which in view of Proposition 4 implies that f is convex.

Now suppose the Hessian is not positive semidefinite at some point $\mathbf{x} \in \Omega$. By continuity of the Hessian it can be assumed, without loss of generality, that $\mathbf{x}$ is an interior point of Ω. There is a $\mathbf{y} \in \Omega$ such that $(\mathbf{y} - \mathbf{x})'\mathbf{F}(\mathbf{x})(\mathbf{y} - \mathbf{x}) < 0$. Again by the continuity of the Hessian, $\mathbf{y}$ may be selected so that for all α, $0 \leqslant \alpha \leqslant 1$,

$$(\mathbf{y} - \mathbf{x})'\mathbf{F}(\mathbf{x} + \alpha(\mathbf{y} - \mathbf{x}))(\mathbf{y} - \mathbf{x}) < 0.$$

This in view of (12) implies that (13) does not hold; which in view of Proposition 4 implies that f is not convex. ∎

The Hessian matrix is the generalization to E^n of the concept of the curvature of a function, and correspondingly, positive definiteness of the Hessian is the generalization of positive curvature. Convex functions have positive (or at least nonnegative) curvature in every direction. Motivated by these observations, we sometimes refer to a function as being *locally convex* if its Hessian matrix is positive semidefinite in a small region, and *locally strictly convex* if the Hessian is positive definite in the region. In these terms we see that the second-order sufficiency result of the last section requires that the function be locally strictly convex at the point x^*. Thus, even the local theory, derived solely in terms of the elementary calculus, is actually intimately related to convexity—at least locally. For this reason we can view the two theories, local and global, not as disjoint parallel developments but as complementary and interactive. Results that are based on convexity apply even to nonconvex problems in a region near the solution, and conversely, local results apply to a global minimum point.

6.4 MINIMIZATION AND MAXIMIZATION OF CONVEX FUNCTIONS

We turn now to the three classic results concerning minimization or maximization of convex functions.

Theorem 1. *Let f be a convex function defined on the convex set Ω. Then the set Γ where f achieves its minimum is convex, and any relative minimum of f is a global minimum.*

Proof. If f has no relative minima the theorem is valid by default. Assume now that c_0 is the minimum of f. Then clearly $\Gamma = \{\mathbf{x} : f(\mathbf{x}) \leqslant c_0, \mathbf{x} \in \Omega\}$ and this is convex by Proposition 3 of the last section.

Suppose now that $\mathbf{x}^* \in \Omega$ is a relative minimum point of f, but that there is another point $\mathbf{y} \in \Omega$ with $f(\mathbf{y}) < f(\mathbf{x}^*)$. On the line $\alpha\mathbf{y} + (1-\alpha)\mathbf{x}^*$, $0 < \alpha < 1$ we have

$$f(\alpha\mathbf{y} + (1-\alpha)\mathbf{x}^*) \leqslant \alpha f(\mathbf{y}) + (1-\alpha)f(\mathbf{x}^*) < f(\mathbf{x}^*)$$

contradicting the fact that $\mathbf{x}^*$ is a relative minimum point. ▮

We might paraphrase the above theorem as saying that for convex functions, all minimum points are located together (in a convex set) and all relative minima are global minima. The next theorem says that if f is continuously differentiable and convex, then satisfaction of the first order necessary conditions are both necessary and sufficient for a point to be a global minimizing point.

Theorem 2. *Let $f \in C^1$ be convex on the convex set Ω. If there is a point $\mathbf{x}^* \in \Omega$ such that, for all $\mathbf{y} \in \Omega$, $\nabla f(\mathbf{x}^*)(\mathbf{y} - \mathbf{x}^*) \geqslant 0$, then $\mathbf{x}^*$ is a global minimum point of f over Ω.*

Proof. We note parenthetically that since $\mathbf{y} - \mathbf{x}^*$ is a feasible direction at $\mathbf{x}^*$, the given condition is equivalent to the first-order necessary condition stated in Section 6.1. The proof of the proposition is immediate, since by Proposition 4 of the last section

$$f(\mathbf{y}) \geqslant f(\mathbf{x}^*) + \nabla f(\mathbf{x}^*)(\mathbf{y} - \mathbf{x}^*) \geqslant f(\mathbf{x}^*). \quad ▮$$

Next we turn to the question of maximizing a convex function over a convex set. There is, however, no analog of Theorem 1 for maximization; indeed, the tendency is for the occurrence of numerous non-global relative maximum points. Nevertheless, it is possible to prove one important result. It is not used in subsequent chapters, but it is useful for some areas of optimization.

Theorem 3. *Let f be a convex function defined on the bounded, closed convex set Ω. If f has a maximum over Ω it is achieved at an extreme point of Ω.*

Proof. Suppose f achieves a global maximum at $\mathbf{x}^* \in \Omega$. We show first that this maximum is achieved at some boundary point of Ω. If $\mathbf{x}^*$ is itself a boundary point, then there is nothing to prove, so assume $\mathbf{x}^*$ is not a boundary point. Let L be any line passing through the point $\mathbf{x}^*$. The intersection of this line with Ω is an interval of the line L having end points $\mathbf{y}_1, \mathbf{y}_2$ which are boundary points of Ω, and we have $\mathbf{x}^* = \alpha\mathbf{y}_1 + (1 - \alpha)\mathbf{y}_2$ for some α, $0 < \alpha < 1$. By convexity of f

$$f(\mathbf{x}^*) \leqslant \alpha f(\mathbf{y}_1) + (1 - \alpha)f(\mathbf{y}_2) \leqslant \max\{f(\mathbf{y}_1), f(\mathbf{y}_2)\}.$$

Thus either $f(\mathbf{y}_1)$ or $f(\mathbf{y}_2)$ must be at least as great as $f(\mathbf{x}^*)$. Since $\mathbf{x}^*$ is a maximum point, so is either $\mathbf{y}_1$ or $\mathbf{y}_2$.

We have shown that the maximum, if achieved, must be achieved at a boundary point of Ω. If this boundary point, $\mathbf{x}^*$, is an extreme point of Ω there is nothing more to prove. If it is not an extreme point, consider the intersection of Ω with a supporting hyperplane H at $\mathbf{x}^*$. This intersection, T_1, is of dimension $n - 1$ or less and the global maximum of f over T_1 is equal to $f(\mathbf{x}^*)$ and must be achieved at a boundary point $\mathbf{x}_1$ of T_1. If this boundary point is an extreme point of T_1, it is also an extreme point of Ω by Lemma 1, Section B.4, and hence the theorem is proved. If $\mathbf{x}_1$ is not an extreme point of T_1, we form T_2, the intersection of T_1 with a hyperplane in E^{n-1} supporting T_1 at $\mathbf{x}_1$. This process can continue at most a total of n times when a set T_n of dimension zero, consisting of a single point, is obtained. This single point is an extreme point of T_n and also, by repeated application of Lemma 1, Section B.4, an extreme point of Ω. ∎

6.5 GLOBAL CONVERGENCE OF DESCENT ALGORITHMS

A good portion of the remainder of this book is devoted to presentation and analysis of various algorithms designed to solve nonlinear programming problems. Although these algorithms vary substantially in their motivation, application, and detailed analysis, ranging from the simple to the highly complex, they have the common heritage of all being iterative descent algorithms. By *iterative*, we mean, roughly, that the algorithm generates a series of points, each point being calculated on the basis of the points preceding it. By *descent*, we mean that as each new point is generated by the algorithm the corresponding value of some function (evaluated at the most recent point) decreases in value. Ideally, the sequence of points generated by the algorithm in this way converges in a finite or infinite number of steps to a solution of the original problem.

An iterative algorithm is initiated by specifying a starting point. If for arbitrary starting points the algorithm is guaranteed to generate a sequence of points converging to a solution, then the algorithm is said to be *globally convergent*. Quite definitely, not all algorithms have this obviously desirable property. Indeed, many of the most important algorithms for solving nonlinear programming problems are not globally convergent in their purest form and thus occasionally generate sequences that either do not converge at all or converge to points that are not solutions. It is often possible, however, to modify such algorithms, by appending special devices, so as to guarantee global convergence.

Fortunately, the subject of global convergence can be treated in a unified manner through the analysis of a general theory of algorithms developed mainly by Zangwill. From this analysis, which is presented in this section, we derive the Global Convergence Theorem that is applicable to the study of any iterative descent algorithm. Frequent reference to this important result is made in subsequent chapters.

Algorithms

We think of an algorithm as a mapping. Given a point $\mathbf{x}$ in some space X, the output of an algorithm applied to $\mathbf{x}$ is a new point. Operated iteratively, an algorithm is repeatedly reapplied to the new points it generates so as to produce a whole sequence of points. Thus, as a preliminary definition, we might formally define an algorithm $\mathbf{A}$ as a mapping taking points in a space X into (other) points in X. Operated iteratively, the algorithm $\mathbf{A}$ initiated at $\mathbf{x}_0 \in X$ would generate the sequence $\{\mathbf{x}_k\}$ defined by

$$\mathbf{x}_{k+1} = \mathbf{A}(\mathbf{x}_k).$$

In practice, the mapping $\mathbf{A}$ might be defined explicitly by a simple mathematical expression or it might be defined implicitly by, say, a lengthy complex computer program. Given an input vector, both define a corresponding output.

With this intuitive idea of an algorithm in mind, we now generalize the concept somewhat so as to provide greater flexibility in our analyses.

Definition. An *algorithm* $\mathbf{A}$ is a mapping defined on a space X that assigns to every point $\mathbf{x} \in X$ a subset of X.

In this definition the term "space" can be interpreted loosely. Usually X is the vector space E^n but it may be only a subset of E^n or even a more general metric space. The most important aspect of the definition, however, is that the mapping $\mathbf{A}$, rather than being a point-to-point mapping of X, is a *point-to-set mapping* of X.

An algorithm $\mathbf{A}$ generates a sequence of points in the following way. Given $\mathbf{x}_k \in X$ the algorithm yields $\mathbf{A}(\mathbf{x}_k)$ which is a subset of X. From this subset an arbitrary element $\mathbf{x}_{k+1}$ is selected. In this way, given an initial point $\mathbf{x}_0$, the algorithm generates sequences through the iteration

$$\mathbf{x}_{k+1} \in \mathbf{A}(\mathbf{x}_k).$$

It is clear that, unlike the case where $\mathbf{A}$ is a point-to-point mapping, the sequence generated by the algorithm $\mathbf{A}$ cannot, in general, be predicted solely from knowledge of the initial point $\mathbf{x}_0$. This degree of uncertainty is designed to reflect uncertainty that we may have in practice as to specific details of an algorithm.

Example 1. Suppose for x on the real line we define

$$A(x) = [-|x|/2, |x|/2]$$

so that $A(x)$ is an interval of the real line. Starting at $x_0 = 100$, each of the sequences below might be generated from iterative application of this algorithm.

$$100, 50, 25, 12, -6, -2, 1, 1/2, \ldots$$

$$100, -40, 20, -5, -2, 1, 1/4, 1/8, \ldots$$

$$100, 10, -1, 1/16, 1/100, -1/1000, 1/10{,}000, \ldots .$$

The apparent ambiguity that is built into this definition of an algorithm is not meant to imply that actual algorithms are random in character. In actual implementation algorithms are not defined ambiguously. Indeed, a particular computer program executed twice from the same starting point will generate two copies of the same sequence. In other words, in practice algorithms are point-to-point mappings. The utility of the more general definition is that it allows one to analyze, in a single step, the convergence of an infinite family of similar algorithms. Thus, two computer programs, designed from the same basic idea, may differ slightly in some details, and therefore perhaps may not produce identical results when given the same starting point. Both programs may, however, be regarded as implementations of the same point-to-set mappings. In the example above, for instance, it is not necessary to know exactly how x_{k+1} is determined from x_k so long as it is known that its absolute value is no greater than one-half x_k's absolute value. In this manner, the generalized concept of an algorithm sometimes leads to simpler analysis.

Descent

In order to describe the idea of a descent algorithm we first must agree on a subset Γ of the space X, referred to as the *solution set*. The basic idea of a *descent function*, which is defined below, is that for points outside the solution set, a single step of the algorithm yields a decrease in the value of the descent function.

Definition. Let $\Gamma \subset X$ be a given solution set and let $\mathbf{A}$ be an algorithm on X. A continuous real-valued function Z on X is said to be a *descent function* for Γ and $\mathbf{A}$ if it satisfies

i) if $\mathbf{x} \notin \Gamma$ and $\mathbf{y} \in \mathbf{A}(\mathbf{x})$, then $Z(\mathbf{y}) < Z(\mathbf{x})$

ii) if $\mathbf{x} \in \Gamma$ and $\mathbf{y} \in \mathbf{A}(\mathbf{x})$, then $Z(\mathbf{y}) \leqslant Z(\mathbf{x})$.

There are a number of ways a solution set, algorithm, and descent function can be defined. A natural set-up for the problem

$$\begin{aligned} &\text{minimize} \quad f(\mathbf{x}) \\ &\text{subject to} \quad \mathbf{x} \in \Omega \end{aligned} \tag{14}$$

is to let Γ be the set of minimizing points, and define an algorithm $\mathbf{A}$ on Ω in such a way that f decreases at each step and thereby serves as a descent function. Indeed, this is the procedure followed in a majority of cases. Another possibility for unconstrained problems is to let Γ be the set of points $\mathbf{x}$ satisfying $\nabla f(\mathbf{x}) = \mathbf{0}$. In this case we might design an algorithm for which $|\nabla f(\mathbf{x})|$ serves as a descent function or for which $f(\mathbf{x})$ serves as a descent function.

Closed Mappings

An important property possessed by some algorithms is that they are closed. This property, which is a generalization for point-to-set mappings of the concept of continuity for point-to-point mappings, turns out to be the key to establishing a general global convergence theorem. In defining this property we allow the point-to-set mapping to map points in one space X into subsets of another space Y.

Definition. A point-to-set mapping $\mathbf{A}$ from X to Y is said to be *closed* at $\mathbf{x} \in X$ if the assumptions

i) $\mathbf{x}_k \to \mathbf{x}, \quad \mathbf{x}_k \in X,$

ii) $\mathbf{y}_k \to \mathbf{y}, \quad \mathbf{y}_k \in \mathbf{A}(\mathbf{x}_k)$

imply

iii) $\mathbf{y} \in \mathbf{A}(\mathbf{x})$.

The point-to-set map $\mathbf{A}$ is said to be *closed* on X if it is closed at each point of X.

Example 2. As a special case, suppose that the mapping $\mathbf{A}$ is a point-to-point mapping, that is, for each $\mathbf{x} \in X$ the set $\mathbf{A}(\mathbf{x})$ consists of a single point in Y. Suppose also that $\mathbf{A}$ is continuous at $\mathbf{x} \in X$. This means that if $\mathbf{x}_k \to \mathbf{x}$ then $\mathbf{A}(\mathbf{x}_k) \to \mathbf{A}(\mathbf{x})$, and it follows that $\mathbf{A}$ is closed at $\mathbf{x}$. Thus for point-to-point mappings continuity implies closedness. The converse is, however, not true in general.

Example 3. The reader should verify that the point-to-set mapping defined in Example 1 is closed.

Many complex algorithms that we analyze are most conveniently regarded as the composition of two or more simple point-to-set mappings. It is therefore natural to ask whether closedness of the individual maps implies closedness of the composite. The answer is a qualified "yes".

Definition. Let $\mathbf{A}: X \to Y$ and $\mathbf{B}: Y \to Z$ be point-to-set mappings. The composite mapping $\mathbf{C} = \mathbf{BA}$ is defined as the point-to-set mapping $\mathbf{C}: X \to Z$ with

$$\mathbf{C}(\mathbf{x}) = \bigcup_{\mathbf{y} \in \mathbf{A}(\mathbf{x})} \mathbf{B}(\mathbf{y}).$$

This definition is illustrated in Fig. 6.4.

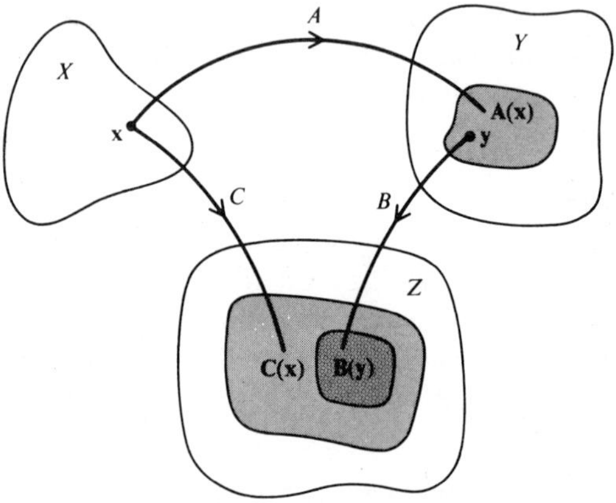

Fig. 6.4 Composition of mappings

Proposition. *Let $\mathbf{A}: X \to Y$ and $\mathbf{B}: Y \to Z$ be point-to-set mappings. Suppose $\mathbf{A}$ is closed at $\mathbf{x}$ and $\mathbf{B}$ is closed on $\mathbf{A}(\mathbf{x})$. Suppose also that if $\mathbf{x}_k \to \mathbf{x}$ and $\mathbf{y}_k \in \mathbf{A}(\mathbf{x}_k)$, there is a $\mathbf{y}$ such that, for some subsequence $\{\mathbf{y}_{k_i}\}$, $\mathbf{y}_{k_i} \to \mathbf{y}$. Then the composite mapping $\mathbf{C} = \mathbf{BA}$ is closed at $\mathbf{x}$.*

Proof. Let $\mathbf{x}_k \to \mathbf{x}$ and $\mathbf{z}_k \to \mathbf{z}$ with $\mathbf{z}_k \in \mathbf{C}(\mathbf{x}_k)$. It must be shown that $\mathbf{z} \in \mathbf{C}(\mathbf{x})$.

Select $\mathbf{y}_k \in \mathbf{A}(\mathbf{x}_k)$ such that $\mathbf{z}_k \in \mathbf{B}(\mathbf{y}_k)$ and according to the hypothesis let $\mathbf{y}$ and $\{\mathbf{y}_{k_i}\}$ be such that $\mathbf{y}_{k_i} \to \mathbf{y}$. Since $\mathbf{A}$ is closed at $\mathbf{x}$ it follows that $\mathbf{y} \in \mathbf{A}(\mathbf{x})$.

Likewise, since $\mathbf{y}_{k_i} \to \mathbf{y}$, $\mathbf{z}_{k_i} \to \mathbf{z}$ and $\mathbf{B}$ is closed at $\mathbf{y}$, it follows that $\mathbf{z} \in \mathbf{B}(\mathbf{y}) \subset \mathbf{BA}(\mathbf{x}) = \mathbf{C}(\mathbf{x})$. ∎

Two important corollaries follow immediately.

Corollary 1. *Let $\mathbf{A}: X \to Y$ and $\mathbf{B}: Y \to Z$ be point-to-set mappings. If $\mathbf{A}$ is closed at $\mathbf{x}$, $\mathbf{B}$ is closed on $\mathbf{A}(\mathbf{x})$ and Y is compact, then the composite map $\mathbf{C} = \mathbf{BA}$ is closed at $\mathbf{x}$.*

Corollary 2. *Let* $\mathbf{A}: X \to Y$ *be a point-to-point mapping and* $\mathbf{B}: Y \to Z$ *a point-to-point mapping. If* $\mathbf{A}$ *is continuous at* $\mathbf{x}$ *and* $\mathbf{B}$ *is closed at* $\mathbf{A}(\mathbf{x})$*, then the composite mapping* $\mathbf{C} = \mathbf{BA}$ *is closed at* $\mathbf{x}$.

Global Convergence Theorem

We now state the main result.

Global Convergence Theorem. *Let* $\mathbf{A}$ *be an algorithm on* X*, and suppose that given* $\mathbf{x}_0$ *the sequence* $\{\mathbf{x}_k\}_{k=0}^{\infty}$ *is generated satisfying*

$$\mathbf{x}_{k+1} \in \mathbf{A}(\mathbf{x}_k).$$

Let a solution set $\Gamma \subset X$ *be given, and suppose*

i) *all points* $\mathbf{x}_k$ *are contained in a compact set* $S \subset X$.

ii) *there is a continuous function* Z *on* X *such that*

(a) *if* $\mathbf{x} \notin \Gamma$*, then* $Z(\mathbf{y}) < Z(\mathbf{x})$ *for all* $\mathbf{y} \in \mathbf{A}(\mathbf{x})$.

(b) *if* $\mathbf{x} \in \Gamma$*, then* $Z(\mathbf{y}) \leqslant Z(\mathbf{x})$ *for all* $\mathbf{y} \in \mathbf{A}(\mathbf{x})$.

iii) *the mapping* $\mathbf{A}$ *is closed at points outside* Γ.

Then the limit of any convergent subsequence of $\{\mathbf{x}_k\}$ *is a solution.*

Proof. Suppose the convergent subsequence $\{\mathbf{x}_k\}, k \in \mathscr{K}$ converges to the limit $\mathbf{x}$. Since Z is continuous, it follows that for $k \in \mathscr{K}$, $Z(\mathbf{x}_k) \to Z(\mathbf{x})$. This means that Z is convergent with respect to the subsequence, and we shall show that it is convergent with respect to the entire sequence. By the monotonicity of Z on the sequence $\{\mathbf{x}_k\}$ we have $Z(\mathbf{x}_k) - Z(\mathbf{x}) \geqslant 0$ for all k. By the convergence of Z on the subsequence, there is, for a given $\varepsilon > 0$, a $K \in \mathscr{K}$ such that $Z(\mathbf{x}_k) - Z(\mathbf{x}) < \varepsilon$ for all $k > K, k \in \mathscr{K}$.

Thus for all $k > K$

$$Z(\mathbf{x}_k) - Z(\mathbf{x}) = Z(\mathbf{x}_k) - Z(\mathbf{x}_K) + Z(\mathbf{x}_K) - Z(\mathbf{x}) < \varepsilon,$$

which shows that $Z(\mathbf{x}_k) \to Z(\mathbf{x})$.

To complete the proof it is only necessary to show that $\mathbf{x}$ is a solution. Suppose $\mathbf{x}$ is not a solution. Consider the subsequence $\{\mathbf{x}_{k+1}\}_{\mathscr{K}}$. Since all members of this sequence are contained in a compact set, there is a $\overline{\mathscr{K}} \subset \mathscr{K}$ such that $\{\mathbf{x}_{k+1}\}_{\overline{\mathscr{K}}}$ converges to some limit $\bar{\mathbf{x}}$. We thus have $\mathbf{x}_k \to \mathbf{x}, k \in \overline{\mathscr{K}}$, and $\mathbf{x}_{k+1} \in \mathbf{A}(\mathbf{x}_k)$ with $\mathbf{x}_{k+1} \to \bar{\mathbf{x}}, k \in \overline{\mathscr{K}}$. Thus since $\mathbf{A}$ is closed at $\mathbf{x}$ it follows that $\bar{\mathbf{x}} \in \mathbf{A}(\mathbf{x})$. But from above, $Z(\bar{\mathbf{x}}) = Z(\mathbf{x})$ which contradicts the fact that Z is a descent function. ∎

In later chapters the Global Convergence Theorem is used to establish the convergence of several standard algorithms. Here we consider some simple examples designed to illustrate the roles of the various conditions of the theorem.

Example 4. In many respects condition (iii) of the theorem, the closedness of **A** outside the solution set, is the most important condition. The failure of many popular algorithms can be traced to nonsatisfaction of this condition. On the real line consider the point-to-point algorithm

$$A(x) = \begin{cases} \frac{1}{2}(x-1)+1 & x > 1 \\ \frac{1}{2}x & x \leqslant 1 \end{cases}$$

and the solution set $\Gamma = \{0\}$. It is easily verified that a descent function for this solution set and this algorithm is $Z(x) = |x|$. However, starting from $x > 1$, the algorithm generates a sequence converging to $x = 1$ which is not a solution. The difficulty is that A is not closed at $x = 1$.

Example 5. On the real line X consider the solution set to be empty, the descent function $Z(x) = e^{-x}$, and the algorithm $A(x) = x + 1$. All conditions of the convergence theorem except (i) hold. The sequence generated from any starting condition diverges to infinity. This is not strictly a violation of the conclusion of the theorem but simply an example illustrating that if no compactness assumption is introduced, the generated sequence may have no convergent subsequence.

Example 6. Consider the point-to-set algorithm A defined by the graph in Fig. 6.5 and given explicitly on $X = [0, 1]$ by

$$A(x) = \begin{cases} [0, x) & 1 \geqslant x > 0 \\ 0 & x = 0. \end{cases}$$

Letting $\Gamma = \{0\}$, the function $Z(x) = x$ serves as a descent function, because for $x \neq 0$ all points in $A(x)$ are less than x.

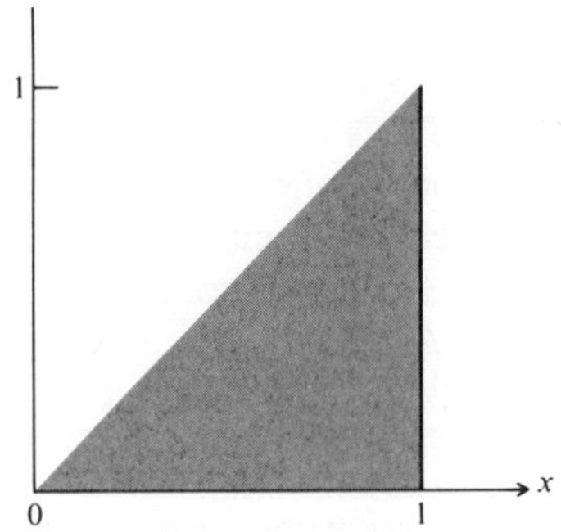

Fig. 6.5 Graph for example

The sequence defined by

$$x_0 = 1$$

$$x_{k+1} = x_k - \frac{1}{2^{k+2}}$$

satisfies $x_{k+1} \in A(x_k)$ but it can easily be seen that $x_k \to \frac{1}{2} \notin \Gamma$. The difficulty here, of course, is that the algorithm A is not closed outside the solution set.

6.6 SPEED OF CONVERGENCE

The study of speed of convergence is substantially deeper than that of global convergence, and the principal results cannot be cast in the form of a single theorem as is possible in the case of global convergence. Nevertheless, there is a rich and yet elementary theory of convergence rates that enables one to predict with confidence the relative effectiveness of a wide class of algorithms. In this section we introduce various concepts designed to measure speed of convergence, and prepare for a study of this most important aspect of nonlinear programming.

Order of Convergence

Consider a sequence of real numbers $\{r_k\}_{k=0}^{\infty}$ converging to the limit r^*. We define several notions related to the speed of convergence of such a sequence.

Definition. Let the sequence $\{r_k\}$ converge to r^*. The *order* of convergence of $\{r_k\}$ is defined as the supremum of the nonnegative numbers p satisfying

$$0 \leqslant \overline{\lim_{k \to \infty}} \frac{|r_{k+1} - r^*|}{|r_k - r^*|^p} < \infty.$$

To ensure that the definition is applicable to any sequence, it is stated in terms of limit superior rather than just limit and 0/0 (which occurs if $r_k = r^*$ for all k) is regarded as finite. But these technicalities are rarely necessary in actual analysis, since the sequences generated by algorithms are generally quite well behaved.

It should be noted that the order of convergence, as with all other notions related to speed of convergence that are introduced, is determined only by the properties of the sequence that hold as $k \to \infty$. Somewhat loosely but picturesquely, we are therefore led to refer to the *tail* of a sequence—that part of the sequence that is arbitrarily far out. In this language we might say that the order of convergence is a measure of how good the worst part of the tail is. Larger values of the order p imply, in a sense, faster convergence, since the distance from the limit r^* is reduced, at least in the tail, by the pth power in a single step. Indeed, if the sequence has order p and (as is the usual case) the limit

$$\beta = \lim_{k\to\infty} \frac{|r_{k+1} - r^*|}{|r_k - r^*|^p}$$

exists, then asymptotically we have

$$|r_{k+1} - r^*| = \beta |r_k - r^*|^p.$$

Example 1. The sequence with $r_k = a^k$ where $0 < a < 1$ converges to zero with order unity, since $r_{k+1}/r_k = a$.

Example 2. The sequence with $r_k = a^{(2^k)}$ for $0 < a < 1$ converges to zero with order two, since $r_{k+1}/r_k^{\,2} = 1$.

Linear Convergence

Most algorithms discussed in this book have an order of convergence equal to unity. It is therefore appropriate to consider this class in greater detail and distinguish certain cases within it.

Definition. If the sequence $\{r_k\}$ converges to r^* in such a way that

$$\lim_{k\to\infty} \frac{|r_{k+1} - r^*|}{|r_k - r^*|} = \beta < 1$$

the sequence is said to converge *linearly* to r^* with *convergence ratio* β.

Linear convergence is, for our purposes, without doubt the most important type of convergence behavior. A linearly convergent sequence, with convergence ratio β, can be said to have a tail that converges at least as fast as the geometric sequence $c\beta^k$ for some constant c. Thus linear convergence is sometimes referred to as *geometric convergence*, although in this book we reserve that phrase for the case when a sequence is exactly geometric.

As a rule, when comparing the relative effectiveness of two competing algorithms both of which produce linearly convergent sequences, the comparison is based on their corresponding convergence ratios—the smaller the ratio the faster the rate. The ultimate case where $\beta = 0$ is referred to as *superlinear convergence*. We note immediately that convergence of any order greater than unity is superlinear, but it is also possible for superlinear convergence to correspond to unity order.

Example 3. The sequence $r_k = 1/k$ converges to zero. The convergence is of order one but it is not linear, since $\lim_{k\to\infty} (r_{k+1}/r_k) = 1$.

Example 4. The sequence $r_k = (1/k)^k$ is of order unity, since $r_{k+1}/r_k{}^p \to \infty$ for $p > 1$. However, $r_{k+1}/r_k \to 0$ as $k \to \infty$ and hence this is superlinear convergence.

Average Rates

All the definitions given above can be referred to as *step-wise* concepts of convergence, since they define bounds on the progress made by going a single step: from k to $k+1$. Another approach is to define concepts related to the average progress per step over a large number of steps. We briefly illustrate how this can be done.

Definition. Let the sequence $\{r_k\}$ converge to r^*. The *average order* of convergence is the infinum of the numbers $p > 1$ such that

$$\overline{\lim_{k\to\infty}} |r_k - r^*|^{1/p^k} = 1.$$

The order is infinity if the equality holds for no $p > 1$.

Example 5. For the sequence $r_k = a^{(2^k)}$, $0 < a < 1$, given in Example 2, we have

$$|r_k|^{1/2^k} = a$$

while

$$|r_k|^{1/p^k} = a^{(2/p)^k} \to 1$$

for $p > 2$. Thus the average order is two.

Example 6. For $r_k = a^k$ with $0 < a < 1$ we have

$$(r_k)^{1/p^k} = a^{k(1/p)^k} \to 1$$

for any $p > 1$. Thus the average order is unity.

As before, the most important case is that of unity order, and in this case we define the *average convergence ratio* as $\overline{\lim_{k\to\infty}} |r_k - r^*|^{1/k}$. Thus for the geometric sequence $r_k = ca^k$, $0 < a < 1$, the average convergence ratio is a. Paralleling the earlier definitions, the reader can then in a similar manner define corresponding notions of average linear and average superlinear convergence.

Although the above array of definitions can be further embellished and expanded, it is quite adequate for our purposes. For the most part we work with the step-wise definitions, since in analyzing iterative algorithms it is natural to compare one step with the next. In most situations, moreover, when the sequences are well-behaved and the limits exist in the definitions, then the step-wise and average concepts of convergence rates coincide.

Convergence of Vectors

Suppose $\{\mathbf{x}_k\}_{k=0}^{\infty}$ is a sequence of vectors in E^n converging to a vector $\mathbf{x}^*$. The convergence properties of such a sequence are defined with respect to some particular function that converts the sequence of vectors into a sequence of numbers. Thus, if f is given continuous function on E^n, the convergence properties of $\{\mathbf{x}_k\}$ can be defined with respect to f by analyzing the convergence of $f(\mathbf{x}_k)$ to $f(\mathbf{x}^*)$. The function f used in this way to measure convergence is called the *error function.*

In optimization theory it is common to choose the error function by which to measure convergence as the same function that defines the objective function of the original optimization problem. This means we measure convergence by how fast the objective converges to its minimum. Alternatively, we sometimes use the function $|\mathbf{x} - \mathbf{x}^*|^2$ and thereby measure convergence by how fast the (squared) distance from the solution point decreases to zero.

Generally, the order of convergence of a sequence is insensitive to the particular error function used; but for step-wise linear convergence the associated convergence ratio is not. Nevertheless, the average convergence ratio is not too sensitive, as the following proposition demonstrates, and hence the particular error function used to measure convergence is not really very important.

Proposition. *Let f and g be two error functions satisfying $f(\mathbf{x}^*) = g(\mathbf{x}^*) = 0$ and, for all $\mathbf{x}$, a relation of the form*

$$0 \leqslant a_1 g(\mathbf{x}) \leqslant f(\mathbf{x}) \leqslant a_2 g(\mathbf{x})$$

for some fixed $a_1 > 0$, $a_2 > 0$. If the sequence $\{\mathbf{x}_k\}_{k=0}^{\infty}$ converges to $\mathbf{x}^*$ linearly with average ratio β with respect to one of these functions, it also does so with respect to the other.

Proof. The statement is easily seen to be symmetric in f and g. Thus we assume $\{\mathbf{x}_k\}$ is linearly convergent with average convergence ratio β with respect to f, and will prove that the same is true with respect to g. We have

$$\beta = \overline{\lim_{k\to\infty}}\, f(\mathbf{x}_k)^{1/k} \leqslant \overline{\lim_{k\to\infty}}\, {a_2}^{1/k} g(\mathbf{x}_k)^{1/k} = \overline{\lim_{k\to\infty}}\, g(\mathbf{x}_k)^{1/k}$$

and

$$\beta = \overline{\lim_{k\to\infty}}\, f(\mathbf{x}_k)^{1/k} \geqslant \overline{\lim_{k\to\infty}}\, {a_1}^{1/k} g(\mathbf{x}_k)^{1/k} = \overline{\lim_{k\to\infty}}\, g(\mathbf{x}_k)^{1/k}.$$

Thus

$$\beta = \overline{\lim_{k\to\infty}}\, g(\mathbf{x}_k)^{1/k}. \blacksquare$$

As an example of an application of the above proposition, consider the case where $g(\mathbf{x}) = |\mathbf{x} - \mathbf{x}^*|^2$ and $f(\mathbf{x}) = (\mathbf{x} - \mathbf{x}^*)'\mathbf{Q}(\mathbf{x} - \mathbf{x}^*)$, where $\mathbf{Q}$ is a positive definite symmetric matrix. Then a_1 and a_2 correspond, respectively, to the smallest and largest eigenvalues of $\mathbf{Q}$. Thus, average linear convergence is identical with respect to any error function constructed from a positive definite quadratic form.

6.7 SUMMARY

There are two different but complementary ways to characterize the solution to unconstrained optimization problems. In the local approach, one examines the relation of a given point to its neighbors. This leads to the conclusion that, at an unconstrained relative minimum point of a smooth function, the gradient of the function vanishes and the Hessian is positive semidefinite; and conversely, if at a point the gradient vanishes and the Hessian is positive definite, that point is a relative minimum point. This characterization has a natural extension to the global approach where convexity ensures that if the gradient vanishes at a point, that point is a global minimum point.

In considering iterative algorithms for finding either local or global minimum points, there are two distinct issues: global convergence properties and local convergence properties. The first is concerned with whether starting at an arbitrary point the sequence generated will converge to a solution. This is insured if the algorithm is closed, has a descent function, and generates a bounded sequence. Local convergence properties are a measure of the ultimate speed of convergence and generally determine the relative advantage of one algorithm to another.

6.8 EXERCISES

1. a) Using the first-order necessary conditions, find a minimum point of the function

$$f(x, y, z) = 2x^2 + xy + y^2 + yz + z^2 - 6x - 7y - 8z + 9.$$

   b) Verify that the point is a relative minimum point by verifying that the second-order sufficiency conditions hold.

   c) Prove that the point is a global minimum point.

2. In this problem and the next we develop a method for determining whether a given symmetric matrix is positive definite. Given an $n \times n$ matrix $\mathbf{A}$ let $\mathbf{A}_k$ denote the principle minor matrix made up of the first k rows and columns. Show (by induction) that if the first $n - 1$ principle minors are nonsingular, then there is a unique lower triangular matrix $\mathbf{L}$ with unit diagonal and a unique upper triangular matrix $\mathbf{U}$ such that $\mathbf{A} = \mathbf{LU}$. (See Appendix C.)

3. A symmetric matrix is positive definite if and only if the determinant of each of its principle minor matrices is positive. Using this fact and the considerations of Exercise 2, show that an $n \times n$ symmetric matrix $\mathbf{A}$ is positive definite if and only

if it has an **LU** decomposition (without interchange of rows) and the diagonal elements of **U** are all positive.

4. Using Exercise 3 show that a $n \times n$ matrix $\mathbf{A}$ is symmetric and positive definite if and only if it can be written as $\mathbf{A} = \mathbf{GG}'$ where $\mathbf{G}$ is a lower triangular matrix with positive diagonal elements.
5. Let f_i, $i \in I$ be a collection of convex functions defined on a convex set Ω. Show that the function f defined by $f(\mathbf{x}) = \sup_{i \in I} f_i(\mathbf{x})$ is convex on the region where it is finite.
6. Let γ be a monotone nondecreasing function of a single variable (that is, $\gamma(r) \leqslant \gamma(r')$ for $r' > r$) which is also convex; and let f be a convex function defined on a convex set Ω. Show that the function $\gamma(f)$ defined by $\gamma(f)(\mathbf{x}) = \gamma[f(\mathbf{x})]$ is convex on Ω.
7. Let f be twice continuously differentiable on a region $\Omega \subset E^n$. Show that a sufficient condition for a point $\mathbf{x}^*$ in the interior of Ω to be a relative minimum point of f is that $\nabla f(\mathbf{x}^*) = \mathbf{0}$ and that f be locally convex at $\mathbf{x}^*$.
8. Define the point-to-set mapping on E^n by
$$\mathbf{A}(\mathbf{x}) = \{\mathbf{y} : \mathbf{y}'\mathbf{x} \leqslant b\}$$
where b is a fixed constant. Is $\mathbf{A}$ closed?
9. Prove the two corollaries in Section 6.5 on the closedness of composite mappings.
10. Show that if $\mathbf{A}$ is a continuous point-to-point mapping, the Global Convergence Theorem is valid even without assumption (i). Compare with Example 2, Section 6.5.
11. Let $\{r_k\}_{k=0}^{\infty}$ and $\{c_k\}_{k=0}^{\infty}$ be sequences of real numbers. Suppose $r_k \to 0$ average linearly and that there are constants $c > 0$ and C such that $c \leqslant c_k \leqslant C$ for all k. Show that $c_k r_k \to 0$ average linearly.
12. Prove a proposition, similar to the one in Section 6.6, showing that the order of convergence is insensitive to the error function.
13. Show that if $r_k \to r^*$ (step-wise) linearly with convergence ratio β, then $r_k \to r^*$ (average) linearly with average convergence ratio no greater than β.

REFERENCES

6.1–6.4. For alternative discussion of the material in these sections see Hadley [H2], Fiacco and McCormick [F3], Zangwill [Z2] or Luenberger [L6].
6.5. The idea of using a descent function (usually the objective itself) in order to guarantee convergence of minimization algorithms is an old one that runs through most literature on optimization, and has long been used to establish global convergence. Formulation of the general global convergence theorem, which captures the essence of many previously diverse arguments, and the idea of representing an algorithm as a point-to-set mapping are both due to Zangwill [Z2].
6.6. Most of the definitions given in this section have been standard for quite some time. A recent thorough discussion which contributes substantially to the unification of these concepts is contained on Ortega and Rheinboldt [O5].

Chapter 7 BASIC DESCENT METHODS

We turn now to a description of the basic techniques used for iteratively solving unconstrained minimization problems. These techniques are, of course, important for practical application since they often offer the simplest, most direct alternatives for obtaining solutions; but perhaps their greatest importance is that they establish certain reference plateaus with respect to difficulty of implementation and speed of convergence. Thus, in later chapters as more efficient techniques and techniques capable of handling constraints are developed, reference is continually made to the basic techniques of this chapter both for guidance and as points of comparison.

There is a fundamental underlying structure for almost all the descent algorithms we discuss. One starts at an initial point; determines, according to a fixed rule, a direction of movement; and then moves in that direction to a (relative) minimum of the objective function on that line. At the new point a new direction is determined and the process is repeated. The primary differences between algorithms (steepest descent, Newton's method, etc.) rest with the rule by which successive directions of movement are selected. Once the selection is made, all algorithms call for movement to the minimum point on the corresponding line.

The process of determining the minimum point on a given line is called *line search.* For general nonlinear functions that cannot be minimized analytically, this process actually is accomplished by searching, in an intelligent manner, along the line for the minimum point. These line search techniques, which are really procedures for solving one-dimensional minimization problems, form the backbone of nonlinear programming algorithms, since higher dimensional problems are ultimately solved by executing a sequence of successive line searches. There are a number of different approaches to this important phase of minimization and the first half of this chapter is devoted to their discussion.

The last sections of the chapter are devoted to a description and analysis of the basic descent algorithms for unconstrained problems: steepest descent,

Newton's method, and coordinate descent. These algorithms serve as primary models for the development and analysis of all others discussed in the book.

7.1 FIBONACCI AND GOLDEN SECTION SEARCH

A very popular method for resolving the line search problem is the Fibonacci search method described in this section. The method has a certain degree of theoretical elegance, which no doubt partially accounts for its popularity, but on the whole the procedure has only marginal value since, as we shall see, there are others which in most circumstances are vastly superior.

Let f be a function of a single (scalar) variable x on an interval $[c_1, c_2]$. The only property that is assumed of f is that it is *unimodal*, that is, it has a single relative minimum (see Fig. 7.1). The minimum point of f is to be determined, at least approximately, by measuring the value of f at a certain number of points. It should be imagined, as is indeed the case in the setting of nonlinear programming, that each measurement of f is somewhat costly—of time if nothing more.

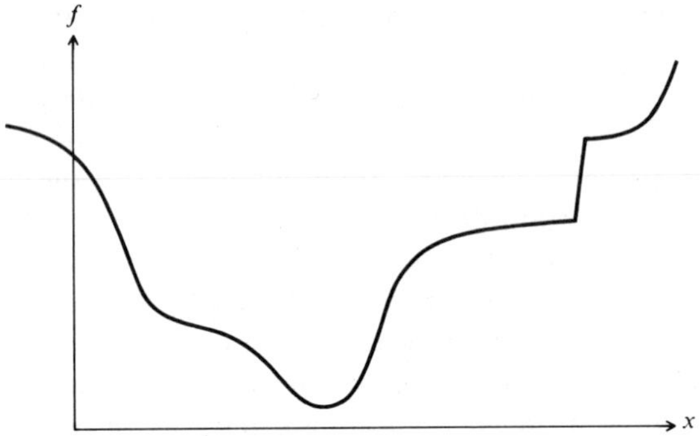

Fig. 7.1 A unimodal function

To develop an appropriate search strategy, that is, a strategy for selecting measurement points based on the previously obtained values, we pose the following problem: Find how to successively select N measurement points so that, without explicit knowledge of f, we can determine the smallest possible region of uncertainty in which the minimum must lie. In this problem the region of uncertainty is determined in any particular case by the relative values of the measured points in conjunction with our assumption that f is unimodal. Thus, after values are known at N points $x_1, x_2, \ldots, x_N$ with

$$c_1 \leqslant x_1 < x_2 \cdots < x_{N-1} < x_N \leqslant c_2,$$

the region of uncertainty is the interval $[x_{k-1}, x_{k+1}]$ where x_k is the minimum point among the N, and we define $x_0 = c_1, x_{N+1} = c_2$ for consistency. The minimum of f must lie somewhere in this interval.

The derivation of the optimal strategy for successively selecting measurement points to obtain the smallest region of uncertainty is fairly straightforward but somewhat tedious. We simply state the result and give an example.

Let

$$d_1 = c_2 - c_1, \text{ the initial width of uncertainty}$$
$$d_k = \text{width of uncertainty after } k \text{ measurements.}$$

Then, if a total of N measurements are to be made, we have

$$d_k = \left(\frac{F_{N-k+1}}{F_N}\right) d_1 \tag{1}$$

where the integers F_k are members of the Fibonacci sequence generated by the recurrence relation

$$F_N = F_{N-1} + F_{N-2}, \qquad F_0 = F_1 = 1. \tag{2}$$

The procedure for reducing the width of uncertainty to d_N is this: The first two measurements are made symmetrically at a distance of $(F_{N-1}/F_N)d_1$ from the ends of the initial intervals; according to which of these is of lesser value, an uncertainty interval of width $d_2 = (F_{N-1}/F_N)d_1$ is determined. The third measurement point is placed symmetrically in this new interval of uncertainty with respect to the measurement already in the interval. The result of this third measurement gives an interval of uncertainty $d_3 = (F_{N-2}/F_N)d_1$. In general, each successive measurement point is placed in the current interval of uncertainty symmetrically with the point already existing in that interval.

Some examples are shown in Fig. 7.2. In these examples the sequence of measurement points is determined in accordance with the assumption that each measurement is of lower value than its predecessors. Note that the procedure always calls for the last two measurements to be made at the midpoint of the semifinal interval of uncertainty. We are to imagine that these two points are actually separated a small distance so that a comparison of their respective values will reduce the interval to nearly half. This terminal anomaly of the Fibonacci search process is, of course, of no great practical consequence.

Search by Golden Section

If the number N of allowed measurement points in a Fibonacci search is made to approach infinity, we obtain the golden section method. It can be argued, based on the optimal property of the finite Fibonacci method, that the corresponding infinite version yields a sequence of intervals of uncertainty whose widths tend to zero faster than that which would be obtained by other methods.

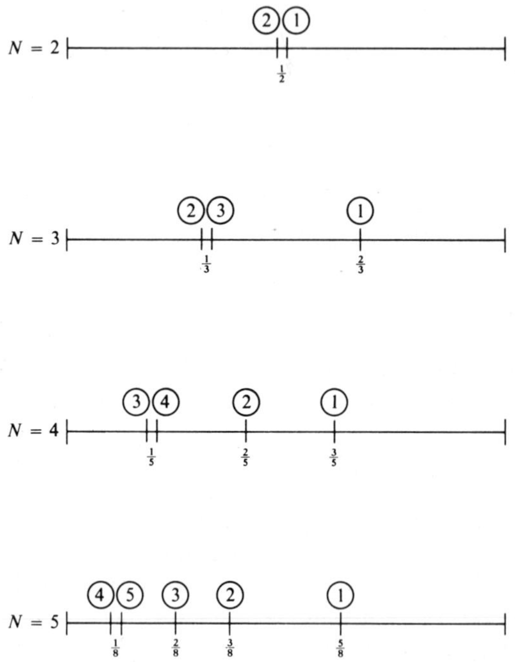

Fig. 7.2 Fibonacci search

The solution to the Fibonacci difference equation

$$F_N = F_{N-1} + F_{N-2} \tag{3}$$

is of the form

$$F_N = A\tau_1{}^N + B\tau_2{}^N \tag{4}$$

where τ_1 and τ_2 are roots of the characteristic equation

$$\tau^2 = \tau + 1.$$

Explicitly,

$$\tau_1 = \frac{1+\sqrt{5}}{2}, \qquad \tau_2 = \frac{1-\sqrt{5}}{2}.$$

(The number $\tau_1 \simeq 1.618$ is known as the *golden section* ratio and was considered by early Greeks to be the most aesthetic value for the ratio of two adjacent sides of a rectangle.)

For large N the first term on the right side of (4) dominates the second, and hence

$$\lim_{N\to\infty} \frac{F_{N-1}}{F_N} = \frac{1}{\tau_1} \simeq 0.618.$$

It follows from (1) that the interval of uncertainty at any point in the process has width

$$d_k = \left(\frac{1}{\tau_1}\right)^{k-1} d_1 \tag{5}$$

and from this it follows that

$$\frac{d_{k+1}}{d_k} = \frac{1}{\tau_1} = 0.618. \tag{6}$$

Therefore, we conclude that, with respect to the width of the uncertainty interval, the search by golden section converges linearly (see Section 6.6) to the overall minimum of the function f with convergence ratio $1/\tau_1 = 0.618$.

7.2 LINE SEARCH BY CURVE FITTING

The Fibonacci search method has a certain amount of theoretical appeal, since it assumes only that the function being searched is unimodal and with respect to this broad class of functions the method is, in some sense, optimal. In most problems, however, it can be safely assumed that the function being searched, as well as being unimodal, possesses a certain degree of smoothness, and one might, therefore, expect that more efficient search techniques exploiting this smoothness can be devised; and indeed they can. Techniques of this nature are usually based on curve fitting procedures where a smooth curve is passed through the previously measured points in order to determine an estimate of the minimum point. A variety of such techniques can be devised depending on whether or not derivatives of the function as well as the values can be measured, how many previous points are used to determine the fit, and the criterion used to determine the fit. In this section a number of possibilities are outlined and analyzed. All of them have orders of convergence greater than unity.

Newton's Method

Suppose that the function f of a single variable x is to be minimized, and suppose that at a point x_k where a measurement is made it is possible to evaluate the three numbers $f(x_k)$, $f'(x_k)$, $f''(x_k)$. It is then possible to construct a quadratic function q which at x_k agrees with f up to second derivatives, that is

$$q(x) = f(x_k) + f'(x_k)(x - x_k) + \tfrac{1}{2}f''(x_k)(x - x_k)^2. \tag{7}$$

We may then calculate an estimate x_{k+1} of the minimum point of f by finding the point where the derivative of q vanishes. Thus setting

$$0 = q'(x_{k+1}) = f'(x_k) + f''(x_k)(x_{k+1} - x_k)$$

we find

$$x_{k+1} = x_k - \frac{f'(x_k)}{f''(x_k)}. \tag{8}$$

This process, which is illustrated in Fig. 7.3, can then be repeated at x_{k+1}.

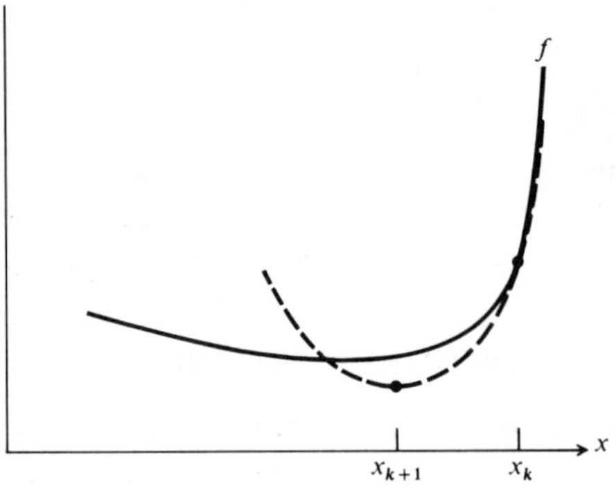

Fig. 7.3 Newton's method for minimization

We note immediately that the new point x_{k+1} resulting from Newton's method does not depend on the value $f(x_k)$. The method can more simply be viewed as a technique for iteratively solving equations of the form

$$g(x) = 0$$

where, when applied to minimization, we put $g(x) \triangleq f'(x)$. In this notation Newton's method takes the form

$$x_{k+1} = x_k - \frac{g(x_k)}{g'(x_k)}. \tag{9}$$

This form is illustrated in Fig. 7.4.

We now show that Newton's method has order two convergence:

Proposition. *Let the function g have a continuous second derivative, and let x^* satisfy $g(x^*) = 0$, $g'(x^*) \neq 0$. Then, provided x_0 is sufficiently close to x^*, the sequence $\{x_k\}_{k=0}^{\infty}$ generated by Newton's method* (9) *converges to x^* with an order of convergence at least two.*

Proof. For points ξ in a region near x^* there is a k_1 such that $|g''(\xi)| < k_1$ and a k_2 such that $|g'(\xi)| > k_2$. Then since $g(x^*) = 0$ we can write

$$x_{k+1} - x^* = x_k - x^* - \frac{g(x_k) - g(x^*)}{g'(x_k)}$$
$$= -[g(x_k) - g(x^*) + g'(x_k)(x^* - x_k)]/g'(x_k)$$

The term in brackets is, by Taylor's theorem, zero to first-order. In fact, using the remainder term in a Taylor series expansion about x_k, we obtain

$$x_{k+1} - x^* = -\frac{1}{2}\frac{g''(\xi)}{g'(x_k)}(x_k - x^*)^2$$

for some ξ between x^* and x_k. Thus in the region near x^*,

$$|x_{k+1} - x^*| \leqslant \frac{k_1}{2k_2}|x_k - x^*|^2.$$

We see that if $|x_k - x^*|k_1/2k_2 < 1$, then $|x_{k+1} - x^*| < |x_k - x^*|$ and thus we conclude that if started close enough to the solution, the method will converge to x^* with an order of convergence at least two. ∎

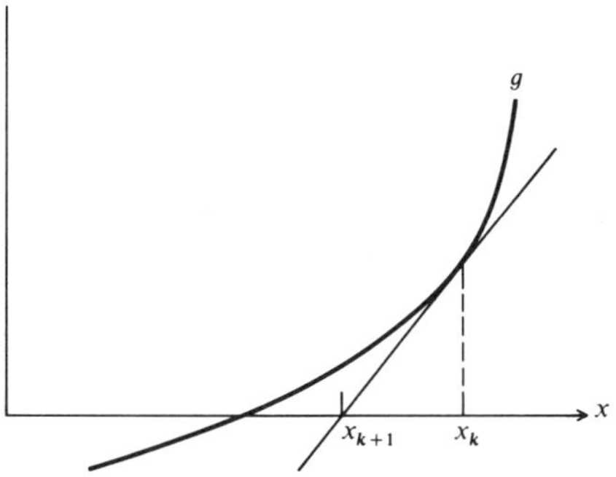

Fig. 7.4 Newton's method for solving equations

Method of False Position

Newton's method for minimization is based on fitting a quadratic on the basis of information at a single point; by using more points, less information is required at each of them. Thus, using $f(x_k), f'(x_k), f'(x_{k-1})$ it is possible to fit the quadratic

$$q(x) = f(x_k) + f'(x_k)(x - x_k) + \frac{f'(x_{k-1}) - f'(x_k)}{x_{k-1} - x_k} \cdot \frac{(x - x_k)^2}{2}$$

which has the same corresponding values. An estimate x_{k+1} can then be determined by finding the point where the derivative of q vanishes; thus

$$x_{k+1} = x_k - f'(x_k)\left[\frac{x_{k-1} - x_k}{f'(x_{k-1}) - f'(x_k)}\right]. \tag{10}$$

(See Fig. 7.5.) Comparing this formula with Newton's method, we see again that the value $f(x_k)$ does not enter; hence, our fit could have been passed through either $f(x_k)$ or $f(x_{k-1})$. Also the formula can be regarded as an approximation to Newton's method where the second derivative is replaced by the difference of two first derivatives.

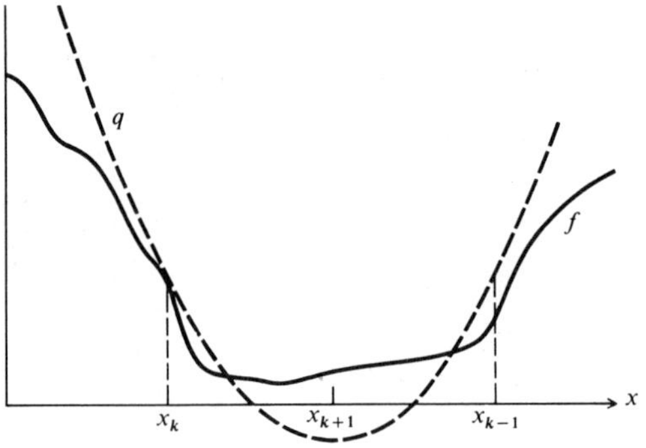

Fig. 7.5 False position for minimization

Again, since this method does not depend on values of f directly, it can be regarded as a method for solving $f'(x) \triangleq g(x) = 0$. Viewed in this way the method, which is illustrated in Fig. 7.6, takes the form

$$x_{k+1} = x_k - g(x_k)\left[\frac{x_k - x_{k-1}}{g(x_k) - g(x_{k-1})}\right]. \tag{11}$$

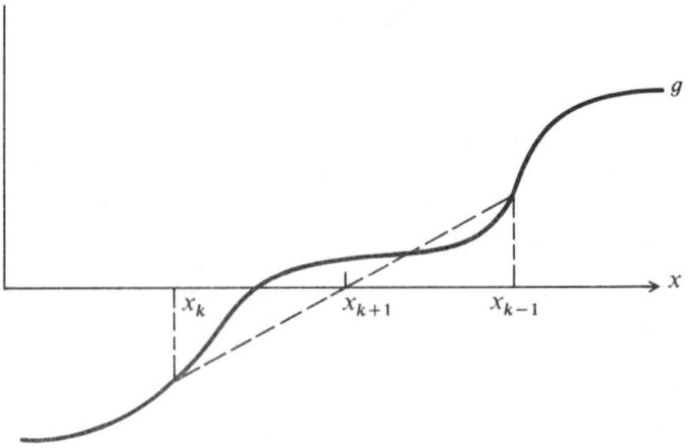

Fig. 7.6 False position for solving equations

We next investigate the order of convergence of the method of false position and discover that it is order $\tau_1 \simeq 1.618$, the golden mean.

Proposition. *Let g have a continuous second derivative and suppose x^* is such that $g(x^*) = 0, g'(x^*) \neq 0$. Then for x_0 sufficiently close to x^*, the sequence $\{x_k\}_{k=0}^{\infty}$ generated by the method of false position* (11) *converges to x^* with order $\tau_1 \simeq 1.618$.*

Proof. Introducing the notation

$$g[a, b] = \frac{g(b) - g(a)}{b - a} \tag{12}$$

we have

$$\begin{aligned} x_{k+1} - x^* &= x_k - x^* - g(x_k)\left[\frac{x_k - x_{k-1}}{g(x_k) - g(x_{k-1})}\right] \\ &= (x_k - x^*)\left\{\frac{g[x_{k-1}, x_k] - g[x_k, x^*]}{g[x_{k-1}, x_k]}\right\}. \end{aligned} \tag{13}$$

Further, upon the introduction of the notation

$$g[a, b, c] = \frac{g[a, b] - g[b, c]}{a - c} \tag{14}$$

we may write (13) as

$$x_{k+1} - x^* = (x_k - x^*)(x_{k-1} - x^*)\left\{\frac{g[x_{k-1}, x_k, x^*]}{g[x_{k-1}, x_k]}\right\}. \tag{15}$$

Now, by the mean value theorem with remainder, we have (see Exercise 2)

$$g[x_{k-1}, x_k] = g'(\xi_k) \tag{16}$$

and

$$g[x_{k-1}, x_k, x^*] = \tfrac{1}{2}g''(\eta_k) \tag{17}$$

where ξ_k and η_k are convex combinations of x_k, x_{k-1} and x_k, x_{k-1}, x^*, respectively. Thus

$$x_{k+1} - x^* = \frac{g''(\eta_k)}{2g'(\xi_k)}(x_k - x^*)(x_{k-1} - x^*). \tag{18}$$

It follows immediately that the process converges if it is started sufficiently close to x^*.

To determine the order of convergence, we note that for large k Eq. (18) becomes approximately

$$x_{k+1} - x^* = M(x_k - x^*)(x_{k-1} - x^*)$$

where

$$M = \frac{g''(x^*)}{2g'(x^*)}.$$

Thus defining $\varepsilon_k = M(x_k - x^*)$ we have, in the limit,

$$\varepsilon_{k+1} = \varepsilon_k \varepsilon_{k-1}.$$

Taking the logarithm of this equation we have, with $y_k = \log \varepsilon_k$,

$$y_{k+1} = y_k + y_{k-1}$$

which is the Fibonacci difference equation discussed in Section 1. A solution to this equation will satisfy

$$y_{k+1} - \tau_1 y_k \to 0.$$

Thus

$$\log \varepsilon_{k+1} - \tau_1 \log \varepsilon_k \to 0 \qquad \text{or} \qquad \log \frac{\varepsilon_{k+1}}{\varepsilon_k^{\tau_1}} \to 0$$

and hence

$$\frac{\varepsilon_{k+1}}{\varepsilon_k^{\tau_1}} \to 1. \blacksquare$$

Cubic Fit

Given the points x_{k-1} and x_k together with the values $f(x_{k-1}), f'(x_{k-1})$, $f(x_k), f'(x_k)$, it is possible to fit a cubic equation to the points having corresponding values. The next point x_{k+1} can then be determined as the relative minimum point of this cubic. This leads to

$$x_{k+1} = x_k - (x_k - x_{k-1}) \left[\frac{f'(x_k) + u_2 - u_1}{f'(x_k) - f'(x_{k-1}) + 2u_2} \right] \tag{19}$$

where

$$u_1 = f'(x_{k-1}) + f'(x_k) - 3\frac{f(x_{k-1}) - f(x_k)}{x_{k-1} - x_k}$$

$$u_2 = [u_1{}^2 - f'(x_{k-1})f'(x_k)]^{\frac{1}{2}}$$

which is easily implementable for computations.

It can be shown (see Exercise 3) that the order of convergence of the cubic fit method is 2.0. Thus, although the method is exact for cubic functions indicating that its order might be three, its order is actually only two.

Quadratic Fit

The scheme that is often most useful in line searching is that of fitting a quadratic through three given points. This has the advantage of not requiring any derivative information. Given x_1, x_2, x_3 and corresponding values $f(x_1) = f_1, f(x_2) = f_2, f(x_3) = f_3$ we construct the quadratic passing through these points

$$q(x) = \sum_{i=1}^{3} f_i \frac{\prod_{j \neq i} (x - x_j)}{\prod_{j \neq i} (x_i - x_j)} \tag{20}$$

and determine a new point x_4 as the point where the derivative of q vanishes. Thus,

$$x_4 = \frac{1}{2} \frac{b_{23} f_1 + b_{31} f_2 + b_{12} f_3}{a_{23} f_1 + a_{31} f_2 + a_{12} f_3} \tag{21}$$

where $a_{ij} = x_i - x_j$, $b_{ij} = x_i^2 - x_j^2$.

Defining the errors $\varepsilon_i = x^* - x_i$, $i = 1, 2, 3, 4$, it can be argued on the basis of symmetry, and it can be shown directly by detailed analysis, that near a minimum point x^* of f, the errors are related approximately by

$$\varepsilon_4 = M(\varepsilon_1 \varepsilon_2 + \varepsilon_2 \varepsilon_3 + \varepsilon_1 \varepsilon_3) \tag{22}$$

where M depends on the values of the second and third derivatives of f at x^*.

If we assume that $\varepsilon_k \to 0$ with an order greater than unity, then for large k the error is governed approximately by

$$\varepsilon_{k+2} = M \varepsilon_k \varepsilon_{k-1}.$$

Letting $y_k = \log M\varepsilon_k$ this becomes

$$y_{k+2} = y_k + y_{k-1}$$

with characteristic equation

$$\lambda^3 - \lambda - 1 = 0.$$

The largest root of this equation is $\lambda \simeq 1.3$ which thus determines the rate of growth of y_k and is the order of convergence of the quadratic fit method.

7.3 GLOBAL CONVERGENCE OF CURVE FITTING

Above, we analyzed the convergence of various curve fitting procedures in the neighborhood of the solution point. If, however, any of these procedures were applied in pure form to search a line for a minimum, there is the danger—alas, the most likely possibility—that the process would diverge or wander about meaninglessly. In other words, the process may never get close enough to the solution for our detailed local convergence analysis to be applicable. It is therefore important to artfully combine our knowledge of the local behavior with conditions guaranteeing global convergence to yield a workable and effective procedure.

The key to guaranteeing global convergence is the Global Convergence Theorem of Chapter 6. Application of this theorem in turn hinges on the construction of a suitable descent function and minor modifications of a pure curve fitting algorithm. We offer below a particular blend of this kind of construction and analysis, taking as departure point the quadratic fit procedure discussed in Section 2 above.

Let us assume that the function f that we wish to minimize is strictly unimodal and has continuous second partial derivatives. We initiate our search procedure by searching along the line until we find three points x_1, x_2, x_3 with $x_1 < x_2 < x_3$ such that $f(x_1) \geqslant f(x_2) \leqslant f(x_3)$. In other words, the value at the middle of these three points is less than that at either end. Such a sequence of points can be determined in a number of ways—see Exercise 6.

The main reason for using points having this pattern is that a quadratic fit to these points will have a minimum (rather than a maximum) and the minimum point will lie in the interval $[x_1, x_3]$. See Fig. 7.7. We modify the pure quadratic fit algorithm so that it always works with points in this basic *three-point pattern.*

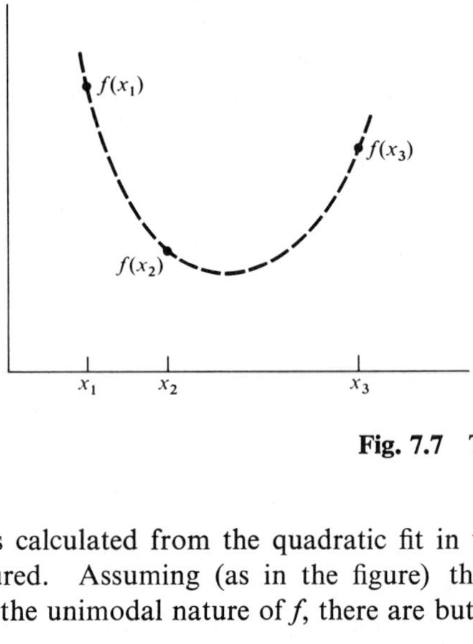

Fig. 7.7 Three-point pattern

The point x_4 is calculated from the quadratic fit in the standard way and $f(x_4)$ is measured. Assuming (as in the figure) that $x_2 < x_4 < x_3$, and accounting for the unimodal nature of f, there are but two possibilities:

1. $f(x_4) \leqslant f(x_2)$
2. $f(x_2) < f(x_4) \leqslant f(x_3)$.

In either case a new three-point pattern, $\bar{x}_1, \bar{x}_2, \bar{x}_3$, involving x_4 and two of the old points, can be determined: In case (1) it is

$$(\bar{x}_1, \bar{x}_2, \bar{x}_3) = (x_2, x_4, x_3),$$

while in case (2) it is

$$(\bar{x}_1, \bar{x}_2, \bar{x}_3) = (x_1, x_2, x_4).$$

We then use this three-point pattern to fit another quadratic and continue. The pure quadratic fit procedure determines the next point from the current point and the previous two points. In the modification above, the next point is determined from the current point and the two out of three last points that form a three-point pattern with it. This simple modification leads to global convergence.

To prove convergence, we note that each three-point pattern can be thought of as defining a vector $\mathbf{x}$ in E^3. Corresponding to an $\mathbf{x} = (x_1, x_2, x_3)$ such that (x_1, x_2, x_3) form a three-point pattern with respect to f, we define $\mathbf{A}(\mathbf{x}) = (\bar{x}_1, \bar{x}_2, \bar{x}_3)$ as discussed above. For completeness we must consider the case where two or more of the $x_i, i = 1, 2, 3$ are equal, since this may occur. The appropriate definitions are simply limiting cases of the earlier ones. For example, if $x_1 = x_2$, then (x_1, x_2, x_3) form a three-point pattern if $f(x_2) \leqslant f(x_3)$ and $f'(x_2) < 0$ (which is the limiting case of $f(x_2) < f(x_1)$). A quadratic is fit in this case by using the values at the two distinct points and the derivative at the duplicated point. In case $x_1 = x_2 = x_3$, (x_1, x_2, x_3) forms a three-point pattern if $f'(x_2) = 0$ and $f''(x_2) \geqslant 0$. With these definitions, the map $\mathbf{A}$ is well-defined. It is also continuous, since curve fitting depends continuously on the data.

We next define the solution set $\Omega \subset E^3$ as the point $\mathbf{x}^* = (x^*, x^*, x^*)$ where x^* is the minimum point of f.

Finally, we let $Z(\mathbf{x}) = f(x_1) + f(x_2) + f(x_3)$. It is easy to see that Z is a descent function for $\mathbf{A}$. After application of $\mathbf{A}$ one of the values $f(x_1)$, $f(x_2), f(x_3)$ will be replaced by $f(x_4)$, and by construction, and the assumption that f is unimodal, it will replace a strictly larger value. Of course, at $\mathbf{x}^* = (x^*, x^*, x^*)$ we have $\mathbf{A}(\mathbf{x}^*) = \mathbf{x}^*$ and hence $Z(\mathbf{A}(\mathbf{x}^*)) = Z(\mathbf{x}^*)$.

Since all points are contained in the initial interval, we have all the requirements for the Global Convergence Theorem. Thus the process converges to the solution. The order of convergence has not been destroyed by this modification, since near the solution the three-point pattern will always be formed from the previous three points. Hence we still have convergence of order 1.3.

It has often been implicitly suggested, and accepted, that when using the quadratic fit technique one should require

$$f(x_{k+1}) < f(x_k)$$

so as to guarantee convergence. If the inequality is not satisfied at some cycle, then a special local search is used to find a better x_{k+1} that does satisfy it. This philosophy amounts to taking $Z(\mathbf{x}) = f(x_3)$ in our general framework and, unfortunately, this is not a descent function even for unimodal functions, and hence the special local search is likely to be necessary several times. It is true, of course, that a similar special local search may, occasionally, be required for the technique we suggest in regions of multiple minima, but it is never required in a unimodal region.

The above construction, based on the pure quadratic fit technique, can be emulated to produce effective procedures based on other curve fitting techniques. For application to smooth functions these techniques seem to be the best available in terms of flexibility to accommodate as much derivative information as is available, fast convergence, and a guarantee of global convergence.

7.4 CLOSEDNESS OF LINE SEARCH ALGORITHMS

Since searching along a line for a minimum point is a component part of most nonlinear programming algorithms, it is desirable to establish at once that this procedure is closed; that is, that the end product of the iterative procedures outlined above, when viewed as a single algorithmic step finding a minimum along a line, define closed algorithms. That is the objective of this section.

To initiate a line search with respect to a function f, two vectors must be specified, the initial point $\mathbf{x}$ and the direction $\mathbf{d}$ in which the search is to be made. The result of the search is a new point. Thus we define the search algorithm S as a mapping from E^{2n} to E^n.

We assume that the search is to be made over the semi-infinite line emanating from $\mathbf{x}$ in the direction $\mathbf{d}$. We also assume, for simplicity, that the search is not made in vain, that is, we assume that there is a minimum point along the line. This will be the case, for instance, if f is continuous and increases without bound as $\mathbf{x}$ tends toward infinity.

Definition. The algorithm $\mathbf{S} : E^{2n} \to E^n$ is defined by

$$\mathbf{S}(\mathbf{x}, \mathbf{d}) = \{\mathbf{y} : \mathbf{y} = \mathbf{x} + \alpha\mathbf{d} \text{ for some } \alpha \geqslant 0, f(\mathbf{y}) = \min_{0 \leqslant \alpha < \infty} f(\mathbf{x} + \alpha\mathbf{d})\}. \tag{23}$$

In some cases there may be many vectors $\mathbf{y}$ yielding the minimum, so S is a set-valued mapping. We must verify that S is closed.

Theorem. *Let f be continuous on E^n. Then the mapping defined by* (23) *is closed at* $\mathbf{x}$ *if* $\mathbf{d} \neq \mathbf{0}$.

Proof. Suppose $\{\mathbf{x}_k\}$ and $\{\mathbf{d}_k\}$ are sequences with $\mathbf{x}_k \to \mathbf{x}, \mathbf{d}_k \to \mathbf{d} \neq \mathbf{0}$. Suppose also that $\mathbf{y}_k \in S(\mathbf{x}_k, \mathbf{d}_k)$ and that $\mathbf{y}_k \to \mathbf{y}$. We must show that $\mathbf{y} \in S(\mathbf{x}, \mathbf{d})$.

For each k we have $\mathbf{y}_k = \mathbf{x}_k + \alpha_k \mathbf{d}_k$ for some α_k. From this we may write

$$\alpha_k = \frac{|\mathbf{y}_k - \mathbf{x}_k|}{|\mathbf{d}_k|}. \tag{24}$$

Taking the limit of the right-hand side of (24), we see that

$$\alpha_k \to \bar{\alpha} \triangleq \frac{|\mathbf{y} - \mathbf{x}|}{|\mathbf{d}|}.$$

It then follows that $\mathbf{y} = \mathbf{x} + \bar{\alpha}\mathbf{d}$. It still remains to be shown that $\mathbf{y} \in \mathbf{S}(\mathbf{x}, \mathbf{d})$.

For each k and each α, $0 \leqslant \alpha < \infty$,

$$f(\mathbf{y}_k) \leqslant f(\mathbf{x}_k + \alpha\mathbf{d}_k).$$

Letting $k \to \infty$ we obtain

$$f(\mathbf{y}) \leqslant f(\mathbf{x} + \alpha\mathbf{d}).$$

Thus

$$f(\mathbf{y}) \leqslant \min_{0 \leqslant \alpha < \infty} f(\mathbf{x} + \alpha\mathbf{d})$$

and hence $\mathbf{y} \in \mathbf{S}(\mathbf{x}, \mathbf{d})$. ∎

The requirement that $\mathbf{d} \neq \mathbf{0}$ is natural both theoretically and practically. From a practical point of view this condition implies that, when constructing algorithms, $\mathbf{d} = \mathbf{0}$ had better correspond to the solution set; but it is clear that if $\mathbf{d} = \mathbf{0}$, no search will be made. Theoretically, the map $\mathbf{S}$ can be not closed at $\mathbf{d} = \mathbf{0}$, as illustrated below.

Example On E^1 define $f(x) = (x - 1)^2$. Then $S(x, d)$ is not closed at $x = 0, d = 0$. To see this we note that for any $d > 0$

$$\min_{0 \leqslant \alpha < \infty} f(\alpha d) = f(1)$$

and hence

$$S(0, d) = 1,$$

but

$$\min_{0 \leqslant \alpha < \infty} f(\alpha \cdot 0) = f(0)$$

so that

$$S(0, 0) = 0.$$

Thus as $d \to 0$, $S(0, d) \nrightarrow S(0, 0)$.

7.5 INACCURATE LINE SEARCH

In practice, of course, it is impossible to obtain the exact minimum point called for by the ideal line search algorithm S described above. As a matter of fact, it is often desirable to sacrifice accuracy in the line search routine in order to conserve overall computation time. Because of these factors we must, to be realistic, be certain, at every stage of development, that our theory does not crumble if inaccurate line searches are introduced.

Inaccuracy generally is introduced in a line search algorithm by simply terminating the search procedure before it has converged. The exact nature of the inaccuracy introduced may therefore depend on the particular search technique employed and the criterion used for terminating the search. We cannot develop a theory that simultaneously covers every important version of inaccuracy without seriously detracting from the underlying simplicity of the algorithms discussed later. For this reason our general approach, which is admittedly more free-wheeling in spirit than necessary but thereby more transparent and less encumbered than a detailed account of inaccuracy, will be to analyze algorithms as if an accurate line search were made at every step, and then point out in side remarks and exercises the effect of inaccuracy.

One important inaccurate line search algorithm is the one that determines the search parameter α to within a fixed percentage of its true value. Specifically, a constant $c, 0 < c < 1$ is selected ($c = 0.10$ is reasonable) and the parameter α in the line search is found so as to satisfy $|\alpha - \bar{\alpha}| \leqslant c\bar{\alpha}$ where $\bar{\alpha}$ is the true minimizing value of the parameter. This criterion is easy to incorporate in conjunction with the standard iterative search techniques described in the first sections of this chapter. For example, in the case of the quadratic fit technique using three-point patterns applied to a unimodal function, at each stage it is known that the true minimum point lies in the interval spanned by the three-point pattern, and hence a bound on the maximum possible fractional error at that stage is easily deduced. One iterates until this bound is no greater than c. It can be shown (see Exercise 11) that this algorithm is closed. Various other modified search algorithms are discussed in the exercises at the end of the chapter.

7.6 THE METHOD OF STEEPEST DESCENT

One of the oldest and most widely known methods for minimizing a function of several variables is the method of steepest descent (often referred to as the gradient method). The method is extremely important from a theoretical viewpoint, since it is one of the simplest for which a satisfactory analysis exists. More advanced algorithms are often motivated by an attempt to modify the basic steepest descent technique in such a way that the new algorithm will have superior convergence properties. The method of steepest descent remains, therefore, not only the technique most often first tried on a new problem but also the standard of reference against which other techniques are measured.

The Method

Let f have continuous first partial derivatives on E^n. We will frequently have need for the gradient vector of f and therefore we introduce some simplifying notation. The gradient $\nabla f(\mathbf{x})$ is, according to our conventions, defined as a n-dimensional *row* vector. For convenience we define the n-dimensional *column* vector $\mathbf{g}(\mathbf{x}) = \nabla f(\mathbf{x})'$. When there is no chance for ambiguity, we sometimes suppress the argument $\mathbf{x}$ and, for example, write $\mathbf{g}_k$ for $\mathbf{g}(\mathbf{x}_k) = \nabla f(\mathbf{x}_k)'$.

The method of steepest descent is defined by the iterative algorithm

$$\mathbf{x}_{k+1} = \mathbf{x}_k - \alpha_k \mathbf{g}_k \tag{25}$$

where α_k is a nonnegative scalar minimizing $f(\mathbf{x}_k - \alpha \mathbf{g}_k)$. In words, from the point $\mathbf{x}_k$ we search along the direction of the negative gradient $-\mathbf{g}_k$ to a minimum point on this line; this minimum point is taken to be $\mathbf{x}_{k+1}$.

In formal terms, the overall algorithm $\mathbf{A} : E^n \to E^n$ which gives $\mathbf{x}_{k+1} \in \mathbf{A}(\mathbf{x}_k)$ can be decomposed in the form $\mathbf{A} = \mathbf{SG}$. Here $\mathbf{G} : E^n \to E^{2n}$ is defined by $\mathbf{G}(\mathbf{x}) = (\mathbf{x}, -\mathbf{g}(\mathbf{x}))$, giving the initial point and direction of a line search. This is followed by the line search $\mathbf{S} : E^{2n} \to E^n$ defined in Section 7.4.

Global Convergence

It was shown in Section 7.4 that $\mathbf{S}$ is closed if $\nabla f(\mathbf{x}) \neq \mathbf{0}$, and it is clear that $\mathbf{G}$ is continuous. Therefore, by Corollary 2 on p. 125 $\mathbf{A}$ is closed.

We define the solution set to be the points $\mathbf{x}$ where $\nabla f(\mathbf{x}) = \mathbf{0}$. Then $Z(\mathbf{x}) = f(\mathbf{x})$ is a descent function for $\mathbf{A}$, since for $\nabla f(\mathbf{x}) \neq \mathbf{0}$

$$\min_{0 \leqslant \alpha < \infty} f(\mathbf{x} - \alpha \mathbf{g}(\mathbf{x})) < f(\mathbf{x}).$$

Thus by the Global Convergence Theorem, if the sequence $\{\mathbf{x}_k\}$ is bounded, it will have limit points and each of these is a solution.

The Quadratic Case

Essentially all of the important local convergence characteristics of the method of steepest descent are revealed by an investigation of the method when applied to quadratic problems. Consider

$$f(\mathbf{x}) = \tfrac{1}{2}\mathbf{x}'\mathbf{Q}\mathbf{x} - \mathbf{x}'\mathbf{b} \tag{26}$$

where $\mathbf{Q}$ is a positive definite symmetric $n \times n$ matrix. Since $\mathbf{Q}$ is positive definite, all of its eigenvalues are positive. We assume that these eigenvalues are ordered: $0 < a = \lambda_1 \leqslant \lambda_2 \ldots \leqslant \lambda_n = A$. With $\mathbf{Q}$ positive definite, it follows (from Proposition 5, Section 6.3) that f is strictly convex.

The unique minimum point of f can be found directly, by setting the gradient to zero, as the vector $\mathbf{x}^*$ satisfying

$$\mathbf{Q}\mathbf{x}^* = \mathbf{b}. \tag{27}$$

Moreover, introducing the function

$$E(\mathbf{x}) = \tfrac{1}{2}(\mathbf{x} - \mathbf{x}^*)'\mathbf{Q}(\mathbf{x} - \mathbf{x}^*) \tag{28}$$

we have $E(\mathbf{x}) = f(\mathbf{x}) + \frac{1}{2}\mathbf{x}^{*\prime}\mathbf{Q}\mathbf{x}^*$, which shows that the function E differs from f only by a constant. For many purposes then, it will be convenient to consider that we are minimizing E rather than f.

The gradient (of both f and E) is given explicitly by

$$\mathbf{g}(\mathbf{x}) = \mathbf{Q}\mathbf{x} - \mathbf{b}. \tag{29}$$

Thus, the method of steepest descent can be expressed as

$$\mathbf{x}_{k+1} = \mathbf{x}_k - \alpha_k \mathbf{g}_k \tag{30}$$

where $\mathbf{g}_k = \mathbf{Q}\mathbf{x}_k - \mathbf{b}$ and where α_k minimizes $f(\mathbf{x}_k - \alpha \mathbf{g}_k)$. We can, however, in this special case, determine the value of α_k explicitly. We have, by definition (26),

$$f(\mathbf{x}_k - \alpha\mathbf{g}_k) = \tfrac{1}{2}(\mathbf{x}_k - \alpha\mathbf{g}_k)'\mathbf{Q}(\mathbf{x}_k - \alpha\mathbf{g}_k) - (\mathbf{x}_k - \alpha\mathbf{g}_k)'\mathbf{b}$$

which (as can be found by differentiating with respect to α) is minimized at

$$\alpha_k = \frac{\mathbf{g}_k{}'\mathbf{g}_k}{\mathbf{g}_k{}'\mathbf{Q}\mathbf{g}_k}. \tag{31}$$

Hence the method of steepest descent (30) takes the explicit form

$$\mathbf{x}_{k+1} = \mathbf{x}_k - \left(\frac{\mathbf{g}_k{}'\mathbf{g}_k}{\mathbf{g}_k{}'\mathbf{Q}\mathbf{g}_k}\right)\mathbf{g}_k \tag{32}$$

where $\mathbf{g}_k = \mathbf{Q}\mathbf{x}_k - \mathbf{b}$.

The function f and the steepest descent process can be illustrated as in Fig. 7.8 by showing contours of constant values of f and a typical sequence developed by the process. The contours of f are n-dimensional ellipsoids with axes in the directions of the n-mutually orthogonal eigenvectors of $\mathbf{Q}$. The axis corresponding to the ith eigenvector has length proportional to $1/\lambda_i$. We now analyze this process and show that the rate of convergence depends on the ratio of the lengths of the axes of the elliptical contours of f, that is, on the eccentricity of the ellipsoids.

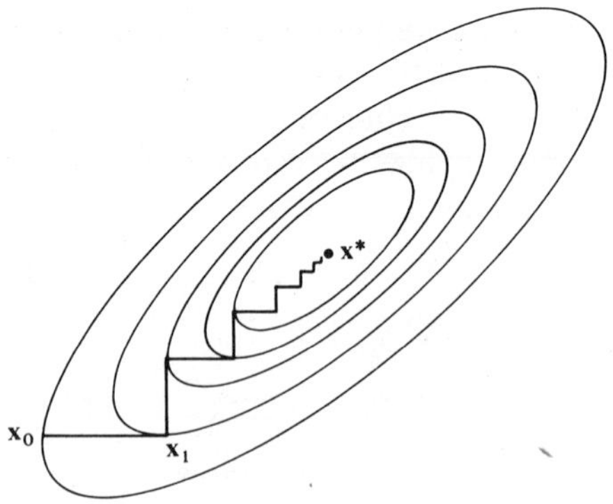

Fig. 7.8 Steepest descent

Lemma 1. *The iterative process* (32) *satisfies*

$$E(\mathbf{x}_{k+1}) = \left\{1 - \frac{(\mathbf{g}_k{}'\mathbf{g}_k)^2}{(\mathbf{g}_k{}'\mathbf{Q}\mathbf{g}_k)(\mathbf{g}_k{}'\mathbf{Q}^{-1}\mathbf{g}_k)}\right\} E(\mathbf{x}_k). \tag{33}$$

Proof. The proof is by direct computation. We have, setting $\mathbf{y}_k = \mathbf{x}_k - \mathbf{x}^*$,

$$\frac{E(\mathbf{x}_k) - E(\mathbf{x}_{k+1})}{E(\mathbf{x}_k)} = \frac{2\alpha_k \mathbf{g}_k{}'\mathbf{Q}\mathbf{y}_k - \alpha_k{}^2 \mathbf{g}_k{}'\mathbf{Q}\mathbf{g}_k}{\mathbf{y}_k{}'\mathbf{Q}\mathbf{y}_k}.$$

Using $\mathbf{g}_k = \mathbf{Q}\mathbf{y}_k$ we have

$$\frac{E(\mathbf{x}_k) - E(\mathbf{x}_{k+1})}{E(\mathbf{x}_k)} = \frac{\dfrac{2(\mathbf{g}_k{}'\mathbf{g}_k)^2}{(\mathbf{g}_k{}'\mathbf{Q}\mathbf{g}_k)} - \dfrac{(\mathbf{g}_k{}'\mathbf{g}_k)^2}{(\mathbf{g}_k{}'\mathbf{Q}\mathbf{g}_k)}}{\mathbf{g}_k{}'\mathbf{Q}^{-1}\mathbf{g}_k}$$

$$= \frac{(\mathbf{g}_k{}'\mathbf{g}_k)^2}{(\mathbf{g}_k{}'\mathbf{Q}\mathbf{g}_k)(\mathbf{g}_k{}'\mathbf{Q}^{-1}\mathbf{g}_k)}. \quad \blacksquare$$

In order to obtain a bound on the rate of convergence, we need a bound on the right-hand side of (33). The best bound is due to Kantorovich and his lemma, stated below, is a useful general tool in convergence analysis.

***Kantorovich inequality*:** *Let* $\mathbf{Q}$ *be a positive definite symmetric* $n \times n$ *matrix. For any vector* $\mathbf{x}$ *there holds*

$$\frac{(\mathbf{x}'\mathbf{x})^2}{(\mathbf{x}'\mathbf{Q}\mathbf{x})(\mathbf{x}'\mathbf{Q}^{-1}\mathbf{x})} \geqslant \frac{4aA}{(a + A)^2} \tag{34}$$

where a *and* A *are, respectively, the smallest and largest eigenvalues of* $\mathbf{Q}$.

Proof. Let the eigenvalues $\lambda_1, \lambda_2, \ldots, \lambda_n$ of $\mathbf{Q}$ satisfy

$$0 < a = \lambda_1 \leqslant \lambda_2 \ldots \leqslant \lambda_n = A.$$

By an appropriate change of coordinates the matrix $\mathbf{Q}$ becomes diagonal with diagonal $(\lambda_1, \lambda_2, \ldots, \lambda_n)$. In this coordinate system we have

$$\frac{(\mathbf{x}'\mathbf{x})^2}{(\mathbf{x}'\mathbf{Q}\mathbf{x})(\mathbf{x}'\mathbf{Q}^{-1}\mathbf{x})} = \frac{(\sum_{i=1}^n x_i^2)^2}{(\sum_{i=1}^n \lambda_i x_i^2)(\sum_{i=1}^n (x_i^2/\lambda_i))}$$

which can be written as

$$\frac{(\mathbf{x}'\mathbf{x})^2}{(\mathbf{x}'\mathbf{Q}\mathbf{x})(\mathbf{x}'\mathbf{Q}^{-1}\mathbf{x})} = \frac{1/\sum_{i=1}^n \xi_i\lambda_i}{\sum_{i=1}^n(\xi_i/\lambda_i)} \triangleq \frac{\phi(\boldsymbol{\xi})}{\psi(\boldsymbol{\xi})}$$

where $\xi_i = x_i^2/\sum_{i=1}^n x_i^2$. We have converted the expression to the ratio of two functions involving convex combinations; one a combination of λ_i's; the other a combination of $1/\lambda_i$'s. The situation is shown pictorially in Fig. 7.9. The curve in the figure represents the function $1/\lambda$. Since $\sum_{i=1}^n \xi_i\lambda_i$ is a point between λ_1 and λ_n, the value of $\phi(\boldsymbol{\xi})$ is a point on the curve. On the other hand, the value of $\psi(\xi)$ is a convex combination of points on the curve and its value corresponds to a point in the shaded region. For the same vector ξ both functions are represented by points on the same vertical line. Thus an appropriate bound is

$$\frac{\phi(\xi)}{\psi(\xi)} \geqslant \min_{\lambda_1 \leqslant \lambda \leqslant \lambda_n} \frac{(1/\lambda)}{(\lambda_1 + \lambda_n - \lambda)/(\lambda_1 \lambda_n)}$$

The minimum is achieved at $\lambda = (\lambda_1 + \lambda_n)/2$, yielding

$$\frac{\phi(\xi)}{\psi(\xi)} \geqslant \frac{4\lambda_1 \lambda_n}{(\lambda_1 + \lambda_n)^2}. \quad \blacksquare$$

Combining the above two lemmas, we obtain the central result on the convergence of the method of steepest descent.

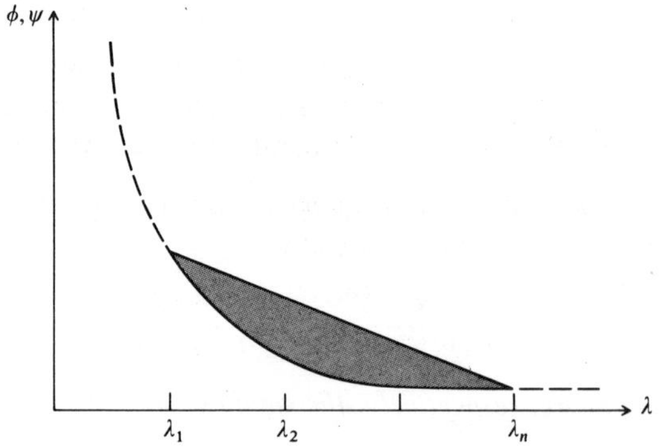

Fig. 7.9 Kantorovich inequality

Theorem (Steepest descent—quadratic case). *For any* $\mathbf{x}_0 \in E^n$ *the method of steepest descent* (32) *converges to the unique minimum point* $\mathbf{x}^*$ *of* f. *Furthermore, with* $E(\mathbf{x}) = \frac{1}{2}(\mathbf{x} - \mathbf{x}^*)'\mathbf{Q}(\mathbf{x} - \mathbf{x}^*)$, *there holds at every step* k

$$E(\mathbf{x}_{k+1}) \leqslant \left(\frac{A - a}{A + a}\right)^2 E(\mathbf{x}_k). \tag{35}$$

Proof. By Lemma 1 and the Kantorovich inequality

$$E(\mathbf{x}_{k+1}) \leqslant \left\{1 - \frac{4aA}{(A + a)^2}\right\} E(\mathbf{x}_k) = \left(\frac{A - a}{A + a}\right)^2 E(\mathbf{x}_k).$$

It follows immediately that $E(\mathbf{x}_k) \to 0$ and hence, since $\mathbf{Q}$ is positive definite, that $\mathbf{x}_k \to \mathbf{x}^*$. $\blacksquare$

Roughly speaking, the above theorem says that the convergence rate of steepest descent is slowed as the contours of f become more eccentric. If $a = A$, corresponding to circular contours, convergence occurs in a single step. Note, however, that even if $n - 1$ of the n eigenvalues are equal and the remaining one is a great distance from these, convergence will be slow, and hence a single abnormal eigenvalue can destroy the effectiveness of steepest descent.

In the terminology introduced in Section 6.6, the above theorem states that with respect to the error function E (or equivalently f) the method of steepest descent converges linearly with a ratio no greater than $[(A - a)/(A + a)]^2$. It also has been shown, by Akaike, that barring certain degenerate starting points, this bound on the rate of convergence is exact. The proof of this fact is fairly complex and we do not give it here; but because of that fact, we say that the convergence ratio of steepest descent *is* $[(A - a)/(A + a)]^2$.

It should be noted that the convergence rate actually depends only on the ratio $r = A/a$ of the largest to the smallest eigenvalue. Thus the convergence ratio is

$$\left(\frac{A - a}{A + a}\right)^2 = \left(\frac{r - 1}{r + 1}\right)^2$$

which clearly shows that convergence is slowed as r increases. The ratio r, which is the single number associated with the matrix $\mathbf{Q}$ that characterizes convergence, is often called the *condition number* of the matrix.

Example. Let us take

$$\mathbf{Q} = \begin{bmatrix} 0.78 & -0.02 & -0.12 & -0.14 \\ -0.02 & 0.86 & -0.04 & 0.06 \\ -0.12 & -0.04 & 0.72 & -0.08 \\ -0.14 & 0.06 & -0.08 & 0.74 \end{bmatrix}$$

$$\mathbf{b} = (0.76, 0.08, -0.12, 0.68).$$

For this matrix it can be calculated that $a = 0.52$, $A = 0.94$ and hence $r = 1.8$. This is a very favorable condition number and leads to the convergence ratio $[(A - a)/(A + a)]^2 = 0.081$. Thus each iteration will reduce the error in the objective by more than a factor of ten; or, equivalently, each iteration will add about one more digit of accuracy. Indeed, starting from the origin the sequence of values obtained by steepest descent as shown in Table 1 is consistent with this estimate.

Table 1. Solution to Example

Step k	$f(\mathbf{x}_k)$
0	0
1	−2.1563625
2	−2.1744062
3	−2.1746440
4	−2.1746585
5	−2.1746595
6	−2.1746595

Solution point $\mathbf{x}^* = (1.534965, 0.1220097, 1.975156, 1.412954)$

The Nonquadratic Case

The result stated here illustrates how, in general, a convergence property derived for quadratic problems translates into a similar one for nonquadratic problems. The general procedure, which is applicable to most methods having unity order of convergence, is to use the Hessian of the objective at the solution point as if it were the $\mathbf{Q}$ matrix of a quadratic problem. The particular theorem stated below is a special case of a theorem in Section 11.5 so we do not prove it here; but it illustrates the generalizability of an analysis of quadratic problems.

Theorem. *Suppose f is defined on E^n, has continuous second partial derivatives, and has a relative minimum at $\mathbf{x}^*$. Suppose further that the Hessian matrix of $f, \mathbf{F}(\mathbf{x}^*)$, has smallest eigenvalue $a > 0$ and largest eigenvalue $A > 0$. If $\{\mathbf{x}_k\}$ is a sequence generated by the method of steepest descent that converges to $\mathbf{x}^*$, then the sequence of objective values $\{f(\mathbf{x}_k)\}$ converges to $f(\mathbf{x}^*)$ linearly with a convergence ratio no greater than $[(A - a)/(A + a)]^2$.*

Scaling

The performance of the method of steepest descent is dependent on the particular choice of variables $\mathbf{x}$ used to define the problem. A new choice may substantially alter the convergence characteristics.

Suppose that $\mathbf{T}$ is an invertible $n \times n$ matrix. We can then represent points in E^n either by the standard vector $\mathbf{x}$ or by $\mathbf{y}$ where $\mathbf{Ty} = \mathbf{x}$. The problem of finding $\mathbf{x}$ to minimize $f(\mathbf{x})$ is equivalent to that of finding $\mathbf{y}$ to minimize $h(\mathbf{y}) = f(\mathbf{Ty})$. Using $\mathbf{y}$ as the underlying set of variables, we then have

$$\nabla h = \nabla f \mathbf{T} \tag{36}$$

where ∇f is the gradient of f with respect to $\mathbf{x}$. Thus, using steepest descent, the direction of search will be

$$\Delta\mathbf{y} = -\mathbf{T}'\nabla f' \tag{37}$$

which in the original variables is

$$\Delta\mathbf{x} = -\mathbf{T}\mathbf{T}'\nabla f'. \tag{38}$$

Thus, we see that the change of variables changes the direction of search.

The rate of convergence of steepest descent with respect to $\mathbf{y}$ will be determined by the eigenvalues of the Hessian of the objective, taken with respect to $\mathbf{y}$. That Hessian is

$$\nabla^2 h(\mathbf{y}) \triangleq \mathbf{H}(\mathbf{y}) = \mathbf{T}'\mathbf{F}(\mathbf{T}\mathbf{y})\mathbf{T}.$$

Thus, if $\mathbf{x}^* = \mathbf{T}\mathbf{y}^*$ is the solution point, the rate of convergence is governed by the matrix

$$\mathbf{H}(\mathbf{y}^*) = \mathbf{T}'\mathbf{F}(\mathbf{x}^*)\mathbf{T}. \tag{39}$$

Very little can be said in comparison of the convergence ratio associated with $\mathbf{H}$ and that of $\mathbf{F}$. If $\mathbf{T}$ is an orthonormal matrix, corresponding to $\mathbf{y}$ being defined from $\mathbf{x}$ by a simple rotation of coordinates, then $\mathbf{T}'\mathbf{T} = \mathbf{I}$, and we see from (38) that the directions remain unchanged and the eigenvalues of $\mathbf{H}$ are the same as those of $\mathbf{F}$.

In general, before attacking a problem with steepest descent, it is desirable, if it is feasible, to introduce a change of variables that leads to a more favorable eigenvalue structure. Usually the only kind of transformation that is at all practical is one having $\mathbf{T}$ equal to a diagonal matrix; corresponding to the introduction of scale factors on each of the variables. One should strive, in doing this, to make the second derivatives with respect to each variable roughly the same. Although appropriate scaling can potentially lead to substantial payoff in terms of enhanced convergence rate, we largely ignore this possibility in our discussions of steepest descent, since without some knowledge of problem structure it is impossible to give concrete guidelines.

7.7 NEWTON'S METHOD

The idea behind Newton's method is that the function f being minimized is approximated locally by a quadratic function, and this approximate function is minimized exactly. Thus, near $\mathbf{x}_k$ we can approximate f by the truncated Taylor series

$$f(\mathbf{x}) \simeq f(\mathbf{x}_k) + \nabla f(\mathbf{x}_k)(\mathbf{x} - \mathbf{x}_k) + \tfrac{1}{2}(\mathbf{x} - \mathbf{x}_k)'\mathbf{F}(\mathbf{x}_k)(\mathbf{x} - \mathbf{x}_k).$$

The right-hand side is minimized at

$$\mathbf{x}_{k+1} = \mathbf{x}_k - [\mathbf{F}(\mathbf{x}_k)]^{-1}\nabla f(\mathbf{x}_k)' \tag{40}$$

and this equation is the pure form of Newton's method.

In view of the second-order sufficiency conditions for a minimum point, we assume that at a relative minimum point, $\mathbf{x}^*$, the Hessian matrix, $\mathbf{F}(\mathbf{x}^*)$, is positive definite. We can then argue that if f has continuous second partial derivatives, $\mathbf{F}(\mathbf{x})$ is positive definite near $\mathbf{x}^*$ and hence the method is well-defined near the solution.

Order Two Convergence

It is a simple matter to show that if Newton's method converges, the order of convergence is two. Define $\mathbf{A}(\mathbf{x}) = \mathbf{x} - [\mathbf{F}(\mathbf{x})]^{-1}\nabla f(\mathbf{x})'$, and suppose $\mathbf{x}^*$ is a point where $\nabla f(\mathbf{x}^*) = \mathbf{0}$ and $\mathbf{F}(\mathbf{x}^*)$ is nonsingular. Then

$$\mathbf{x}_{k+1} - \mathbf{x}^* = \mathbf{A}(\mathbf{x}_k) - \mathbf{A}(\mathbf{x}^*) \tag{41}$$

and

$$|\mathbf{x}_{k+1} - \mathbf{x}^*| = |\mathbf{A}(\mathbf{x}_k) - \mathbf{A}(\mathbf{x}^*)| \leqslant |\mathbf{A}'(\mathbf{x}^*)(\mathbf{x}_k - \mathbf{x}^*)| + \tfrac{1}{2}|\mathbf{A}''(\overline{\mathbf{x}})|\,|\mathbf{x}_k - \mathbf{x}^*|^2$$

where $\overline{\mathbf{x}}$ lies on the line between $\mathbf{x}^*$ and $\mathbf{x}_k$. Since $\mathbf{A}'(\mathbf{x}^*) = \mathbf{0}$ we have

$$|\mathbf{x}_{k+1} - \mathbf{x}^*| \leqslant c|\mathbf{x}_k - \mathbf{x}^*|^2 \tag{42}$$

where c depends on $|f'''(\mathbf{x})|$ near $\mathbf{x}^*$. Thus convergence is of order at least two.

Global Convergence

In contrast to the study of steepest descent, where global convergence followed almost immediately while analysis of local convergence was quite detailed, the reverse is true for Newton's method, where establishing order two local convergence is easy but analysis of global convergence is somewhat involved. At points remote from the solution the algorithm must be modified in order to guarantee convergence.

The first modification is that usually a search parameter α is introduced so that the method takes the form

$$\mathbf{x}_{k+1} = \mathbf{x}_k - \alpha_k[\mathbf{F}(\mathbf{x}_k)]^{-1}\nabla f(\mathbf{x}_k)'$$

where α_k is selected to minimize f. Near the solution we expect, on the basis of how Newton's method was derived, that $\alpha_k \simeq 1$. Introducing the parameter for general points, however, guards against the possibility that the objective might increase with $\alpha_k = 1$, due to nonquadratic terms in the objective function.

The basic considerations required for developing the second modification can be seen most clearly by a brief examination of the general class of algorithms

$$\mathbf{x}_{k+1} = \mathbf{x}_k - \alpha\mathbf{M}_k\mathbf{g}_k \tag{43}$$

where $\mathbf{M}_k$ is a $n \times n$ matrix, α is a positive search parameter, and $\mathbf{g}_k = \nabla f(\mathbf{x}_k)'$. We note that both steepest descent ($\mathbf{M}_k = \mathbf{I}$) and Newton's method ($\mathbf{M}_k = [\mathbf{F}(\mathbf{x}_k)]^{-1}$) belong to this class. The direction vector $\mathbf{d}_k = -\mathbf{M}_k \mathbf{g}_k$ obtained in this way is a direction of descent if for small α the value of f decreases as α increases from zero. For small α we can say

$$f(\mathbf{x}_{k+1}) = f(\mathbf{x}_k) + \nabla f(\mathbf{x}_k)(\mathbf{x}_{k+1} - \mathbf{x}_k) + O(|\mathbf{x}_{k+1} - \mathbf{x}_k|^2)$$

Employing (43) this can be written as

$$f(\mathbf{x}_{k+1}) = f(\mathbf{x}_k) - \alpha \mathbf{g}_k' \mathbf{M}_k \mathbf{g}_k + O(\alpha^2).$$

As $\alpha \to 0$, the second term on the right dominates the third. Hence if one is to guarantee a decrease in f for small α, we must have $\mathbf{g}_k' \mathbf{M}_k \mathbf{g}_k > 0$. The simplest way to insure this is to require that $\mathbf{M}_k$ be positive definite.

Setting $\mathbf{M}_k = \mathbf{I}$, to obtain the method of steepest descent, is about the simplest way to guarantee descent; but this method converges only linearly. Setting $\mathbf{M}_k = [\mathbf{F}(\mathbf{x}_k)]^{-1}$ yields rapid descent near the solution but for a general point it may not yield a direction of descent, since $[\mathbf{F}(\mathbf{x}_k)]^{-1}$ may not be positive definite or even may not exist. In practice, then, Newton's method must be modified to accommodate the possible non-positive definiteness at regions remote from the solution.

Ideally we would like to modify the method so that the algorithm remains closed, is a descent algorithm, retains order two convergence, and is simple to implement. We offer below a candidate that more or less satisfies all of these requirements. We introduce the algorithm in two steps, first giving an analytically simple version, then developing a version more suitable for computation.

Let $\mathbf{F}_k \triangleq \mathbf{F}(\mathbf{x}_k)$. Fix a constant $\delta > 0$. Given $\mathbf{x}_k$, calculate the eigenvalues of $\mathbf{F}_k$ and let ε_k be the smallest nonnegative constant for which the matrix $\varepsilon_k \mathbf{I} + \mathbf{F}_k$ has eigenvalues greater than or equal to δ. Then define

$$\mathbf{d}_k = -(\varepsilon_k \mathbf{I} + \mathbf{F}_k)^{-1} \mathbf{g}_k \tag{44}$$

and iterate according to

$$\mathbf{x}_{k+1} = \mathbf{x}_k + \alpha_k \mathbf{d}_k \tag{45}$$

where α_k minimizes $f(\mathbf{x}_k + \alpha \mathbf{d}_k)$, $\alpha \geqslant 0$.

This algorithm has the desired global and local properties. First, since the eigenvalues of a matrix depend continuously on its elements, ε_k is a continuous function of $\mathbf{x}_k$ and hence the mapping $\mathbf{D} : E^n \to E^{2n}$ defined by $\mathbf{D}(\mathbf{x}_k) = (\mathbf{x}_k, \mathbf{d}_k)$ is continuous. Thus the algorithm $\mathbf{A} = \mathbf{SD}$ is closed at points outside the solution set $\Omega = \{\mathbf{x} : \nabla f(\mathbf{x}) = \mathbf{0}\}$. Second, since $\varepsilon_k \mathbf{I} + \mathbf{F}_k$ is positive definite, $\mathbf{d}_k$ is a descent direction and thus $Z(\mathbf{x}) \triangleq f(\mathbf{x})$ is a continuous descent function for $\mathbf{A}$. Therefore, assuming the generated sequence is bounded, the Global Convergence Theorem applies. Furthermore, if $\delta > 0$ is smaller than the smallest eigenvalue of $\mathbf{F}(\mathbf{x}^*)$, then for $\mathbf{x}_k$

sufficiently close to $\mathbf{x}^*$ we will have $\varepsilon_k = 0$, and the method reduces to Newton's method. Thus this revised method also has order of convergence equal to two.

The more practical version of the method is to reduce $\mathbf{F}_k$ to the form $\mathbf{F}_k = \mathbf{L}_k\mathbf{D}_k\mathbf{L}_k'$, where $\mathbf{L}_k$ is a lower triangular matrix with unity diagonal terms and $\mathbf{D}_k$ is diagonal. This decomposition is accomplished by Gaussian elimination and is far simpler than calculating the smallest eigenvalue of $\mathbf{F}_k$. One then adds the smallest possible positive constant to each of the diagonal elements of $\mathbf{D}_k$ so that they are all greater than or equal to δ—obtaining the new diagonal matrix $\mathbf{E}_k$. Letting $\mathbf{G}_k = \mathbf{L}_k\mathbf{E}_k\mathbf{L}_k'$, one then solves the equation

$$\mathbf{G}_k\mathbf{d}_k = \mathbf{g}_k,$$

which requires only back substitution since $\mathbf{G}_k$ is already decomposed, to obtain $\mathbf{d}_k$, the direction of descent. The remaining steps and the convergence arguments are similar to the above.

The selection of an appropriate δ is somewhat of an art. A small δ means that nearly singular matrices must be inverted, while a large δ means that the order two convergence may be lost. Experimentation and familiarity with a given class of problems is often required to find the best δ.

7.8 COORDINATE DESCENT METHODS

The algorithms discussed in this section are sometimes attractive because of their easy implementation. Generally, however, their convergence properties are poorer than steepest descent.

Let f be a function on E^n having continuous first partial derivatives. Given a point $\mathbf{x} = (x_1, x_2, \ldots, x_n)$, descent with respect to the coordinate x_i (i fixed) means that one solves

$$\underset{x_i}{\text{minimize}}\, f(x_1, x_2, \ldots, x_n).$$

Thus changes in the single component x_i are allowed in seeking a new and better vector $\mathbf{x}$. In our general terminology, each such descent can be regarded as a descent in the direction $\mathbf{e}_i$ (or $-\mathbf{e}_i$) where $\mathbf{e}_i$ is the ith unit vector. By sequentially minimizing with respect to different components, a relative minimum of f might ultimately be determined.

There are a number of ways that this concept can be developed into a full algorithm. The *cyclic coordinate descent* algorithm minimizes f cyclically with respect to the coordinate variables. Thus x_1 is changed first, then x_2 and so forth through x_n. The process is then repeated starting with x_1 again. A variation of this is the *Aitken double sweep method.* In this procedure one searches over $x_1, x_2, \ldots, x_n$, in that order, and then comes back in the order $x_{n-1}, x_{n-2}, \ldots, x_1$. These cyclic methods have the advantage of not requiring any information about ∇f to determine the descent directions.

If the gradient of f is available, then it is possible to select the order of descent coordinates on the basis of the gradient. A popular technique is the *Gauss–Southwell Method* where at each stage the coordinate corresponding to the largest (in absolute value) component of the gradient vector is selected for descent.

Global Convergence

Proof of global convergence is simple for these algorithms. For cyclic coordinate descent, for example, the algorithmic map $\mathbf{A}$ is the composition of $2n$ maps

$$\mathbf{A} = \mathbf{S}\mathbf{C}^n\mathbf{S}\mathbf{C}^{n-1} \ldots \mathbf{S}\mathbf{C}^1$$

where $\mathbf{C}^i(\mathbf{x}) = (\mathbf{x}, \mathbf{e}_i)$ with $\mathbf{e}_i$ equal to the ith unit vector, and $\mathbf{S}$ is the usual line search algorithm but over the doubly infinite line rather than the semi-infinite line. The map $\mathbf{C}^i$ is obviously continuous and $\mathbf{S}$ is closed. If we assume that points are restricted to a compact set, then $\mathbf{A}$ is closed by Corollary 4, page 124. We define the solution set $\Omega = \{\mathbf{x} : \nabla f(\mathbf{x}) = \mathbf{0}\}$. The function $Z(\mathbf{x}) \triangleq f(\mathbf{x})$ serves as a continuous descent for $\mathbf{A}$ with respect to Ω, for if at a point $\mathbf{x}$ we have $\nabla f(\mathbf{x}) \neq \mathbf{0}$, then at least one component of $\nabla f(\mathbf{x})$ does not vanish and a search along the corresponding coordinate direction must yield a decrease.

It should be noted, however, that as with many algorithms, we only proved convergence to a point satisfying $\nabla f(\mathbf{x}^*) = \mathbf{0}$. As illustrated by Example 3, Section 6.1, a coordinate descent algorithm does not necessarily converge to a local minimum point.

Local Convergence Rate

It is difficult to compare the rates of convergence of these algorithms with the rates of others that we analyze. This is partly because coordinate descent algorithms are from an entirely different general class of algorithms than, for example, steepest descent and Newton's method, since coordinate descent algorithms are unaffected by (diagonal) scale factor changes but are affected by rotation of coordinates—the opposite being true for steepest descent. Nevertheless, some comparison is possible.

It can be shown (see Exercise 17) that for the same quadratic problem as treated in Section 7.6, there holds for the Gauss–Southwell method

$$E(\mathbf{x}_{k+1}) \leqslant \left(1 - \frac{a}{A(n-1)}\right) E(\mathbf{x}_k) \tag{46}$$

where a, A are as in Section 7.6 and n is the dimension of the problem. Since

$$\left(\frac{A-a}{A+a}\right)^2 \leqslant \left(1 - \frac{a}{A}\right) \leqslant \left(1 - \frac{a}{A(n-1)}\right)^{n-1} \tag{47}$$

we see that the bound we have for steepest descent is better than the bound we have for $n - 1$ applications of the Gauss–Southwell scheme. Hence, we might argue that it takes essentially $n - 1$ coordinate searches to be as effective as a single gradient search. This is admittedly a crude guess, since (46) is generally not a tight bound, but the overall conclusion is consistent with the results of many experiments. Indeed, unless the variables of a problem are essentially uncoupled from each other (corresponding to a nearly diagonal Hessian matrix) coordinate descent methods seem to require about n line searches to equal the effect of one step of steepest descent.

As an aside, we point out that the above discussion, although crude, illustrates the general objective that we seek in convergence analysis. By comparing the formula giving the rate of convergence for steepest descent with a bound for coordinate descent, we are able to draw some general conclusions on the relative performance of the two methods that are not

Table 2. Solutions to Example

Iteration No.	Value of f for various methods: Gauss–Southwell	Cyclic	Double Sweep
0	0.0	0.0	0.0
1	−0.871111	−0.370256	−0.370256
2	−1.445584	−0.376011	−0.376011
3	−2.087054	−1.446460	−1.446460
4	−2.130796	−2.052949	−2.052949
5	−2.163586	−2.149690	−2.060234
6	−2.170272	−2.149693	−2.060237
7	−2.172786	−2.167983	−2.165641
8	−2.174279	−2.173169	−2.165704
9	−2.174583	−2.174392	−2.168440
10	−2.174638	−2.174397	−2.173981
11	−2.174651	−2.174582	−2.174048
12	−2.174655	−2.174643	−2.174054
13	−2.174658	−2.174656	−2.174608
14	−2.174659	−2.174656	−2.174608
15	−2.174659	−2.174658	−2.174622
16		−2.174659	−2.174655
17		−2.174659	−2.174656
18			−2.174656
19			−2.174659
20			−2.174659

dependent on specific values of a and A. Our analyses of local convergence properties, which usually involve specific formulae, are always guided by this objective of obtaining general qualitative comparisons.

Example. The quadratic problem considered in the Section 7.6 with

$$\mathbf{Q} = \begin{bmatrix} 0.78 & -0.02 & -0.12 & -0.14 \\ -0.02 & 0.86 & -0.04 & 0.06 \\ -0.12 & -0.04 & 0.72 & -0.08 \\ -0.14 & 0.06 & -0.08 & 0.74 \end{bmatrix}$$

$$\mathbf{b} = (0.76, 0.08, -0.12, 0.68)$$

was solved by the various coordinate search methods. The corresponding values of the objective function are shown in Table 2. Observe that the convergence rates of the three coordinate search methods are approximately equal but that they all converge about three times slower than steepest descent. This is in accord with the estimate given above for the Gauss–Southwell method, since in this case $n - 1 = 3$.

7.9 SPACER STEPS

In some of the more complex algorithms presented in later chapters, the rule used to determine a succeeding point in an iteration may depend on several previous points rather than just the current point, or it may depend on the iteration index k. Such features are generally introduced in order to obtain a rapid rate of convergence but they can grossly complicate the analysis of global convergence.

If in such a complex sequence of steps there is inserted, perhaps irregularly but infinitely often, a step of an algorithm such as steepest descent that is known to converge, then it is not difficult to insure that the entire complex process converges. The step which is repeated infinitely often and guarantees convergence is called a *spacer step*, since it separates disjoint portions of the complex sequence. Essentially the only requirement imposed on the other steps of the process is that they do not increase the value of the descent function.

This type of situation can be analyzed easily from the following viewpoint. Suppose $\mathbf{B}$ is an algorithm which together with the descent function Z and solution set Γ, satisfies all the requirements of the Global Convergence Theorem. Define the algorithm $\mathbf{C}$ by $\mathbf{C}(\mathbf{x}) = \{\mathbf{y} : Z(\mathbf{y}) \leqslant Z(\mathbf{x})\}$. In other words, **C** applied to **x** can give any point so long as it does not increase the value of Z. It is easy to verify that **C** is closed. We imagine that **B** represents the spacer step and the complex process between spacer steps is just some realization of **C**. Thus the overall process amounts merely to repeated applications of the composite algorithm **CB**. With this viewpoint we may state the Spacer Step Theorem.

Spacer Step Theorem. *Suppose* **B** *is an algorithm on* X *which is closed outside the solution set* Γ. *Let* Z *be a descent function corresponding to* **B** *and* Γ.

Suppose that the sequence $\{\mathbf{x}_k\}_{k=0}^{\infty}$ *is generated satisfying*

$$\mathbf{x}_{k+1} \in \mathbf{B}(\mathbf{x}_k)$$

for k *in an infinite index set* $\mathscr{K}$, *and that*

$$Z(\mathbf{x}_{k+1}) \leqslant Z(\mathbf{x}_k)$$

for all k. *Suppose also that the set* $S = \{\mathbf{x} : Z(\mathbf{x}) \leqslant Z(\mathbf{x}_0)\}$ *is compact. Then the limit of any convergent subsequence of* $\{\mathbf{x}_k\}_{\mathscr{K}}$ *is a solution.*

Proof. We first define for any $\mathbf{x} \in X$, $\overline{\mathbf{B}}(\mathbf{x}) = S \cap \mathbf{B}(\mathbf{x})$ and then observe that $\mathbf{A} = \mathbf{C}\overline{\mathbf{B}}$ is closed outside the solution set by Corollary 2, p. 125. The Global Convergence Theorem can then be applied to **A**. Since S is compact, there is a subsequence of $\{\mathbf{x}_k\}_{k \in \mathscr{K}}$ converging to a limit **x**. In view of the above we conclude that $\mathbf{x} \in \Gamma$. ▌

7.10 SUMMARY

Most iterative algorithms for minimization require a line search at every stage of the process. By employing any one of a variety of curve fitting techniques, however, the order of convergence of the line search process can be made greater than unity; which means that as compared to the linear convergence that accompanies most full descent algorithms (such as steepest descent) the individual line searches are rapid. Indeed, in common practice, only about three search points are required in any one line search.

It was shown in Sections 7.4, 7.5 and the exercises that line search algorithms of varying degrees of accuracy are all closed. Thus line searching is not only rapid enough to be practical but also behaves in such a way as to make analysis of global convergence simple.

The most important result of this chapter is the fact that the method of steepest descent converges linearly with a convergence ratio equal to $[(A - a)/(A + a)]^2$, where a and A are, respectively, the smallest and largest eigenvalues of the Hessian of the objective function evaluated at the solution point. This formula, which arises frequently throughout the remainder of the book, serves as a fundamental reference point for other algorithms. It is, however, important to understand that it is the *formula* and not its *value* that serves as the reference. We rarely advocate that the formula be evaluated since it involves quantities (namely eigenvalues) that are generally not computable until after the optimal solution is known. The formula itself, however, even though its value is unknown, can be used to make significant comparisons of the effectiveness of steepest descent versus other algorithms.

Newton's method has order two convergence. But unfortunately this speed is rarely realized in practice, since Newton's method must be modified to insure global convergence, and evaluation of the Hessian at every point is usually not worth the trouble. Nevertheless, Newton's method provides another valuable reference point in the study of algorithms.

Coordinate descent algorithms are valuable only in the special situation where the variables are essentially uncoupled or there is special structure that makes searching in the coordinate directions particularly easy. Otherwise steepest descent can be expected to be faster. Even if the gradient is not directly available, it would probably be better to evaluate a finite-difference approximation to the gradient, by taking a single step in each coordinate direction, and use this approximation in a steepest descent algorithm; rather than executing a full line search in each coordinate direction.

Finally, Section 7.9 explains that global convergence is guaranteed simply by the inclusion, in a complex algorithm, of spacer steps. This result is called upon frequently in what follows.

7.11 EXERCISES

1. Show that $g[a, b, c]$ defined by (14) is symmetric, that is, interchange of the arguments does not affect its value.
2. Prove (16) and (17).
   Hint: To prove (17) expand it, and substract and add $g'(x_k)$ to the numerator.
3. Argue using symmetry that the error in the cubic fit method approximately satisfies an equation of the form
   $$\varepsilon_{k+1} = M(\varepsilon_k^2\, \varepsilon_{k-1} + \varepsilon_k \varepsilon_{k-1}^2)$$
   and then find the order of convergence.
4. What conditions on the values and derivatives at two points guarantee that a cubic polynomial fit to this data will have a minimum between the two points? Use your answer to develop a search scheme, based on cubic fit, that is globally convergent for unimodal functions.

5. Consider the iterative process

$$x_{k+1} = \frac{1}{2}\left(x_k + \frac{a}{x_k}\right)$$

where $a > 0$. Assuming the process converges, to what does it converge? What is the order of convergence?

6. Suppose the continuous real-valued function f of a single variable satisfies

$$\min_{x \geqslant 0} f(x) < f(0).$$

Starting at any $x > 0$ show that, through a series of halvings and doublings of x and evaluation of the corresponding $f(x)$'s, a three-point pattern can be determined.

7. For $\delta > 0$ define the map $\mathbf{S}^\delta$ by

$$\mathbf{S}^\delta(\mathbf{x}, \mathbf{d}) = \{\mathbf{y} : \mathbf{y} = \mathbf{x} + \alpha\mathbf{d}, \quad 0 \leqslant \alpha \leqslant \delta; \quad f(\mathbf{y}) = \min_{0 \leqslant \beta \leqslant \delta} f(\mathbf{x} + \beta\mathbf{d})\}.$$

Thus $\mathbf{S}^\delta$ searches the interval $[0, \delta]$ for a minimum of $f(\mathbf{x} + \alpha\mathbf{d})$; representing a "limited range" line search. Show that if f is continuous, $\mathbf{S}^\delta$ is closed at all $(\mathbf{x}, \mathbf{d})$.

8. For $\varepsilon > 0$ define the map ${}^\varepsilon\mathbf{S}$ by

$${}^\varepsilon\mathbf{S}(\mathbf{x}, \mathbf{d}) = \{\mathbf{y} : \mathbf{y} = \mathbf{x} + \alpha\mathbf{d}, \ \alpha \geqslant 0, \ f(\mathbf{y}) \leqslant \min_{0 \leqslant \beta} f(\mathbf{x} + \beta\mathbf{d}) + \varepsilon\}.$$

Show that if f is continuous, ${}^\varepsilon\mathbf{S}$ is closed at $(\mathbf{x}, \mathbf{d})$ if $\mathbf{d} \neq \mathbf{0}$. This map corresponds to an "inaccurate" line search.

9. Referring to the previous two exercises, define and prove a result for ${}^\varepsilon\mathbf{S}^\delta$.

10. Define $\bar{\mathbf{S}}$ as the line search algorithm that finds the first relative minimum of $f(\mathbf{x} + \alpha\mathbf{d})$ for $\alpha \geqslant 0$. If f is continuous and $\mathbf{d} \neq \mathbf{0}$, is $\bar{\mathbf{S}}$ closed?

11. Define the search mapping $\mathbf{F}$ that determines the parameter α to within a given fraction c, $0 \leqslant c \leqslant 1$, by

$$\mathbf{F}(\mathbf{x}, \mathbf{d}) = \{\mathbf{y} : \mathbf{y} = \mathbf{x} + \alpha\mathbf{d}, \ 0 \leqslant \alpha < \infty, \ |\alpha - \bar{\alpha}| \leqslant c\bar{\alpha} \text{ where }$$
$$\frac{d}{d\alpha} f(\mathbf{x} + \bar{\alpha}\mathbf{d}) = 0\}$$

Show that if $\mathbf{d} \neq \mathbf{0}$ and $(d/d\alpha)f(\mathbf{x} + \alpha\mathbf{d})$ is continuous, then $\mathbf{F}$ is closed at $(\mathbf{x}, \mathbf{d})$.

12. Let $\mathbf{e}_1, \mathbf{e}_2, \ldots, \mathbf{e}_n$ denote the eigenvectors of the symmetric positive definite $n \times n$ matrix $\mathbf{Q}$. For the quadratic problem considered in Section 7.6, suppose $\mathbf{x}_0$ is chosen so that $\mathbf{g}_0$ belongs to a subspace M spanned by a subset of the $\mathbf{e}_i$'s. Show that for the method of steepest descent $\mathbf{g}_k \in M$ for all k. Find the rate of convergence in this case.

13. Suppose we use the method of steepest descent to minimize the quadratic function $f(\mathbf{x}) = \frac{1}{2}(\mathbf{x} - \mathbf{x}^*)'\mathbf{Q}(\mathbf{x} - \mathbf{x}^*)$ but we allow a tolerance $\pm\delta\alpha_k$ $(\delta \geqslant 0)$ in the line search, that is

$$\mathbf{x}_{k+1} = \mathbf{x}_k - \alpha_k \mathbf{g}_k$$

where

$$(1 - \delta)\bar{\alpha}_k \leqslant \alpha_k \leqslant (1 + \delta)\bar{\alpha}_k$$

and $\bar{\alpha}_k$ minimizes $f(\mathbf{x}_k - \alpha \mathbf{g}_k)$ over α.

a) Find the convergence rate of the algorithm in terms of a and A, the smallest and largest eigenvalues of $\mathbf{Q}$, and the tolerance δ.
Hint: Assume the extreme case $\alpha_k = (1 + \delta)\bar{\alpha}_k$

b) What is the largest δ that guarantees convergence of the algorithm? Explain this result geometrically.

c) Does the sign of δ make any difference?

14. Suppose in the method of steepest descent for the quadratic problem, the value of α_k is not determined to exactly minimize $E(\mathbf{x}_{k+1})$ but instead only satisfies

$$\frac{E(\mathbf{x}_k) - E(\mathbf{x}_{k+1})}{E(\mathbf{x}_k)} \geqslant \beta \frac{E(\mathbf{x}_k) - \bar{E}}{E(\mathbf{x}_k)}$$

for some β, $0 < \beta < 1$, where $\bar{E}$ is the value that corresponds to the best α_k. Find the best estimate for the rate of convergence in this case.

15. Suppose an iterative algorithm of the form

$$\mathbf{x}_{k+1} = \mathbf{x}_k + \alpha_k \mathbf{d}_k$$

is applied to the quadratic problem with matrix $\mathbf{Q}$, where α_k as usual is chosen as the minimum point of the line search and where $\mathbf{d}_k$ is a vector satisfying $\mathbf{d}_k'\mathbf{g}_k < 0$ and $(\mathbf{d}_k'\mathbf{g}_k)^2 \geqslant \beta(\mathbf{d}_k'\mathbf{Q}\mathbf{d}_k)(\mathbf{g}_k'\mathbf{Q}^{-1}\mathbf{g}_k)$, where $0 < \beta \leqslant 1$. This corresponds to a steepest descent algorithm with "sloppy" choice of direction. Estimate the rate of convergence of this algorithm.

16. Repeat Exercise 15 with the condition on $(\mathbf{d}_k'\mathbf{g}_k)^2$ replaced by

$$(\mathbf{d}_k'\mathbf{g}_k)^2 \geqslant \beta(\mathbf{d}_k'\mathbf{d}_k)(\mathbf{g}_k'\mathbf{g}_k), \qquad 0 < \beta \leqslant 1.$$

17. Use the result of Exercise 16 to derive (46) for the Gauss–Southwell method.

18. Let $f(x, y) = x^2 + y^2 + xy - 3x$

a) Find an unconstrained local minimum point of f.

b) Why is the solution to (a) actually a global minimum point?

c) Find the minimum point of f subject to $x \geqslant 0$, $y \geqslant 0$.

d) If the method of steepest descent were applied to (a), what would be the rate of convergence of the objective function?

19. Find an estimate for the rate of convergence for the modified Newton method

$$\mathbf{x}_{k+1} = \mathbf{x}_k - \alpha_k(\varepsilon_k \mathbf{I} + \mathbf{F}_k)^{-1}\mathbf{g}_k$$

given by (44) and (45) when δ is larger than the smallest eigenvalue of $\mathbf{F}(\mathbf{x}^*)$.

20. Prove global convergence of the Gauss–Southwell method.

21. Consider a problem of the form

$$\text{minimize } f(\mathbf{x})$$
$$\text{subject to } \mathbf{x} \geqslant \mathbf{0}$$

where $\mathbf{x} \in E^n$. A gradient-type procedure has been suggested for this kind of problem that accounts for the constraint. At a given point $\mathbf{x} = (x_1, x_2, \ldots, x_n)$, the direction $\mathbf{d} = (d_1, d_2, \ldots, d_n)$ is determined from the gradient $\nabla f(\mathbf{x})' = \mathbf{g} = (g_1, g_2, \ldots, g_n)$ by

$$d_i = \begin{cases} -g_i & \text{if } \; x_i > 0 \quad \text{or} \quad g_i < 0 \\ 0 & \text{if } \; x_i = 0 \quad \text{and} \quad g_i \geqslant 0. \end{cases}$$

This direction is then used as a direction of search in the usual manner.

a) What are the first-order necessary conditions for a minimum point of this problem?

b) Show that $\mathbf{d}$, as determined by the algorithm, is zero only at a point satisfying the first-order conditions.

c) Show that if $\mathbf{d} \neq \mathbf{0}$, it is possible to decrease the value of f by movement along $\mathbf{d}$.

d) If restricted to a compact region, does the Global Convergence Theorem apply? Why?

22. Consider the quadratic problem and suppose $\mathbf{Q}$ has unity diagonal. Consider a coordinate descent procedure in which the coordinate to be searched is at every stage selected randomly, each coordinate being equally likely. Let $\boldsymbol{\varepsilon}_k = \mathbf{x}_k - \mathbf{x}^*$. Assuming $\boldsymbol{\varepsilon}_k$ is known, show that $\overline{\boldsymbol{\varepsilon}_{k+1}'\mathbf{Q}\boldsymbol{\varepsilon}_{k+1}}$, the expected value of $\boldsymbol{\varepsilon}_{k+1}'\mathbf{Q}\boldsymbol{\varepsilon}_{k+1}$, satisfies

$$\overline{\boldsymbol{\varepsilon}_{k+1}'\mathbf{Q}\boldsymbol{\varepsilon}_{k+1}} = \left(1 - \frac{\boldsymbol{\varepsilon}_k'\mathbf{Q}^2\boldsymbol{\varepsilon}_k}{n\boldsymbol{\varepsilon}_k'\mathbf{Q}\boldsymbol{\varepsilon}_k}\right)\boldsymbol{\varepsilon}_k'\mathbf{Q}\boldsymbol{\varepsilon}_k \leqslant \left(1 - \frac{a^2}{nA}\right)\boldsymbol{\varepsilon}_k'\mathbf{Q}\boldsymbol{\varepsilon}_k.$$

23. If the matrix $\mathbf{Q}$ has a condition number of 10, how many iterations of steepest descent would be required to get six place accuracy in the minimum value of the objective function of the corresponding quadratic problem?

24. *Stopping Criterion.* A question that arises in using an algorithm such as steepest descent to minimize an objective function f is when to stop the iterative process, or, in other words, how can one tell when the current point is close to a solution. If, as with steepest descent, it is known that convergence is linear, this knowledge can be used to develop a stopping criterion. Let $\{f_k\}_{k=0}^{\infty}$ be the

sequence of values obtained by the algorithm. We assume that $f_k \to f^*$ linearly, but both f^* and the convergence ratio β are unknown. However we know that, at least approximately,

$$f_{k+1} - f^* = \beta(f_k - f^*)$$

and

$$f_k - f^* = \beta(f_{k-1} - f^*).$$

These two equations can be solved for β and f^*.

a) Show that

$$f^* = \frac{f_k^2 - f_{k-1}f_{k+1}}{2f_k - f_{k-1} - f_{k+1}}$$

$$\beta = \frac{f_{k+1} - f_k}{f_k - f_{k-1}}$$

b) Motivated by the above we form the sequence $\{f_k^*\}$ defined by

$$f_k^* = \frac{f_k^2 - f_{k-1}f_{k+1}}{2f_k - f_{k-1} - f_{k+1}}.$$

as the original sequence is generated. (This procedure of generating $\{f_k^*\}$ from $\{f_k\}$ is called the Aitken δ^2-process.) If $|f_k - f^*| = \beta^k + o(\beta^k)$ show that $|f_k^* - f^*| = o(\beta^k)$ which means that $\{f_k^*\}$ converges to f^* faster than $\{f_k\}$ does. The iterative search for the minimum of f can then be terminated when $f_k - f_k^*$ is smaller than some prescribed tolerance.

REFERENCES

7.1. For a detailed exposition of Fibonacci search techniques see Wilde and Beightler [W1]. For an introductory discussion of difference equations see Lanczos [L1].

7.2. Many of these techniques are standard among numerical analysts. See, for example, Kowalik and Osborne [K3], or Traub [T3].

7.3. The material in this section is new.

7.4. The closedness of line search algorithms was established by Zangwill [Z2].

7.6. For an alternate exposition of this well-known method see Antosiewicz and Rheinboldt [A3] or Luenberger [L6]. For a proof that the estimate (35) is essentially exact see Akaike [A2]. For early work on the nonquadratic case see Curry [C7]. The numerical problem considered in the example is a standard one. See Faddeev and Faddeeva [F1].

7.8. A detailed analysis of coordinate algorithms can be found in Fox [F11] and Isaacson and Keller [I1]. For a discussion of the Gauss–Southwell method see Forsythe and Wasow [F10].

7.9. A version of the Spacer Step Theorem can be found in Zangwill [Z2].

Chapter 8 CONJUGATE DIRECTION METHODS

Conjugate direction methods can be regarded as being somewhat intermediate between the method of steepest descent and Newton's method. They are motivated by the desire to accelerate the typically slow convergence associated with steepest descent while avoiding the information requirements associated with the evaluation, storage, and inversion of the Hessian as required by Newton's method.

Conjugate direction methods invariably are invented and analyzed for the purely quadratic problem

$$\text{minimize } \tfrac{1}{2}\mathbf{x}'\mathbf{Q}\mathbf{x} - \mathbf{b}'\mathbf{x}$$

where $\mathbf{Q}$ is an $n \times n$ symmetric positive definite matrix. The techniques once worked out for this problem are then extended, by approximation, to more general problems; it being argued that, since near the solution point every problem is approximately quadratic, convergence behavior is similar to that for the pure quadratic situation.

The area of conjugate direction algorithms has been one of great creativity in the nonlinear programming field, illustrating that detailed analysis of the pure quadratic problem can lead to significant practical advances. Indeed, conjugate direction methods, especially the method of conjugate gradients, have proved to be extremely effective in dealing with general objective functions and are considered among the best general purpose methods presently available.

8.1 CONJUGATE DIRECTIONS

Definition. Given a symmetric matrix $\mathbf{Q}$, two vectors $\mathbf{d}_1$ and $\mathbf{d}_2$ are said to be $\mathbf{Q}$-*orthogonal, or conjugate, with respect to* $\mathbf{Q}$ *if* $\mathbf{d}_1'\mathbf{Q}\mathbf{d}_2 = 0$.

In the applications that we consider, the matrix $\mathbf{Q}$ will be positive definite but this is not inherent in the basic definition. Thus if $\mathbf{Q} = \mathbf{0}$, any two vectors are conjugate, while if $\mathbf{Q} = \mathbf{I}$, conjugacy is equivalent to the usual notion of orthogonality. A finite set of vectors $\mathbf{d}_0, \mathbf{d}_1, \ldots, \mathbf{d}_k$ is said to be a $\mathbf{Q}$-orthogonal set if $\mathbf{d}_i'\mathbf{Q}\mathbf{d}_j = 0$ for all $i \neq j$.

Proposition. *If* $\mathbf{Q}$ *is positive definite and the set of nonzero vectors* $\mathbf{d}_0$, $\mathbf{d}_1, \mathbf{d}_2 \ldots, \mathbf{d}_k$ *are* $\mathbf{Q}$*-orthogonal, then these vectors are linearly independent.*

Proof. Suppose there are constants α_i, $i = 0, 1, 2, \ldots, k$ such that

$$\alpha_0 \mathbf{d}_0 + \cdots + \alpha_k \mathbf{d}_k = \mathbf{0}.$$

Multiplying by $\mathbf{Q}$ and taking the scalar product with $\mathbf{d}_i$ yields

$$\alpha_i \mathbf{d}_i'\mathbf{Q}\mathbf{d}_i = 0.$$

Or, since $\mathbf{d}_i'\mathbf{Q}\mathbf{d}_i > 0$ in view of the positive definiteness of $\mathbf{Q}$, we have $\alpha_i = 0$. ∎

Before discussing the general conjugate direction algorithm, let us investigate just why the notion of $\mathbf{Q}$-orthogonality is useful in the solution of the quadratic problem

$$\text{minimize } \tfrac{1}{2}\mathbf{x}'\mathbf{Q}\mathbf{x} - \mathbf{b}'\mathbf{x} \tag{1}$$

when $\mathbf{Q}$ is positive definite. Recall that the unique solution to this problem is also the unique solution to the linear equation

$$\mathbf{Q}\mathbf{x} = \mathbf{b} \tag{2}$$

and hence that the quadratic minimization problem is equivalent to a linear equation problem.

Corresponding to the $n \times n$ positive definite matrix $\mathbf{Q}$ let $\mathbf{d}_0, \mathbf{d}_1, \ldots, \mathbf{d}_{n-1}$ be n nonzero $\mathbf{Q}$-orthogonal vectors. By the above proposition they are linearly independent, which implies that the solution $\mathbf{x}^*$ of (1) or (2) can be expanded in terms of them as

$$\mathbf{x}^* = \alpha_0 \mathbf{d}_0 + \cdots + \alpha_{n-1}\mathbf{d}_{n-1} \tag{3}$$

for some set of α_i's. In fact, multiplying by $\mathbf{Q}$ and then taking the scalar product with $\mathbf{d}_i$ yields directly

$$\alpha_i = \frac{\mathbf{d}_i'\mathbf{Q}\mathbf{x}^*}{\mathbf{d}_i'\mathbf{Q}\mathbf{d}_i} = \frac{\mathbf{d}_i'\mathbf{b}}{\mathbf{d}_i'\mathbf{Q}\mathbf{d}_i}. \tag{4}$$

This shows that the α_i's and consequently the solution $\mathbf{x}^*$ can be found by evaluation of simple scalar products. The end result is

$$\mathbf{x}^* = \sum_{i=0}^{n-1} \frac{\mathbf{d}_i'\mathbf{b}}{\mathbf{d}_i'\mathbf{Q}\mathbf{d}_i}\mathbf{d}_i. \tag{5}$$

There are two basic ideas imbedded in (5). The first is the idea of selecting an orthogonal set of $\mathbf{d}_i$'s so that by taking an appropriate scalar product all terms on the right side of (3), except the ith, vanish. This could, of course, have been accomplished by making the $\mathbf{d}_i$'s orthogonal in the ordinary sense instead of making them $\mathbf{Q}$-orthogonal. The second basic observation, however, is that by using $\mathbf{Q}$-orthogonality the resulting equation for α_i can be expressed in terms of the known vector $\mathbf{b}$ rather than the unknown vector $\mathbf{x}^*$; hence the coefficients can be evaluated without knowing $\mathbf{x}^*$.

The expansion for $\mathbf{x}^*$ can be considered to be the result of an iterative process of n steps where at the ith step $\alpha_i \mathbf{d}_i$ is added. Viewing the procedure this way, and allowing for an arbitrary initial point for the iteration, the basic conjugate direction method is obtained.

Conjugate Direction Theorem. *Let $\{\mathbf{d}_i\}_{i=0}^{n-1}$ be a set of nonzero $\mathbf{Q}$-orthogonal vectors. For any $\mathbf{x}_0 \in E^n$ the sequence $\{\mathbf{x}_k\}$ generated according to*

$$\mathbf{x}_{k+1} = \mathbf{x}_k + \alpha_k \mathbf{d}_k, \qquad k \geqslant 0 \tag{6}$$

with

$$\alpha_k = -\frac{\mathbf{g}_k{}'\mathbf{d}_k}{\mathbf{d}_k{}'\mathbf{Q}\mathbf{d}_k} \tag{7}$$

and

$$\mathbf{g}_k = \mathbf{Q}\mathbf{x}_k - \mathbf{b},$$

converges to the unique solution, $\mathbf{x}^*$, of $\mathbf{Q}\mathbf{x} = \mathbf{b}$ after n steps, that is, $\mathbf{x}_n = \mathbf{x}^*$.

Proof. Since the $\mathbf{d}_k$'s are linearly independent, we can write

$$\mathbf{x}^* - \mathbf{x}_0 = \alpha_0 \mathbf{d}_0 + \alpha_1 \mathbf{d}_1 + \cdots + \alpha_{n-1}\mathbf{d}_{n-1}$$

for some set of α_k's. As we did to get (4), we multiply by $\mathbf{Q}$ and take the scalar product with $\mathbf{d}_k$ to find

$$\alpha_k = \frac{\mathbf{d}_k{}'\mathbf{Q}(\mathbf{x}^* - \mathbf{x}_0)}{\mathbf{d}_k{}'\mathbf{Q}\mathbf{d}_k}\,. \tag{8}$$

Now following the iterative process (6) from $\mathbf{x}_0$ up to $\mathbf{x}_k$ gives

$$\mathbf{x}_k - \mathbf{x}_0 = \alpha_0 \mathbf{d}_0 + \alpha_1 \mathbf{d}_1 + \cdots + \alpha_{k-1}\mathbf{d}_{k-1} \tag{9}$$

and hence by the $\mathbf{Q}$-orthogonality of the $\mathbf{d}_k$'s it follows that

$$\mathbf{d}_k{}'\mathbf{Q}(\mathbf{x}_k - \mathbf{x}_0) = 0. \tag{10}$$

Substituting (10) in (8) produces

$$\alpha_k = \frac{\mathbf{d}_k{}'\mathbf{Q}(\mathbf{x}^* - \mathbf{x}_k)}{\mathbf{d}_k{}'\mathbf{Q}\mathbf{d}_k} = -\frac{\mathbf{g}_k{}'\mathbf{d}_k}{\mathbf{d}_k{}'\mathbf{Q}\mathbf{d}_k}$$

which is identical with (7). ▌

To this point the conjugate direction method has been derived essentially through the observation that solving (1) is equivalent to solving (2).

The conjugate direction method has been viewed simply as a somewhat special, but nevertheless straightforward, orthogonal expansion for the solution to (2). This viewpoint, although important because of its underlying simplicity, ignores some of the most important aspects of the algorithm; especially those aspects that are important when extending the method to nonquadratic problems. These additional properties are discussed in the next section.

Also, methods for selecting or generating sequences of conjugate directions have not yet been presented. Some methods for doing this are discussed in the exercises; while the most important method, that of conjugate gradients, is discussed in Section 8.3.

8.2 DESCENT PROPERTIES OF THE CONJUGATE DIRECTION METHOD

We define $\mathcal{B}_k$ as the subspace of E^n spanned by $\{\mathbf{d}_0, \mathbf{d}_1, \ldots, \mathbf{d}_{k-1}\}$. We shall show that as the method of conjugate directions progresses each $\mathbf{x}_k$ minimizes the objective over the k dimensional linear variety $\mathbf{x}_0 + \mathcal{B}_k$.

Expanding Subspace Theorem. *Let $\{\mathbf{d}_i\}_{i=0}^{n-1}$ be a sequence of nonzero* **Q**-*orthogonal vectors in E^n. Then for any $\mathbf{x}_0 \in E^n$ the sequence $\{\mathbf{x}_k\}$ generated according to*

$$\mathbf{x}_{k+1} = \mathbf{x}_k + \alpha_k \mathbf{d}_k \tag{11}$$

$$\alpha_k = -\frac{\mathbf{g}_k'\mathbf{d}_k}{\mathbf{d}_k'\mathbf{Q}\mathbf{d}_k} \tag{12}$$

has the property that $\mathbf{x}_k$ minimizes $f(\mathbf{x}) = \frac{1}{2}\mathbf{x}'\mathbf{Q}\mathbf{x} - \mathbf{b}'\mathbf{x}$ on the line $\mathbf{x} = \mathbf{x}_{k-1} + \alpha\mathbf{d}_{k-1}$, $-\infty < \alpha < \infty$, as well as on the linear variety $\mathbf{x}_0 + \mathcal{B}_k$.

Proof. It need only be shown that $\mathbf{x}_k$ minimizes f on the linear variety $\mathbf{x}_0 + \mathcal{B}_k$, since it contains the line $\mathbf{x} = \mathbf{x}_{k-1} + \alpha\mathbf{d}_{k-1}$. Since f is a strictly convex function, the conclusion will hold if it can be shown that $\mathbf{g}_k$ is orthogonal to $\mathcal{B}_k$ (that is, the gradient of f at $\mathbf{x}_k$ is orthogonal to the subspace $\mathcal{B}_k$). The situation is illustrated in Fig. 8.1. (Compare Theorem 2, page 119.)

We prove $\mathbf{g}_k \perp \mathcal{B}_k$ by induction. Since $\mathcal{B}_0$ is empty that hypothesis is true for $k = 0$. Assuming that it is true for k, that is, assuming $\mathbf{g}_k \perp \mathcal{B}_k$, we show that $\mathbf{g}_{k+1} \perp \mathcal{B}_{k+1}$. We have

$$\mathbf{g}_{k+1} = \mathbf{g}_k + \alpha_k \mathbf{Q}\mathbf{d}_k \tag{13}$$

and hence

$$\mathbf{d}_k'\mathbf{g}_{k+1} = \mathbf{d}_k'\mathbf{g}_k + \alpha_k \mathbf{d}_k'\mathbf{Q}\mathbf{d}_k = 0 \tag{14}$$

by definition of α_k. Also for $i < k$

$$\mathbf{d}_i'\mathbf{g}_{k+1} = \mathbf{d}_i'\mathbf{g}_k + \alpha_k \mathbf{d}_i'\mathbf{Q}\mathbf{d}_k. \tag{15}$$

The first term on the right-hand side of (15) vanishes because of the induction hypothesis, while the second vanishes by the $\mathbf{Q}$-orthogonality of the $\mathbf{d}_i$'s. Thus $\mathbf{g}_{k+1} \perp \mathcal{B}_{k+1}$. ∎

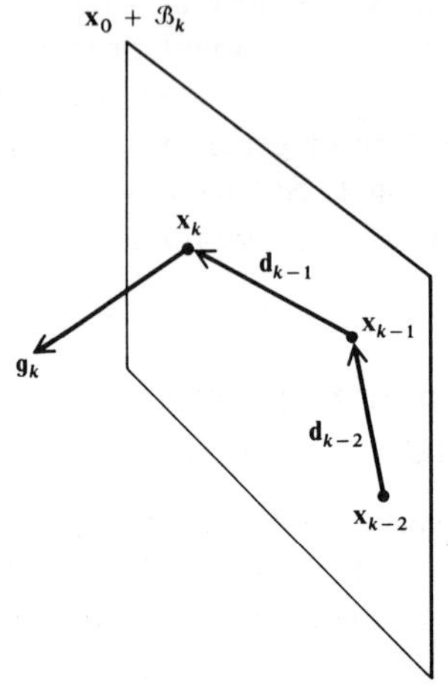

Fig. 8.1 Conjugate direction method

Corollary. *In the method of conjugate directions the gradients* $\mathbf{g}_k$, $k = 0, 1, \ldots, n$ *satisfy*

$$\mathbf{g}_k{}'\mathbf{d}_i = 0 \qquad \text{for} \qquad i < k.$$

The above theorem is referred to as the Expanding Subspace Theorem, since the $\mathcal{B}_k$'s form a sequence of subspaces with $\mathcal{B}_{k+1} \supset \mathcal{B}_k$. Since $\mathbf{x}_k$ minimizes f over $\mathbf{x}_0 + \mathcal{B}_k$, it is clear that $\mathbf{x}_n$ must be the overall minimum of f.

To obtain another interpretation of this result we again introduce the function

$$E(\mathbf{x}) = \tfrac{1}{2}(\mathbf{x} - \mathbf{x}^*)'\mathbf{Q}(\mathbf{x} - \mathbf{x}^*) \tag{16}$$

as a measure of how close the vector $\mathbf{x}$ is to the solution $\mathbf{x}^*$. Since $E(\mathbf{x}) = f(\mathbf{x}) + \frac{1}{2}\mathbf{x}^{*\prime}\mathbf{Q}\mathbf{x}^*$ the function E can be regarded as the objective that we seek to minimize.

By considering the minimization of E we can regard the original problem as one of minimizing a generalized distance from the point $\mathbf{x}^*$. Indeed, if we had $\mathbf{Q} = \mathbf{I}$, the generalized notion of distance would correspond (within a factor of two) to the usual Euclidean distance. For an arbitrary positive–definite $\mathbf{Q}$ we say E is a generalized Euclidean metric or distance function.

Vectors $\mathbf{d}_i$, $i = 0, 1, \ldots, n-1$ that are $\mathbf{Q}$-orthogonal may be regarded as orthogonal in this generalized Euclidean space and this leads to the simple interpretation of the Expanding Subspace Theorem illlustrated in Fig. 8.2. For simplicity we assume $\mathbf{x}_0 = \mathbf{0}$. In the figure $\mathbf{d}_k$ is shown as being orthogonal to $\mathcal{B}_k$ with respect to the geneıalized metric. The point $\mathbf{x}_k$ minimizes E over $\mathcal{B}_k$ while $\mathbf{x}_{k+1}$ minimizes E over $\mathcal{B}_{k+1}$. The basic property is that, since $\mathbf{d}_k$ is orthogonal to $\mathcal{B}_k$, the point $\mathbf{x}_{k+1}$ can be found by minimizing E along $\mathbf{d}_k$ and adding the result to $\mathbf{x}_k$.

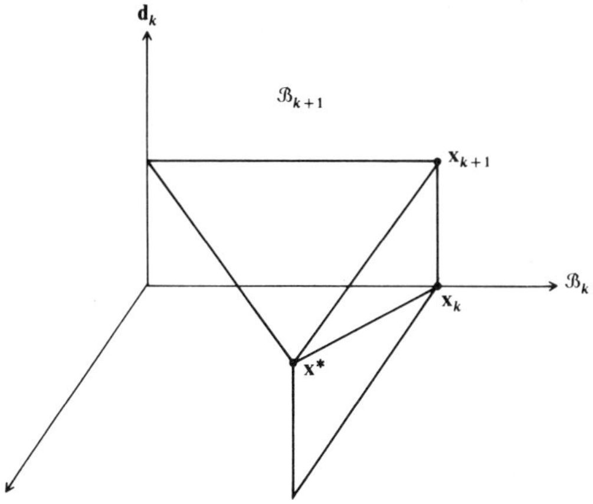

Fig. 8.2 Interpretation of expanding subspace theorem

8.3 THE CONJUGATE GRADIENT METHOD

The conjugate gradient method is the conjugate direction method that is obtained by selecting the successive direction vectors as a conjugate version of the successive gradients obtained as the method progresses. Thus the directions are not specified beforehand, but rather are determined sequentially at each step of the iteration. At step k one evaluates the current negative gradient vector and adds to it a linear combination of the previous direction vectors to obtain a new conjugate direction vector along which to move.

There are three primary advantages to this method of direction selection. First, unless the solution is attained in less than n steps, the gradient is always nonzero and linearly independent of all previous direction vectors. Indeed, the gradient $\mathbf{g}_k$ is orthogonal to the subspace $\mathcal{B}_k$ generated by $\mathbf{d}_0, \mathbf{d}_1, \ldots, \mathbf{d}_{k-1}$. If the solution is reached before n steps are taken, the gradient vanishes and the process terminates—it being unnecessary, in this case, to find additional directions.

Second, a more important advantage of the conjugate gradient method is the especially simple formula that is used to determine the new direction vector. This simplicity makes the method only slightly more complicated than steepest descent.

Third, because the directions are based on the gradients, the process makes good uniform progress toward the solution at every step. This is in contrast to the situation for arbitrary sequences of conjugate directions in which progress may be slight until the final few steps. Although for the pure quadratic problem uniform progress is of no great importance, it is important for generalizations to nonquadratic problems.

Conjugate Gradient Algorithm

Starting at any $\mathbf{x}_0 \in E^n$ define $\mathbf{d}_0 = -\mathbf{g}_0 = \mathbf{b} - \mathbf{Q}\mathbf{x}_0$ and

$$\mathbf{x}_{k+1} = \mathbf{x}_k + \alpha_k \mathbf{d}_k \tag{17}$$

$$\alpha_k = -\frac{\mathbf{g}_k{}'\mathbf{d}_k}{\mathbf{d}_k{}'\mathbf{Q}\mathbf{d}_k} \tag{18}$$

$$\mathbf{d}_{k+1} = -\mathbf{g}_{k+1} + \beta_k \mathbf{d}_k \tag{19}$$

$$\beta_k = \frac{\mathbf{g}_{k+1}'\mathbf{Q}\mathbf{d}_k}{\mathbf{d}_k{}'\mathbf{Q}\mathbf{d}_k} \tag{20}$$

where $\mathbf{g}_k = \mathbf{Q}\mathbf{x}_k - \mathbf{b}$.

In the algorithm the first step is identical to a steepest descent step; each succeeding step moves in a direction that is a linear combination of the current gradient and the preceding direction vector. The attractive feature of the algorithm is the simple formulae, (19) and (20), for updating the direction vector. The method is only slightly more complicated to implement than the method of steepest descent but converges in a finite number of steps.

Verification of the Algorithm

To verify that the algorithm is a conjugate direction algorithm, it is necessary to verify that the vectors $\{\mathbf{d}_k\}$ are $\mathbf{Q}$-orthogonal. It is easiest to prove this by simultaneously proving a number of other properties of the algorithm. This is done in the theorem below where the notation $[\mathbf{d}_0, \mathbf{d}_1, \ldots, \mathbf{d}_k]$ is ısed to denote the subspace spanned by the vectors $\mathbf{d}_0, \mathbf{d}_1, \ldots, \mathbf{d}_k$.

Conjugate Gradient Theorem. *The conjugate gradient algorithm* (17)–(20) *is a conjugate direction method. If it does not terminate at* $\mathbf{x}_k$, *then*

a) $[\mathbf{g}_0, \mathbf{g}_1, \ldots, \mathbf{g}_k] = [\mathbf{g}_0, \mathbf{Q}\mathbf{g}_0, \ldots, \mathbf{Q}^k\mathbf{g}_0]$

b) $[\mathbf{d}_0, \mathbf{d}_1, \ldots, \mathbf{d}_k] = [\mathbf{g}_0, \mathbf{Q}\mathbf{g}_0, \ldots, \mathbf{Q}^k\mathbf{g}_0]$

c) $\mathbf{d}_k{}'\mathbf{Q}\mathbf{d}_i = 0 \quad \text{for} \quad i \leqslant k-1$

d) $\alpha_k = \mathbf{g}_k{}'\mathbf{g}_k / \mathbf{d}_k{}'\mathbf{Q}\mathbf{d}_k$

e) $\beta_k = \mathbf{g}_{k+1}'\mathbf{g}_{k+1} / \mathbf{g}_k{}'\mathbf{g}_k$.

Proof. We first prove (a), (b) and (c) simultaneously by induction. Clearly, they are true for $k = 0$. Now suppose they are true for k, we show that they are true for $k + 1$. We have

$$\mathbf{g}_{k+1} = \mathbf{g}_k + \alpha_k \mathbf{Q}\mathbf{d}_k.$$

By the induction hypothesis both $\mathbf{g}_k$ and $\mathbf{Q}\mathbf{d}_k$ belong to $[\mathbf{g}_0, \mathbf{Q}\mathbf{g}_0, \ldots, \mathbf{Q}^{k+1}\mathbf{g}_0]$, the first by (a) and the second by (b). Thus $\mathbf{g}_{k+1} \in [\mathbf{g}_0, \mathbf{Q}\mathbf{g}_0, \ldots, \mathbf{Q}^{k+1}\mathbf{g}_0]$. Furthermore $\mathbf{g}_{k+1} \notin [\mathbf{g}_0, \mathbf{Q}\mathbf{g}_0, \ldots, \mathbf{Q}^{k}\mathbf{g}_0] = [\mathbf{d}_0, \mathbf{d}_1, \ldots, \mathbf{d}_k]$ since otherwise $\mathbf{g}_{k+1} = \mathbf{0}$, because for any conjugate direction method $\mathbf{g}_{k+1}$ is orthogonal to $[\mathbf{d}_0, \mathbf{d}_1, \ldots, \mathbf{d}_k]$. (The induction hypothesis on (c) guarantees that the method is a conjugate direction method up to $\mathbf{x}_{k+1}$.) Thus, finally we conclude that

$$[\mathbf{g}_0, \mathbf{g}_1, \ldots, \mathbf{g}_{k+1}] = [\mathbf{g}_0, \mathbf{Q}\mathbf{g}_0, \ldots, \mathbf{Q}^{k+1}\mathbf{g}_0]$$

which proves (a).

To prove (b) we write

$$\mathbf{d}_{k+1} = -\mathbf{g}_{k+1} + \beta_k \mathbf{d}_k$$

and (b) immediately follows from (a) and the induction hypothesis on (b).

Next, to prove (c) we have

$$\mathbf{d}'_{k+1}\mathbf{Q}\mathbf{d}_i = -\mathbf{g}'_{k+1}\mathbf{Q}\mathbf{d}_i + \beta_k \mathbf{d}_k'\mathbf{Q}\mathbf{d}_i.$$

For $i = k$ the right side is zero by definition of β_k. For $i < k$ both terms vanish. The first term vanishes since $\mathbf{Q}\mathbf{d}_i \in [\mathbf{d}_1, \mathbf{d}_2, \ldots, \mathbf{d}_{i+1}]$, the induction hypothesis which guarantees the method is a conjugate direction method up to $\mathbf{x}_{k+1}$, and by the Expanding Subspace Theorem that guarantees that $\mathbf{g}_{k+1}$ is orthogonal to $[\mathbf{d}_0, \mathbf{d}_1, \ldots, \mathbf{d}_{i+1}]$. The second term vanishes by the induction hypothesis on (c). This proves (c), which also proves that the method is a conjugate direction method.

To prove (d) we have

$$-\mathbf{g}_k'\mathbf{d}_k = \mathbf{g}_k'\mathbf{g}_k - \beta_{k-1}\mathbf{g}_k'\mathbf{d}_{k-1}$$

and the second term is zero by the Expanding Subspace Theorem.

Finally, to prove (e) we note that $\mathbf{g}_{k+1}'\mathbf{g}_k = 0$, because $\mathbf{g}_k \in [\mathbf{d}_0, \ldots, \mathbf{d}_k]$ and $\mathbf{g}_{k+1}$ is orthogonal to $[\mathbf{d}_0, \ldots, \mathbf{d}_k]$. Thus since

$$\mathbf{Q}\mathbf{d}_k = \frac{1}{\alpha_k}(\mathbf{g}_{k+1} - \mathbf{g}_k)$$

we have

$$\mathbf{g}'_{k+1}\mathbf{Q}\mathbf{d}_k = \frac{1}{\alpha_k}\mathbf{g}'_{k+1}\mathbf{g}_{k+1}. \quad \blacksquare$$

Parts (a) and (b) of this theorem are a formal statement of the interrelation between the direction vectors and the gradient vectors. Part (c) is the equation that verifies that the method is a conjugate direction method. Parts (d) and (e) are identities yielding alternative formulae for α_k and β_k that are often more convenient than the original ones.

8.4 THE C–G METHOD AS AN OPTIMAL PROCESS

We turn now to the description of a special viewpoint that leads quickly to some very profound convergence results for the method of conjugate gradients. The basis of the viewpoint is part (b) of the Conjugate Gradient Theorem. This result tells us the spaces $\mathcal{B}_k$ over which we successively minimize are determined by the original gradient $\mathbf{g}_0$ and multiplications of it by $\mathbf{Q}$. Each step of the method brings into consideration an additional power of $\mathbf{Q}$ times $\mathbf{g}_0$. It is this observation we exploit.

Let us consider a new general approach for solving the quadratic minimization problem. Given an arbitrary starting point $\mathbf{x}_0$, let

$$\mathbf{x}_{k+1} = \mathbf{x}_0 + P_k(\mathbf{Q})\mathbf{g}_0 \tag{21}$$

where P_k is a polynomial of degree k. Selection of a set of coefficients for each of the polynomials P_k determines a sequence of $\mathbf{x}_k$'s. We have

$$\begin{aligned}\mathbf{x}_{k+1} - \mathbf{x}^* &= \mathbf{x}_0 - \mathbf{x}^* + P_k(\mathbf{Q})\mathbf{Q}(\mathbf{x}_0 - \mathbf{x}^*)\\ &= [\mathbf{I} + \mathbf{Q}P_k(\mathbf{Q})](\mathbf{x}_0 - \mathbf{x}^*)\end{aligned} \tag{22}$$

and hence

$$\begin{aligned}E(\mathbf{x}_{k+1}) &= \tfrac{1}{2}(\mathbf{x}_{k+1} - \mathbf{x}^*)'\mathbf{Q}(\mathbf{x}_{k+1} - \mathbf{x}^*)\\ &= \tfrac{1}{2}(\mathbf{x}_0 - \mathbf{x}^*)'\mathbf{Q}[\mathbf{I} + \mathbf{Q}P_k(\mathbf{Q})]^2(\mathbf{x}_0 - \mathbf{x}^*).\end{aligned} \tag{23}$$

We may now pose the problem of selecting the polynomial P_k in such a way as to minimize $E(\mathbf{x}_{k+1})$ with respect to all possible polynomials of degree k. Expanding (21), however, we obtain

$$\mathbf{x}_{k+1} = \mathbf{x}_0 + \gamma_0\mathbf{g}_0 + \gamma_1\mathbf{Q}\mathbf{g}_0 + \cdots + \gamma_k\mathbf{Q}^k\mathbf{g}_0 \tag{24}$$

where the γ_i's are the coefficients of P_k. In view of

$$\mathcal{B}_{k+1} = [\mathbf{d}_0, \mathbf{d}_1, \ldots, \mathbf{d}_k] = [\mathbf{g}_0, \mathbf{Q}\mathbf{g}_0, \ldots, \mathbf{Q}^k\mathbf{g}_0]$$

the vector $\mathbf{x}_{k+1} = \mathbf{x}_0 + \alpha_0\mathbf{d}_0 + \alpha_1\mathbf{d}_1 + \cdots + \alpha_k\mathbf{d}_k$ generated by the method of conjugate gradients has precisely this form; moreover, according to the Expanding Subspace Theorem, the coefficients γ_i determined by the conjugate gradient process are such as to minimize $E(\mathbf{x}_{k+1})$. Therefore, the problem posed of selecting the optimal P_k is solved by the conjugate gradient procedure.

The explicit relation between the optimal coefficients γ_i of P_k and the constants α_i, β_i associated with the conjugate gradient method is, of course, somewhat complicated, as is the relation between the coefficients of P_k and those of P_{k+1}. The power of the conjugate gradient method is that as it progresses it successively solves each of the optimal polynomial problems while updating only a small amount of information.

We summarize the above development by the following very useful theorem.

Theorem 1. *The point* $\mathbf{x}_{k+1}$ *generated by the conjugate gradient method satisfies*

$$E(\mathbf{x}_{k+1}) = \min_{P_k} \tfrac{1}{2}(\mathbf{x}_0 - \mathbf{x}^*)'\mathbf{Q}[\mathbf{I} + \mathbf{Q}P_k(\mathbf{Q})]^2(\mathbf{x}_0 - \mathbf{x}^*) \tag{25}$$

where the minimum is taken with respect to all polynomials P_k *of degree* k.

Bounds on Convergence

To use Theorem 1 most effectively it is convenient to recast it in terms of eigenvectors and eigenvalues of the matrix $\mathbf{Q}$. Suppose that the vector $\mathbf{x}_0 - \mathbf{x}^*$ is written in the eigenvector expansion

$$\mathbf{x}_0 - \mathbf{x}^* = \xi_1\mathbf{e}_1 + \xi_2\mathbf{e}_2 + \cdots + \xi_n\mathbf{e}_n$$

where the $\mathbf{e}_i$'s are normalized eigenvectors of $\mathbf{Q}$. Then since $\mathbf{Q}(\mathbf{x}_0 - \mathbf{x}^*) = \lambda_1\xi_1\mathbf{e}_1 + \lambda_2\xi_2\mathbf{e}_2 + \cdots + \lambda_n\xi_n\mathbf{e}_n$ and since the eigenvectors are mutually orthogonal, we have

$$E(\mathbf{x}_0) = \tfrac{1}{2}(\mathbf{x}_0 - \mathbf{x}^*)'\mathbf{Q}(\mathbf{x}_0 - \mathbf{x}^*) = \tfrac{1}{2}\sum_{i=1}^{n} \lambda_i\xi_i^2 \tag{26}$$

where the λ_i's are the corresponding eigenvalues of $\mathbf{Q}$. Applying the same manipulations to (25), we find that for *any* polynomial P_k of degree k there holds

$$E(\mathbf{x}_{k+1}) \leqslant \tfrac{1}{2}\sum_{i=1}^{n} [1 + \lambda_i P_k(\lambda_i)]^2\lambda_i\xi_i^2$$

It then follows that

$$E(\mathbf{x}_{k+1}) \leqslant \max_{\lambda_i} [1 + \lambda_i P_k(\lambda_i)]^2 \tfrac{1}{2}\sum_{i=1}^{n} \lambda_i\xi_i^2$$

and hence finally

$$E(\mathbf{x}_{k+1}) \leqslant \max_{\lambda_i} (1 + \lambda_i P_k(\lambda_i)]^2 E(\mathbf{x}_0).$$

We summarize this result by the following theorem.

Theorem 2. *In the method of conjugate gradients we have*

$$E(\mathbf{x}_{k+1}) \leqslant \max_{\lambda_i} [1 + \lambda_i P_k(\lambda_i)]^2 E(\mathbf{x}_0) \tag{27}$$

for any polynomial P_k *of degree* k, *where the maximum is taken over all eigenvalues* λ_i *of* $\mathbf{Q}$.

This way of viewing the conjugate gradient method as an optimal process is exploited in the next section. We note here that it implies the far from obvious fact that every step of the conjugate gradient method is at least as good as a steepest descent step would be from the same point. To see this, suppose $\mathbf{x}_k$ has been computed by the conjugate gradient method. From (24) we know $\mathbf{x}_k$ has the form

$$\mathbf{x}_k = \mathbf{x}_0 + \bar{\gamma}_0\mathbf{g}_0 + \bar{\gamma}_1\mathbf{Q}\mathbf{g}_0 + \cdots + \bar{\gamma}_{k-1}\mathbf{Q}^{k-1}\mathbf{g}_0.$$

Now if $\mathbf{x}_{k+1}$ is computed from $\mathbf{x}_k$ by steepest descent, we have $\mathbf{x}_{k+1} = \mathbf{x}_k - \alpha_k\mathbf{g}_k$ for some α_k. In view of part (a) of the Conjugate Gradient Theorem $\mathbf{x}_{k+1}$ will have the form (24). Since for the conjugate direction method $E(\mathbf{x}_{k+1})$ is lower than any other $\mathbf{x}_{k+1}$ of the form (24), we obtain the desired conclusion.

Typically when some information about the eigenvalue structure of $\mathbf{Q}$ is known, that information can be exploited by construction of a suitable polynomial P_k to use in (27). Suppose, for example, it were known that $\mathbf{Q}$ had only $m < n$ distinct eigenvalues. Then it is clear that by suitable choice of P_{m-1} it would be possible to make the mth degree polynomial $1 + \lambda P_{m-1}(\lambda)$ have its m zeros at the m eigenvalues. Using that particular polynomial in (27) shows that $E(\mathbf{x}_m) = 0$. Thus the optimal solution will be obtained in at most m, rather than n, steps. More sophisticated examples of this type of reasoning are contained in the next section and in the exercises at the end of the chapter.

8.5 THE PARTIAL CONJUGATE GRADIENT METHOD

A collection of procedures that are natural to consider at this point are those in which the conjugate gradient procedure is carried out for $m + 1 < n$ steps and then, rather than continuing, the process is restarted from the current point and $m + 1$ more conjugate gradient steps are taken. The special case of $m = 0$ corresponds to the standard method of steepest descent, while $m = n - 1$ corresponds to the full conjugate gradient method. These *partial conjugate gradient* methods are of extreme theoretical and practical importance, and their analysis yields additional insight into the method of conjugate gradients. The development of the least section forms the basis of our analysis.

As before, given the problem

$$\text{minimize } \tfrac{1}{2}\mathbf{x}'\mathbf{Q}\mathbf{x} - \mathbf{b}'\mathbf{x} \tag{28}$$

we define for any point $\mathbf{x}_k$ the gradient $\mathbf{g}_k = \mathbf{Q}\mathbf{x}_k - \mathbf{b}$. We consider an iteration scheme of the form

$$\mathbf{x}_{k+1} = \mathbf{x}_k + P^k(\mathbf{Q})\mathbf{g}_k, \tag{29}$$

where P^k is a polynomial of degree m. We select the coefficients of the polynomial P^k so as to minimize

$$E(\mathbf{x}_{k+1}) = \tfrac{1}{2}(\mathbf{x}_{k+1} - \mathbf{x}^*)'\mathbf{Q}(\mathbf{x}_{k+1} - \mathbf{x}^*) \tag{30}$$

where $\mathbf{x}^*$ is the solution to (28). In view of the development of the last section, it is clear that $\mathbf{x}_{k+1}$ can be found by taking $m + 1$ conjugate gradient steps rather than explicitly determining the appropriate polynomial directly. (The sequence indexing is slightly different here than in the previous section, since now we do not give separate indices to the intermediate steps of this process. Going from $\mathbf{x}_k$ to $\mathbf{x}_{k+1}$ by the partial conjugate gradient method involves m other points.)

The results of the previous section provide a tool for convergence analysis of this method. In this case, however, we develop a result that is of particular interest for $\mathbf{Q}$'s having a special eigenvalue structure that occurs frequently in optimization problems, especially, as shown below and in Chapter 12, in the context of penalty function methods for solving problems with constraints. We imagine that the eigenvalues of $\mathbf{Q}$ are of two kinds: there are m large eigenvalues that may or may not be located near each other, and $n - m$ smaller eigenvalues located within an interval [a, b]. Such a distribution of eigenvalues is shown in Fig. 8.3.

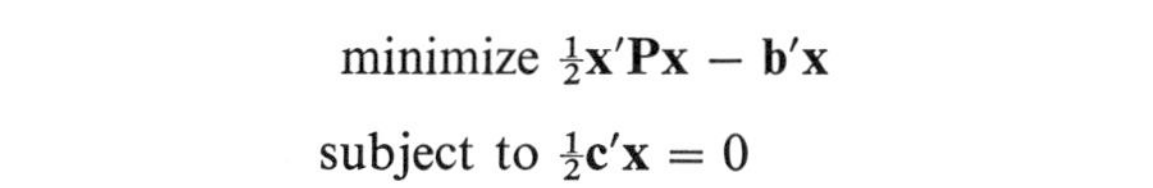

Fig. 8.3 Eigenvalue distribution

As an example, consider the problem on E^n

$$\text{minimize } \tfrac{1}{2}\mathbf{x}'\mathbf{P}\mathbf{x} - \mathbf{b}'\mathbf{x}$$

$$\text{subject to } \tfrac{1}{2}\mathbf{c}'\mathbf{x} = 0$$

where $\mathbf{P}$ is a symmetric positive definite matrix with eigenvalues in the interval $[a, A]$ and $\mathbf{b}$ and $\mathbf{c}$ are vectors in E^n. This is a constrained problem but it can be approximated by the unconstrained problem

$$\text{minimize } \tfrac{1}{2}\mathbf{x}'\mathbf{P}\mathbf{x} - \mathbf{b}'\mathbf{x} + \tfrac{1}{2}\mu(\mathbf{c}'\mathbf{x})^2$$

where μ is a large positive constant. The last term in the objective function is called a *penalty term*; for large μ minimization with respect to $\mathbf{x}$ will tend to make $\mathbf{c}'\mathbf{x}$ small.

The total quadratic term in the objective is $\frac{1}{2}\mathbf{x}'(\mathbf{P} + \mu\mathbf{c}\mathbf{c}')\mathbf{x}$, and thus it is appropriate to consider the eigenvalues of the matrix $\mathbf{Q} = \mathbf{P} + \mu\mathbf{c}\mathbf{c}'$. As μ tends to infinity it can be shown (see Chapter 12) that one eigenvalue of this matrix tends to infinity and the other $n-1$ eigenvalues remain bounded within the original interval [a, A].

Notice that, if the gradient method were applied to a $\mathbf{Q}$ with such a structure, convergence would be governed by the ratio of the smallest to largest eigenvalue, which in this case would be quite unfavorable. In the theorem below it is stated that by successively repeating $m+1$ conjugate gradient steps the effects of the m largest eigenvalues are eliminated and the rate of convergence is determined as if they were not present. A computational example of this phenomenon is presented in Section 12.6. The reader may find it interesting to read that section right after this one.

Theorem (Partial conjugate gradient method). *Suppose the symmetric positive definite matrix* $\mathbf{Q}$ *has* $n-m$ *eigenvalues in the interval* $[a, b]$, $a > 0$ *and the remaining* m *eigenvalues are greater than* b. *Then the method of partial conjugate gradients, restarted every* $m+1$ *steps, satisfies*

$$E(\mathbf{x}_{k+1}) \leqslant \left(\frac{b-a}{b+a}\right)^2 E(\mathbf{x}_k). \tag{31}$$

(The point $\mathbf{x}_{k+1}$ *is found from* $\mathbf{x}_k$ *by taking* $m+1$ *conjugate gradient steps so that each increment in* k *is a composite of several simple steps.)*

Proof. Application of (27) yields

$$E(\mathbf{x}_{k+1}) \leqslant \max_{\lambda_i} [1 + \lambda_i P(\lambda_i)]^2 E(\mathbf{x}_k) \tag{32}$$

for any mth order polynomial P, where the λ_i's are the eigenvalues of $\mathbf{Q}$. Let us select P so that the $m+1$th degree polynomial $q(\lambda) = 1 + \lambda P(\lambda)$ vanishes at $(a+b)/2$ and at the m large eigenvalues of $\mathbf{Q}$. This is illustrated in Fig. 8.4. For this choice of P we may write (32) as

$$E(\mathbf{x}_{k+1}) \leqslant \max_{a \leqslant \lambda_i \leqslant b} [1 + \lambda_i P(\lambda_i)]^2 E(\mathbf{x}_k).$$

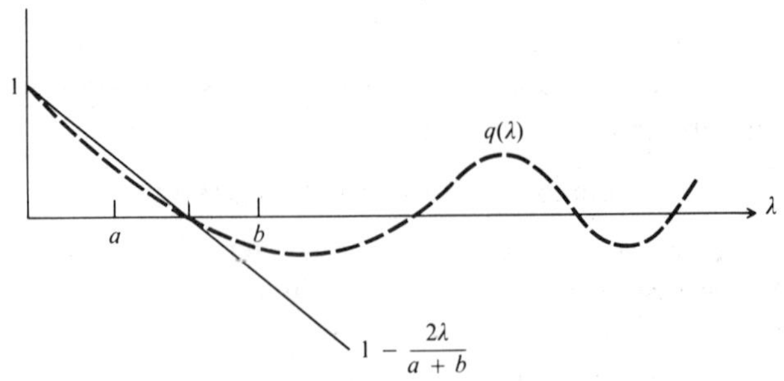

Fig. 8.4 Construction for proof

Since the polynomial $q(\lambda) = 1 + \lambda P(\lambda)$ has $m + 1$ real roots, $q'(\lambda)$ will have m real roots which alternate between the roots of $q(\lambda)$ on the real axis. Likewise, $q''(\lambda)$ will have $m - 1$ real roots which alternate between the roots of $q'(\lambda)$. Thus, since $q(\lambda)$ has no root in the interval $(-\infty, (a + b)/2)$, we see that $q''(\lambda)$ does not change sign in that interval and since it is easily verified that $q''(0) > 0$ it follows that $q(\lambda)$ is convex for $\lambda < (a + b)/2$. Therefore, on $[0, (a + b)/2]$, $q(\lambda)$ lies below the straight line $1 - [2\lambda/(a + b)]$. Thus we conclude that

$$q(\lambda) \leqslant 1 - \frac{2\lambda}{a + b}$$

on $[0, (a + b)/2]$ and that

$$q'\left(\frac{a + b}{2}\right) \geqslant -1.$$

We can see that on $[(a + b)/2, b]$

$$q(\lambda) \geqslant 1 - \frac{2\lambda}{a + b},$$

since for $q(\lambda)$ to cross first the line $1 - [2\lambda/(a + b)]$ and then the λ-axis would require at least two changes in sign of $q''(\lambda)$, whereas, at most one root of $q''(\lambda)$ exists to the left of the second root of $q(\lambda)$. We see then that the inequality

$$|1 + \lambda P(\lambda)| \leqslant 1 - \frac{2\lambda}{a + b}$$

is valid on the interval $[a, b]$. The final result (31) follows immediately. ∎

In view of this theorem. the method of partial conjugate gradients can be regarded as a generalization of steepest descent, not only in its philosophy and implementation, but also in its behaviour. Its rate of convergence is bounded by exactly the same formula as that of steepest descent but with the largest eigenvalues removed from consideration.

8.6 EXTENSION TO NONQUADRATIC PROBLEMS

The general unconstrained minimization problem on E^n

$$\text{minimize } f(\mathbf{x})$$

can be attacked by making suitable approximations to the conjugate gradient algorithm. There are a number of ways that this might be accomplished; the choice depends partially on what properties of f are easily computable. We look at two methods in this section and another in the following section.

Quadratic Approximation

In the quadratic approximation method we make the following associations at $\mathbf{x}_k$:

$$\mathbf{g}_k \leftrightarrow \nabla f(\mathbf{x}_k)', \qquad \mathbf{Q} \leftrightarrow \mathbf{F}(\mathbf{x}_k)$$

and using associations, reevaluated at each step, all quantities necessary to implement the basic conjugate gradient algorithm can be evaluated. If f is quadratic, these associations are identities, so that the general algorithm obtained by using them is a generalization of the conjugate gradient scheme. This is similar to the philosophy underlying Newton's method where at each step the solution of a general problem is approximated by the solution of a purely quadratic problem through these same associations.

Since $\mathbf{Q}$-conjugacy of the direction vectors in the pure conjugate gradient algorithm is dependent on the initial direction being the negative gradient, it is important that the general algorithm be reinitialized every n or $n+1$ steps by a pure gradient step; otherwise, the advantage of conjugate directions near the solution (where the problem is nearly quadratic) will likely be lost. The general conjugate gradient algorithm is then defined as below.

Step 1. Starting at $\mathbf{x}_0$ compute $\mathbf{g}_0 = \nabla f(\mathbf{x}_0)'$ and set $\mathbf{d}_0 = -\mathbf{g}_0$.

Step 2. For $k = 0, 1, \ldots, n-1$

a) set $\mathbf{x}_{k+1} = \mathbf{x}_k + \alpha_k \mathbf{d}_k$ where $\alpha_k = \dfrac{-\mathbf{g}_k'\mathbf{d}_k}{\mathbf{d}_k'\mathbf{F}(\mathbf{x}_k)\mathbf{d}_k}$

b) compute $\mathbf{g}_{k+1} = \nabla f(\mathbf{x}_{k+1})'$

c) unless $k = n-1$, set $\mathbf{d}_{k+1} = -\mathbf{g}_{k+1} + \beta_k \mathbf{d}_k$ where

$$\beta_k = \frac{\mathbf{g}_{k+1}'\mathbf{F}(\mathbf{x}_k)\mathbf{d}_k}{\mathbf{d}_k'\mathbf{F}(\mathbf{x}_k)\mathbf{d}_k}$$

and repeat (a).

Step 3. Replace $\mathbf{x}_0$ by $\mathbf{x}_n$ and go back to step 1.

An attractive feature of the algorithm is that, just as in the pure form of Newton's method, no line searching is required at any stage. Also, the algorithm converges in a finite number of steps for a quadratic problem. The undesirable features are that $\mathbf{F}(\mathbf{x}_k)$ must be evaluated at each point, which is often impractical, and that the algorithm is not, in this form, globally convergent.

Fletcher–Reeves Method

The *Fletcher–Reeves Method* for implementing the conjugate gradient algorithm for nonquadratic problems does not require the evaluation of $\mathbf{F}(\mathbf{x}_k)$ but does require a line search. The algorithm is based on the association $\mathbf{g}_k \leftrightarrow \nabla f(\mathbf{x}_k)'$, part (e) of the Conjugate Gradient Theorem, and the fact that α_k as computed by the pure conjugate gradient algorithm minimizes the objective function in the direction $\mathbf{d}_k$. The complete algorithm is:

Step 1. Given $\mathbf{x}_0$ compute $\mathbf{g}_0 = \nabla f(\mathbf{x}_0)'$ and set $\mathbf{d}_0 = -\mathbf{g}_0$.

Step 2. For $k = 0, 1, \ldots, n-1$

a) Set $\mathbf{x}_{k+1} = \mathbf{x}_k + \alpha_k \mathbf{d}_k$ where α_k minimizes $f(\mathbf{x}_k + \alpha \mathbf{d}_k)$

b) Compute $\mathbf{g}_{k+1} = \nabla f(\mathbf{x}_{k+1})'$

c) Unless $k = n-1$, set $\mathbf{d}_{k+1} = -\mathbf{g}_{k+1} + \beta_k \mathbf{d}_k$ where

$$\beta_k = \frac{\mathbf{g}_{k+1}' \mathbf{g}_{k+1}}{\mathbf{g}_k' \mathbf{g}_k}$$

Step 3. Replace $\mathbf{x}_0$ by $\mathbf{x}_n$ and go back to step 1.

Convergence

Global convergence of the Fletcher–Reeves method is established by noting that a pure steepest descent step is taken every n steps and serves as a spacer step. Since the other steps do not increase the objective, and in fact hopefully they decrease it, global convergence is assured. Thus the restarting aspect of the algorithm is important for global convergence analysis, since in general one cannot guarantee that the directions $\mathbf{d}_k$ generated by the method are descent directions.

The local convergence properties of both of the above, and most other, nonquadratic extensions of the conjugate gradient method can be inferred from the quadratic analysis. Assuming that at the solution, $\mathbf{x}^*$, the matrix $\mathbf{F}(\mathbf{x}^*)$ is positive definite, we expect the asymptotic convergence rate per step to be at least as good as steepest descent, since this is true in the quadratic case. In addition to this bound on the single step rate we expect that the method is of order two with respect to each complete cycle of n steps. In other words, since one complete cycle solves a quadratic problem exactly just as Newton's method does in one step, we expect that for general nonquadratic problems there will hold $|\mathbf{x}_{k+n} - \mathbf{x}^*| \leqslant c|\mathbf{x}_k - \mathbf{x}^*|^2$ for some c and $k = 0, n, 2n, 3n, \ldots$. This can indeed be proved, and of course underlies the original motivation for the method. For problems with large n, however, a result of this type is in itself of little comfort, since we probably hope to terminate in fewer than n steps. Further discussions on this general topic is contained in Section 9.4.

If one of these extended conjugate gradient methods were restarted every $m+1$ steps where $m < n$, then by the results of Section 8.5 the convergence of a complete cycle would approach that given by (31) but with the eigenvalues taken as those of $\mathbf{F}(\mathbf{x}^*)$. Thus, the partial conjugate gradient method will, for general problems, possess linear convergence.

8.7 PARALLEL TANGENTS

In early experiments with the method of steepest descent the path of descent was noticed to be highly zig-zag in character, making slow indirect progress toward the solution. (This phenomenon is now quite well understood and is predicted by the convergence analysis of Section 7.6.) It was also noticed that in two dimensions the solution point often lies close to the line that connects the zig-zag points, as illustrated in Fig. 8.5. This observation motivated the *accelerated gradient method* in which a complete cycle consists of taking two steepest descent steps and then searching along the line connecting the initial point and the point obtained after the two gradient steps. The method of parallel tangents (PARTAN) was developed through an attempt to extend this idea to an acceleration scheme involving all previous steps. The original development was based largely on a special geometric property of the tangents to the contours of a quadratic function, but the method is now recognized as a particular implementation of the method of conjugate gradients, and this is the context in which it is treated here.

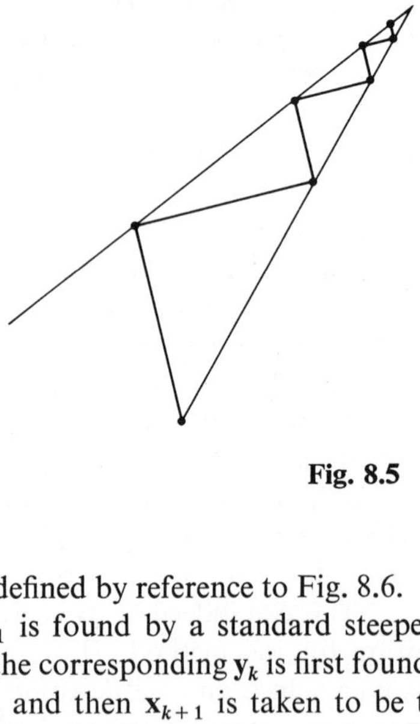

Fig. 8.5 Path of gradient method

The algorithm is defined by reference to Fig. 8.6. Starting at an arbitrary point $\mathbf{x}_0$ the point $\mathbf{x}_1$ is found by a standard steepest descent step. After that, from a point $\mathbf{x}_k$ the corresponding $\mathbf{y}_k$ is first found by a standard steepest descent step from $\mathbf{x}_k$, and then $\mathbf{x}_{k+1}$ is taken to be the minimum point on the line connecting $\mathbf{x}_{k-1}$ and $\mathbf{y}_k$. The process is continued for n steps and then restarted with a standard steepest descent step.

Notice that except for the first step, $\mathbf{x}_{k+1}$ is determined from $\mathbf{x}_k$, not by searching along a single line, but by searching along two lines. The direction $\mathbf{d}_k$ connecting two successive points (indicated as dotted lines in the figure) is thus determined only indirectly. We shall see, however, that, in the case where the objective function is quadratic, the $\mathbf{d}_k$'s are the same directions, and the $\mathbf{x}_k$'s are the same points, as would be generated by the method of conjugate gradients.

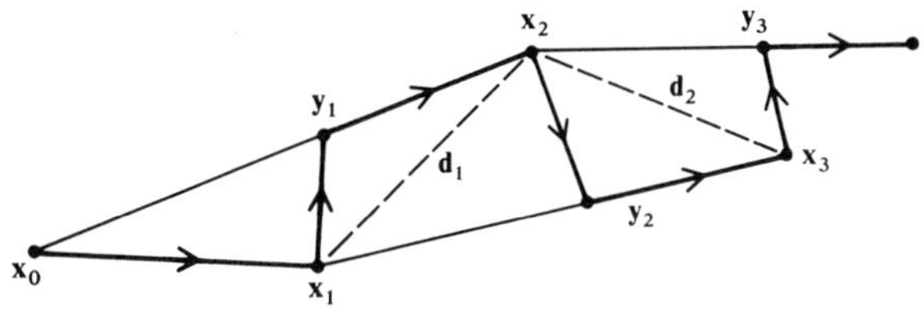

Fig. 8.6 Partan

PARTAN Theorem. *For a quadratic function, PARTAN is equivalent to the method of conjugate gradients.*

Proof. The proof is by induction. It is certainly true of the first step, since it is a steepest descent step. Suppose that $\mathbf{x}_0, \mathbf{x}_1, \ldots, \mathbf{x}_k$ have been generated by the conjugate gradient method and $\mathbf{x}_{k+1}$ is determined according to PARTAN. This single step is shown in Fig. 8.7. We want to show that $\mathbf{x}_{k+1}$ is the same point as would be generated by another step of the conjugate gradient method. For this to be true $\mathbf{x}_{k+1}$ must be that point which mininizes f over the plane defined by $\mathbf{d}_{k-1}$ and $\mathbf{g}_k = \nabla f(\mathbf{x}_k)'$. From the theory of conjugate gradients, this point will also minimize f over the subspace determined by $\mathbf{g}_k$ and all previous $\mathbf{d}_i$'s. Equivalently, we must find the point $\mathbf{x}$ where $\nabla f(\mathbf{x})$ is orthogonal to both $\mathbf{g}_k$ and $\mathbf{d}_{k-1}$. Since $\mathbf{y}_k$ minimizes f along $\mathbf{g}_k$, we see that $\nabla f(\mathbf{y}_k)$ is orthogonal to $\mathbf{g}_k$. Since $\nabla f(\mathbf{x}_{k-1})$ is contained in the subspace $[\mathbf{d}_0, \mathbf{d}_1, \ldots, \mathbf{d}_{k-1}]$ and because $\mathbf{g}_k$ is orthogonal to this subspace by the Expanding Subspace Theorem, we see that $\nabla f(\mathbf{x}_{k-1})$ is also orthogonal to $\mathbf{g}_k$. Since $\nabla f(\mathbf{x})$ is linear in $\mathbf{x}$, it follows that at every point $\mathbf{x}$ on the line through $\mathbf{x}_{k-1}$ and $\mathbf{y}_k$ we have $\nabla f(\mathbf{x})$ orthogonal to $\mathbf{g}_k$. By minimizing f along this line, a point $\mathbf{x}_{k+1}$ is obtained where in addition $\nabla f(\mathbf{x}_{k+1})$ is orthogonal to the line. Thus $\nabla f(\mathbf{x}_{k+1})$ is orthogonal to both $\mathbf{g}_k$ and the line joining $\mathbf{x}_{k-1}$ and $\mathbf{y}_k$. It follows that $\nabla f(\mathbf{x}_{k+1})$ is orthogonal to the plane. ∎

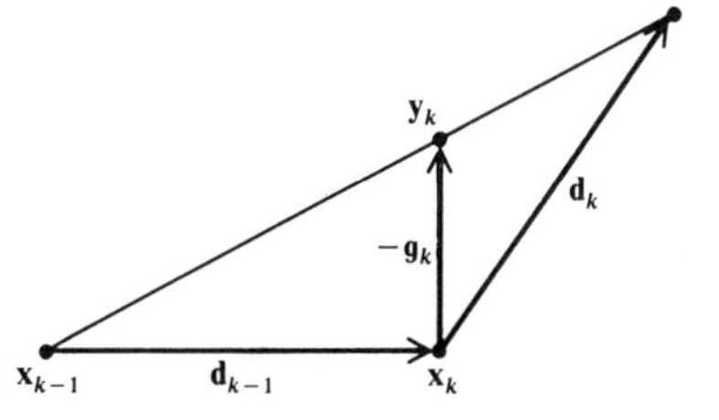

Fig. 8.7 One step of Partan

There are advantages and disadvantages of PARTAN relative to other methods when applied to nonquadratic problems. One attractive feature of the algorithm is its simplicity and ease of implementation. Probably its most desirable property, however, is its strong global convergence characteristics. Each step of the process is at least as good as steepest descent; since going from $\mathbf{x}_k$ to $\mathbf{y}_k$ is exactly steepest descent, and the additional move to $\mathbf{x}_{k+1}$ provides further decrease of the objective function. Thus global convergence is not tied to the fact that the process is restarted every n steps. It should be emphasized, however, that PARTAN should be restarted every n steps (or $n+1$ steps) so that it will behave like the conjugate gradient method near the solution. Running PARTAN indefinitely without restarting will converge—but slowly relative to the restarted method.

An undesirable feature of the algorithm is that two line searches are required at each step, except the first, rather than one as is required by, say, the Fletcher–Reeves method. This is at least partially compensated by the fact that searches need not be as accurate for PARTAN, for while inaccurate searches in the Fletcher–Reeves method may yield nonsensical successive search directions, PARTAN will at least do as well as steepest descent.

8.8 EXERCISES

1. Let $\mathbf{Q}$ be a positive definite symmetric matrix and suppose $\mathbf{p}_0, \mathbf{p}_1, \ldots, \mathbf{p}_{n-1}$ are linearly independent vectors in E^n. Show that a Gram–Schmidt procedure can be used to generate a sequence of $\mathbf{Q}$-conjugate directions from the $\mathbf{p}_i$'s. Specifically, show that $\mathbf{d}_0, \mathbf{d}_1, \ldots, \mathbf{d}_{n-1}$ defined recursively by

$$\mathbf{d}_0 = \mathbf{p}_0$$

$$\mathbf{d}_{k+1} = \mathbf{p}_{k+1} - \sum_{i=0}^{k} \frac{\mathbf{p}_{k+1}'\mathbf{Q}\mathbf{d}_i}{\mathbf{d}_i'\mathbf{Q}\mathbf{d}_i}\mathbf{d}_i$$

forms a $\mathbf{Q}$-conjugate set.

2. Suppose the $\mathbf{p}_i$'s in Exercise 1 are generated as *moments* of $\mathbf{Q}$, that is, suppose $\mathbf{p}_k = \mathbf{Q}^k\mathbf{p}_0$, $k = 1, 2, \ldots, n-1$. Show that the corresponding $\mathbf{d}_k$'s can then be generated by a (three-term) recursion formula where $\mathbf{d}_{k+1}$ is defined only in terms of $\mathbf{Q}\mathbf{d}_k$, $\mathbf{d}_k$ and $\mathbf{d}_{k-1}$.

3. Suppose the $\mathbf{p}_k$'s in Exercise 1 are taken as $\mathbf{p}_k = \mathbf{e}_k$ where $\mathbf{e}_k$ is the kth unit coordinate vector and the $\mathbf{d}_k$'s are constructed accordingly. Show that using $\mathbf{d}_k$'s in a conjugate direction method to minimize $\frac{1}{2}\mathbf{x}'\mathbf{Q}\mathbf{x} - \mathbf{b}'\mathbf{x}$ is equivalent to the application of Gaussian elimination to solve $\mathbf{Q}\mathbf{x} = \mathbf{b}$.

4. Let $f(\mathbf{x}) = \frac{1}{2}\mathbf{x}'\mathbf{Q}\mathbf{x} - \mathbf{b}'\mathbf{x}$ be defined on E^n with $\mathbf{Q}$ positive definite. Let $\mathbf{x}_1$ be a minimum point of f over a subspace of E^n containing the vector $\mathbf{d}$ and let $\mathbf{x}_2$ be

the minimum of f over another subspace containing $\mathbf{d}$. Suppose $f(\mathbf{x}_1) < f(\mathbf{x}_2)$. Show that $\mathbf{x}_1 - \mathbf{x}_2$ is $\mathbf{Q}$-conjugate to $\mathbf{d}$.

5. Let $\mathbf{Q}$ be a symmetric matrix. Show that any two eigenvectors of $\mathbf{Q}$, corresponding to distinct eigenvalues, are $\mathbf{Q}$-conjugate.

6. Let $\mathbf{Q}$ be an $n \times n$ symmetric matrix and let $\mathbf{d}_0, \mathbf{d}_1, \ldots, \mathbf{d}_{n-1}$ be $\mathbf{Q}$-conjugate. Show how to find an E such that $\mathbf{E}'\mathbf{Q}\mathbf{E}$ is diagonal.

7. Show that in the conjugate gradient method $\mathbf{Q}\mathbf{d}_{k-1} \in \mathcal{B}_{k+1}$.

8. Derive the rate of convergence of the method of steepest descent by viewing it as a one-step optimal process.

9. Let $P^k(\mathbf{Q}) = c_0 + c_1\mathbf{Q} + c_2\mathbf{Q}^2 + \cdots + c_m\mathbf{Q}^m$ be the optimal polynomial in (29) minimizing (30). Show that the c_i's can be found explicitly by solving the vector equation

$$-\begin{bmatrix} \mathbf{g}_k'\mathbf{Q}\mathbf{g}_k & \mathbf{g}_k'\mathbf{Q}^2\mathbf{g}_k \cdots & \mathbf{g}_k'\mathbf{Q}^{m+1}\mathbf{g}_k \\ \mathbf{g}_k'\mathbf{Q}^2\mathbf{g}_k & \mathbf{g}_k'\mathbf{Q}^3\mathbf{g}_k \cdots & \mathbf{g}_k'\mathbf{Q}^{m+2}\mathbf{g}_k \\ \cdot & & \\ \cdot & & \\ \cdot & & \\ \mathbf{g}_k'\mathbf{Q}^{m+1}\mathbf{g}_k & \cdots & \mathbf{g}_k'\mathbf{Q}^{2m+1}\mathbf{g}_k \end{bmatrix} \begin{bmatrix} c_0 \\ c_1 \\ \cdot \\ \cdot \\ \cdot \\ c_m \end{bmatrix} = \begin{bmatrix} \mathbf{g}_k'\mathbf{g}_k \\ \mathbf{g}_k'\mathbf{Q}\mathbf{g}_k \\ \cdot \\ \cdot \\ \cdot \\ \mathbf{g}_k'\mathbf{Q}^m\mathbf{g}_k \end{bmatrix}$$

Show that this reduces to steepest descent when $m = 0$.

10. Show that for the method of conjugate directions there holds

$$E(\mathbf{x}_k) \leqslant 4\left(\frac{1-\sqrt{\gamma}}{1+\sqrt{\gamma}}\right)^{2k} E(\mathbf{x}_0)$$

where $\gamma = a/A$ and a and A are the smallest and largest eigenvalues of $\mathbf{Q}$. Hint: In (27) select $P_{k-1}(\lambda)$ so that

$$1 + \lambda P_{k-1}(\lambda) = \frac{T_k\left(\dfrac{A + a - 2\lambda}{A - a}\right)}{T_k\left(\dfrac{A + a}{A - a}\right)}$$

where $T_k(\lambda) = \cos(k \arccos \lambda)$ is the kth Chebyshev polynomial. This choice gives the minimum maximum magnitude on $[a, A]$. Verify and use the inequality

$$\frac{(1-\gamma)^k}{(1+\gamma)^{2k} + (1-\gamma)^{2k}} \leqslant \left(\frac{1-\sqrt{\gamma}}{1+\sqrt{\gamma}}\right)^k.$$

11. Suppose it is known that each eigenvalue of $\mathbf{Q}$ lies either in the interval $[a, A]$ or in the interval $[a + \Delta, A + \Delta]$ where a, A and Δ are all positive. Show that the partial conjugate method restarted every two steps will converge with a ratio no greater than $[(A - a)/(A + a)]^2$ no matter how large Δ is.

12. Modify the first method given in Section 8.6 so that it is globally convergent.
13. Show that in the purely quadratic form of the conjugate gradient method $\mathbf{d}_k'\mathbf{Q}\mathbf{d}_k = -\mathbf{d}_k'\mathbf{Q}\mathbf{g}_k$. Using this show that to obtain $\mathbf{x}_{k+1}$ from $\mathbf{x}_k$ it is necessary to use $\mathbf{Q}$ only to evaluate $\mathbf{g}_k$ and $\mathbf{Q}\mathbf{g}_k$.
14. Show that in the quadratic problem $\mathbf{Q}\mathbf{g}_k$ can be evaluated by taking a unit step from $\mathbf{x}_k$ in the direction of the negative gradient and evaluating the gradient there. Specifically, if $\mathbf{y}_k = \mathbf{x}_k - \mathbf{g}_k$ and $\mathbf{p}_k = \nabla f(\mathbf{y}_k)'$, then $\mathbf{Q}\mathbf{g}_k = \mathbf{g}_k - \mathbf{p}_k$.
15. Combine the results of Exercises 13 and 14 to derive a conjugate gradient method for general problems much in the spirit of the first method of Section 8.6 but which does not require knowledge of $\mathbf{F}(\mathbf{x}_k)$ or a line search.

REFERENCES

8.1–8.3. For the original development of conjugate direction methods see Hestenes and Stiefel [H6] and Hestenes [H4]. For another introductory treatment see Beckman [B3]. The method was extended to the case where $\mathbf{Q}$ is not positive definite, which arises in constrained problems, by Luenberger [L7], [L9].

8.4. The idea of viewing the conjugate gradient method as an optimal process was originated by Stiefel [S5]. Also see Daniel [D1] and Faddeev and Faddeeva [F1].

8.5. The partial conjugate gradient method presented here is identical to the so-called s-step gradient method. See Faddeev and Faddeeva [F1] and Forsythe [F8]. The bound on the rate of convergence given in this section in terms of the interval containing the $n - m$ smallest eigenvalues was first given in Luenberger [L10]. Although this bound cannot be expected to be tight, it is a reasonable conjecture that it becomes tight as the m largest eigenvalues tend to infinity with arbitrarily large separation.

8.6. For the first approximate method see Daniel [D1]. For the second method see Fletcher and Reeves [F6]. For proof of the n-step, order two convergence see Cohen [C4].

8.7. PARTAN is due to Shah, Buehler and Kempthorne [S1]. Also see Wolfe [W4].

8.8. The approach indicated in Exercises 1 and 2 can be used as a foundation for the development of conjugate gradients, see Antosiewicz and Rheinboldt [A3], Vorobyev [V2], Faddeev and Faddeeva [F1], and Luenberger [L6]. The result stated in Exercise 3 is due to Hestenes and Stiefel [H6]. Exercise 4 is due to Powell [P1]. For the solution to Exercise 10 see Faddeev and Faddeeva [F1] or Daniel [D1].

Chapter 9 QUASI-NEWTON METHODS

In this chapter we take another approach toward the development of methods lying somewhere intermediate to steepest descent and Newton's method. Again working under the assumption that evaluation and inversion of the Hessian matrix is impractical or costly, the idea underlying quasi-Newton methods is to use an approximation to the inverse Hessian in place of the true inverse that is required in Newton's method. The form of the approximation varies among different methods—ranging from the simplest where it remains fixed throughout the iterative process, to the more advanced where improved approximations are built up on the basis of information gathered during the descent process.

The quasi-Newton methods that build up an approximation to the inverse Hessian are analytically the most sophisticated methods discussed in this book for solving unconstrained problems and represent the culmination of the development of algorithms through detailed analysis of the quadratic problem. As might be expected, the convergence properties of these methods are somewhat more difficult to discover than those of simpler methods. Nevertheless, we are able, by continuing with the same basic techniques as before, to illuminate their most important features.

In the course of our analysis we develop two important generalizations of the method of steepest descent and its corresponding convergence rate theorem. The first, discussed in Section 9.1, modifies steepest descent by taking as the direction vector a positive definite transformation of the negative gradient. The second, discussed in Section 9.6, is a combination of steepest descent and Newton's method. Both of these fundamental methods have convergence properties analogous to those of steepest descent.

9.1 MODIFIED NEWTON METHOD

A very basic iterative process for solving the problem

$$\text{minimize} \quad f(\mathbf{x})$$

which includes as special cases most of our earlier ones is

$$\mathbf{x}_{k+1} = \mathbf{x}_k - \alpha_k \mathbf{S}_k \nabla f(\mathbf{x}_k)' \tag{1}$$

where $\mathbf{S}_k$ is a symmetric $n \times n$ matrix and where, as usual, α_k is chosen to minimize $f(\mathbf{x}_{k+1})$. If $\mathbf{S}_k$ is the inverse of the Hessian of f, we obtain Newton's method, while if $\mathbf{S}_k = \mathbf{I}$ we have steepest descent. It would seem to be a good idea, in general, to select $\mathbf{S}_k$ as an approximation to the inverse of the Hessian. We examine that philosophy in this section.

First, we note, as in Section 7.7, that in order that the process (1) be guaranteed to be a descent method for small values of α, it is necessary in general to require that $\mathbf{S}_k$ be positive definite. We shall therefore always impose this as a requirement.

Because of the similarity of the algorithm (1) with steepest descent† it should not be surprising that its convergence properties are similar in character to our earlier results. We derive the actual rate of convergence by considering, as usual, the standard quadratic problem with

$$f(\mathbf{x}) = \tfrac{1}{2}\mathbf{x}'\mathbf{Q}\mathbf{x} - \mathbf{b}'\mathbf{x} \tag{2}$$

where $\mathbf{Q}$ is symmetric and positive definite. For this case we can find an explicit expression for α_k in (1). The algorithm becomes

$$\mathbf{x}_{k+1} = \mathbf{x}_k - \alpha_k \mathbf{S}_k \mathbf{g}_k \tag{3a}$$

where

$$\mathbf{g}_k = \mathbf{Q}\mathbf{x}_k - \mathbf{b} \tag{3b}$$

$$\alpha_k = \frac{\mathbf{g}_k'\mathbf{S}_k\mathbf{g}_k}{\mathbf{g}_k'\mathbf{S}_k\mathbf{Q}\mathbf{S}_k\mathbf{g}_k}. \tag{3c}$$

We may then derive the convergence rate of this algorithm by slightly extending the analysis carried out for the method of steepest descent.

Modified Newton Method Theorem (Quadratic case). *Let* $\mathbf{x}^*$ *be the unique minimum point of* f, *and define* $E(\mathbf{x}) = \frac{1}{2}(\mathbf{x} - \mathbf{x}^*)'\mathbf{Q}(\mathbf{x} - \mathbf{x}^*)$. *Then for the algorithm* (3) *there holds at every step* k

$$E(\mathbf{x}_{k+1}) \leqslant \left(\frac{B_k - b_k}{B_k + b_k}\right)^2 E(\mathbf{x}_k) \tag{4}$$

where b_k *and* B_k *are, respectively, the smallest and largest eigenvalues of the matrix* $\mathbf{S}_k\mathbf{Q}$.

---

† The algorithm (1) is sometimes referred to as the *method of deflected gradients*, since the direction vector can be thought of as being determined by deflecting the gradient through multiplication by $\mathbf{S}_k$.

Proof. We have by direct substitution

$$\frac{E(\mathbf{x}_k) - E(\mathbf{x}_{k+1})}{E(\mathbf{x}_k)} = \frac{(\mathbf{g}_k{}'\mathbf{S}_k\mathbf{g}_k)^2}{(\mathbf{g}_k{}'\mathbf{S}_k{}'\mathbf{Q}\mathbf{S}_k\mathbf{g}_k)(\mathbf{g}_k{}'\mathbf{Q}^{-1}\mathbf{g}_k)}.$$

Letting $\mathbf{T}_k = \mathbf{S}_k{}^{1/2}\mathbf{Q}\mathbf{S}_k{}^{1/2}$ and $\mathbf{p}_k = \mathbf{S}_k{}^{1/2}\mathbf{g}_k$ we obtain

$$\frac{E(\mathbf{x}_k) - E(\mathbf{x}_{k+1})}{E(\mathbf{x}_k)} = \frac{(\mathbf{p}_k{}'\mathbf{p}_k)^2}{(\mathbf{p}_k{}'\mathbf{T}_k\mathbf{p}_k)(\mathbf{p}_k{}'\mathbf{T}_k{}^{-1}\mathbf{p}_k)}.$$

From the Kantorovich inequality we obtain easily

$$E(\mathbf{x}_{k+1}) \leqslant \left(\frac{B_k - b_k}{B_k + b_k}\right)^2 E(\mathbf{x}_k)$$

where b_k and B_k are the smallest and largest eigenvalues of $\mathbf{T}_k$. Since $\mathbf{S}_k{}^{1/2}\mathbf{T}_k\mathbf{S}_k{}^{-1/2} = \mathbf{S}_k\mathbf{Q}$, we see that $\mathbf{S}_k\mathbf{Q}$ is similar to $\mathbf{T}_k$ and therefore has the same eigenvalues. ∎

This theorem supports the intuitive notion that for the quadratic problem one should strive to make $\mathbf{S}_k$ close to $\mathbf{Q}^{-1}$ since then both b_k and B_k would be close to unity and convergence would be rapid. For a nonquadratic objective function f the analog to $\mathbf{Q}$ is the Hessian $\mathbf{F}(\mathbf{x})$, and hence one should try to make $\mathbf{S}_k$ close to $\mathbf{F}(\mathbf{x}_k)^{-1}$.

Two remarks may help to put the above result in proper perspective. The first remark is that both the algorithm (1) and the theorem stated above are only simple, minor, and natural extensions of the work presented in Chapter 7 on steepest descent. As such the result of this section can be regarded, correspondingly, not as a new idea but as an extension of the basic result on steepest descent. The second remark is that this one simple result when properly applied can quickly characterize the convergence properties of some fairly complex algorithms. Thus, rather than an isolated result concerned with a specific form of algorithm, the theorem above should be regarded as a general tool for convergence analysis. It provides significant insight into various quasi-Newton methods discussed in this chapter.

A Classical Method

We conclude this section by mentioning the *classical modified Newton's method*, a standard method for approximating Newton's method without evaluating $\mathbf{F}(\mathbf{x}_k)^{-1}$ for each k. We set

$$\mathbf{x}_{k+1} = \mathbf{x}_k - \alpha_k[\mathbf{F}(\mathbf{x}_0)]^{-1}\nabla f(\mathbf{x}_k)'. \tag{5}$$

In this method the Hessian at the initial point $\mathbf{x}_0$ is inverted and used throughout the process. The effectiveness of this procedure is governed largely by how fast the Hessian is changing—in other words, by the magnitude of the third derivatives of f.

9.2 CONSTRUCTION OF THE INVERSE

The fundamental idea behind most quasi-Newton methods is to try to construct the inverse Hessian, or an approximation of it, using information gathered as the descent process progresses. The current approximation is then used at each stage to define the next descent direction as in the modified Newton method. Ideally, the approximations converge to the inverse of the Hessian at the solution point and the overall method behaves somewhat like Newton's method. In this section we show how the inverse Hessian can be built up from gradient information obtained at various points.

Let f be a function on E^n that has continuous second partial derivatives. If for two points $\mathbf{x}_{k+1}$, $\mathbf{x}_k$ we define $\mathbf{g}_{k+1} = \nabla f(\mathbf{x}_{k+1})'$, $\mathbf{g}_k = \nabla f(\mathbf{x}_k)'$ and $\mathbf{p}_k = \mathbf{x}_{k+1} - \mathbf{x}_k$, then by the Mean Value Theorem

$$\mathbf{g}_{k+1} - \mathbf{g}_k = \mathbf{F}(\mathbf{x}_k + \theta \mathbf{p}_k)\mathbf{p}_k \tag{6}$$

for some θ, $0 \leqslant \theta \leqslant 1$. If the Hessian, $\mathbf{F}$, is constant, then we have

$$\mathbf{q}_k \triangleq \mathbf{g}_{k+1} - \mathbf{g}_k = \mathbf{F}\mathbf{p}_k, \tag{7}$$

and we see that evaluation of the gradient at two points gives information about $\mathbf{F}$. If n linearly independent directions $\mathbf{p}_0, \mathbf{p}_1, \mathbf{p}_2, \ldots, \mathbf{p}_{n-1}$ and the corresponding $\mathbf{q}_k$'s are known, then $\mathbf{F}$ is uniquely determined. Indeed, letting $\mathbf{P}$ and $\mathbf{Q}$ be the $n \times n$ matrices with columns $\mathbf{p}_k$ and $\mathbf{q}_k$ respectively, we have

$$\mathbf{F} = \mathbf{Q}\mathbf{P}^{-1}. \tag{8}$$

It is natural to attempt to construct successive approximations $\mathbf{S}_k$ to $\mathbf{F}^{-1}$ based on data obtained from the first k steps of a descent process in such a way that if $\mathbf{F}$ were constant the approximation would be consistent with (7) for these steps. Specifically, if $\mathbf{F}$ were constant $\mathbf{S}_{k+1}$ would satisfy

$$\mathbf{S}_{k+1}\mathbf{q}_i = \mathbf{p}_i, \qquad 0 \leqslant i \leqslant k. \tag{9}$$

After n linearly independent steps we would then have $\mathbf{S}_n = \mathbf{F}^{-1}$.

For any $k < n$ the problem of constructing a suitable $\mathbf{S}_k$, which in general serves as an approximation to the inverse Hessian and which in the case of constant $\mathbf{F}$ satisfies (9), admits an infinity of solutions, since there are more degrees of freedom than there are constraints. Thus, a particular method can take into account additional considerations. We discuss below one of the most popular schemes that has been proposed. Another is discussed in the next section.

Rank One Correction

Since $\mathbf{F}$ and $\mathbf{F}^{-1}$ are symmetric, it is natural to require that $\mathbf{S}_k$, the approximation to $\mathbf{F}^{-1}$, be symmetric. We investigate the possibility of defining a recursion of the form

$$\mathbf{S}_{k+1} = \mathbf{S}_k + a_k \mathbf{z}_k \mathbf{z}_k' \tag{10}$$

which preserves symmetry. The vector $\mathbf{z}_k$ and the constant a_k define a matrix of rank one, by which the approximation to the inverse is updated. We select them so that (9) is satisfied. Setting i equal to k in (9) and substituting (10) we obtain

$$\mathbf{p}_k = \mathbf{S}_{k+1}\mathbf{q}_k = \mathbf{S}_k\mathbf{q}_k + a_k\mathbf{z}_k\mathbf{z}_k'\mathbf{q}_k. \tag{11}$$

Taking the inner product with $\mathbf{q}_k$ we have

$$\mathbf{q}_k'\mathbf{p}_k - \mathbf{q}_k'\mathbf{S}_k\mathbf{q}_k = a_k(\mathbf{z}_k'\mathbf{q}_k)^2. \tag{12}$$

On the other hand, using (11) we may write (10) as

$$\mathbf{S}_{k+1} = \mathbf{S}_k + \frac{(\mathbf{p}_k - \mathbf{S}_k\mathbf{q}_k)(\mathbf{p}_k - \mathbf{S}_k\mathbf{q}_k)'}{a_k(\mathbf{z}_k'\mathbf{q}_k)^2}$$

which in view of (12) leads finally to

$$\mathbf{S}_{k+1} = \mathbf{S}_k + \frac{(\mathbf{p}_k - \mathbf{S}_k\mathbf{q}_k)(\mathbf{p}_k - \mathbf{S}_k\mathbf{q}_k)'}{\mathbf{q}_k'(\mathbf{p}_k - \mathbf{S}_k\mathbf{q}_k)}. \tag{13}$$

We have determined what a rank one correction must be if it is to satisfy (9) for $i = k$. It remains to be shown that, for the case where $\mathbf{F}$ is constant, (9) is also satisfied for $i < k$. This in turn will imply that the rank one recursion converges to $\mathbf{F}^{-1}$ after at most n steps.

Theorem. *Let* $\mathbf{F}$ *be a fixed symmetric matrix and suppose that* $\mathbf{p}_0, \mathbf{p}_1, \mathbf{p}_2, \ldots, \mathbf{p}_k$ *are given vectors. Define* $\mathbf{q}_i = \mathbf{F}\mathbf{p}_i$, $i = 0, 1, 2, \ldots, k$. *Starting with any initial symmetric matrix* $\mathbf{S}_0$ *let*

$$\mathbf{S}_{i+1} = \mathbf{S}_i + \frac{(\mathbf{p}_i - \mathbf{S}_i\mathbf{q}_i)(\mathbf{p}_i - \mathbf{S}_i\mathbf{q}_i)'}{\mathbf{q}_i'(\mathbf{p}_i - \mathbf{S}_i\mathbf{q}_i)}. \tag{14}$$

Then

$$\mathbf{p}_i = \mathbf{S}_{k+1}\mathbf{q}_i \qquad \text{for} \qquad i \leqslant k. \tag{15}$$

Proof. The proof is by induction. Suppose it is true for $\mathbf{S}_k$, and $i \leqslant k-1$. The relation was shown above to be true for $\mathbf{S}_{k+1}$ and $i = k$. For $i < k$

$$\mathbf{S}_{k+1}\mathbf{q}_i = \mathbf{S}_k\mathbf{q}_k + \mathbf{y}_k(\mathbf{p}_k'\mathbf{q}_i - \mathbf{q}_k'\mathbf{S}_k\mathbf{q}_k) \tag{16}$$

where

$$\mathbf{y}_k = \frac{(\mathbf{p}_k - \mathbf{S}_k\mathbf{q}_k)}{\mathbf{q}_k'(\mathbf{p}_k - \mathbf{S}_k\mathbf{q}_k)}.$$

By the induction hypothesis, (16) becomes

$$\mathbf{S}_{k+1}\mathbf{q}_i = \mathbf{p}_i + \mathbf{y}_k(\mathbf{p}_k'\mathbf{q}_i - \mathbf{q}_k'\mathbf{p}_i).$$

From the calculation

$$\mathbf{q}_k'\mathbf{p}_i = \mathbf{p}_k'\mathbf{F}\mathbf{p}_i = \mathbf{p}_k'\mathbf{q}_i$$

it follows that the second term vanishes. ▌

To incorporate the approximate inverse Hessian in a descent procedure while simultaneously improving it, we calculate the direction $\mathbf{d}_k$ from

$$\mathbf{d}_k = -\mathbf{S}_k\mathbf{g}_k$$

and then minimize $f(\mathbf{x}_k + \alpha\mathbf{d}_k)$ with respect to $\alpha \geqslant 0$. This determines $\mathbf{x}_{k+1} = \mathbf{x}_k + \alpha_k\mathbf{d}_k$, $\mathbf{p}_k = \alpha_k\mathbf{d}_k$, and $\mathbf{g}_{k+1}$. Then $\mathbf{S}_{k+1}$ can be calculated according to (13). We note that if $\mathbf{S}_k$ is positive definite, then $\mathbf{S}_{k+1}$ will also be positive definite if $\mathbf{q}_k'(\mathbf{p}_k - \mathbf{S}_k\mathbf{q}_k) > 0$. It has been recommended, therefore, that $\mathbf{S}_k$ should be updated only if this condition is satisfied, while letting $\mathbf{S}_{k+1} = \mathbf{S}_k$ if it is not satisfied. This heuristic scheme suffers a severe defeat, however, when subjected to the analysis suggested in Exercise 5. Thus, although an excellent simple example of how information gathered during the descent process can in principle be used to update an approximation to the inverse Hessian, the rank one method possesses some serious limitations.

9.3 DAVIDON–FLETCHER–POWELL METHOD

The earliest, and certainly one of the most clever schemes for constructing the inverse Hessian, was originally proposed by Davidon and later developed by Fletcher and Powell. It has the fascinating and desirable property that, for a quadratic objective, it simultaneously generates the directions of the conjugate gradient method while constructing the inverse Hessian. At each step the inverse Hessian is updated by the sum of two symmetric rank one matrices, and this scheme is therefore often referred to as a *rank two correction procedure*. The method is also often referred to as the *variable metric method*, the name originally suggested by Davidon.

The procedure is this: Start with any symmetric positive matrix $\mathbf{S}_0$, and any point $\mathbf{x}_0$, then starting with $k = 0$,

Step 1. Set $\mathbf{d}_k = -\mathbf{S}_k\mathbf{g}_k$.

Step 2. Minimize $f(\mathbf{x}_k + \alpha\mathbf{d}_k)$ with respect to $\alpha \geqslant 0$ to obtain $\mathbf{x}_{k+1}$, $\mathbf{p}_k = \alpha_k\mathbf{d}_k$, and $\mathbf{g}_{k+1}$.

Step 3. Set $\mathbf{q}_k = \mathbf{g}_{k+1} - \mathbf{g}_k$ and

$$\mathbf{S}_{k+1} = \mathbf{S}_k + \frac{\mathbf{p}_k\mathbf{p}_k'}{\mathbf{p}_k'\mathbf{q}_k} - \frac{\mathbf{S}_k\mathbf{q}_k\mathbf{q}_k'\mathbf{S}_k}{\mathbf{q}_k'\mathbf{S}_k\mathbf{q}_k}. \tag{17}$$

Update k and return to step 1.

Positive Definiteness

We first demonstrate that if $\mathbf{S}_k$ is positive definite, then so is $\mathbf{S}_{k+1}$. For any $\mathbf{x} \in E^n$ we have

$$\mathbf{x}'\mathbf{S}_{k+1}\mathbf{x} = \mathbf{x}'\mathbf{S}_k\mathbf{x} + \frac{(\mathbf{x}'\mathbf{p}_k)^2}{\mathbf{p}_k'\mathbf{q}_k} - \frac{(\mathbf{x}'\mathbf{S}_k\mathbf{q}_k)^2}{\mathbf{q}_k'\mathbf{S}_k\mathbf{q}_k}. \tag{18}$$

Defining $\mathbf{a} = \mathbf{S}_k^{1/2}\mathbf{x}$, $\mathbf{b} = \mathbf{S}_k^{1/2}\mathbf{q}_k$ we may rewrite (18) as

$$\mathbf{x}'\mathbf{S}_{k+1}\mathbf{x} = \frac{(\mathbf{a}'\mathbf{a})(\mathbf{b}'\mathbf{b}) - (\mathbf{a}'\mathbf{b})^2}{(\mathbf{b}'\mathbf{b})} + \frac{(\mathbf{x}'\mathbf{p}_k)^2}{\mathbf{p}_k'\mathbf{q}_k}.$$

We also have

$$\mathbf{p}_k'\mathbf{q}_k = \mathbf{p}_k'\mathbf{g}_{k+1} - \mathbf{p}_k'\mathbf{g}_k = -\mathbf{p}_k'\mathbf{g}_k \tag{19}$$

since

$$\mathbf{p}_k'\mathbf{g}_{k+1} = 0 \tag{20}$$

because $\mathbf{x}_{k+1}$ is the minimum point of f along $\mathbf{p}_k$. Thus by definition of $\mathbf{p}_k$

$$\mathbf{p}_k'\mathbf{q}_k = \alpha_k\mathbf{g}_k'\mathbf{S}_k\mathbf{g}_k \tag{21}$$

and hence

$$\mathbf{x}'\mathbf{S}_{k+1}\mathbf{x} = \frac{(\mathbf{a}'\mathbf{a})(\mathbf{b}'\mathbf{b}) - (\mathbf{a}'\mathbf{b})^2}{(\mathbf{b}'\mathbf{b})} + \frac{(\mathbf{x}'\mathbf{p}_k)^2}{\alpha_k\mathbf{g}_k'\mathbf{S}_k\mathbf{g}_k}. \tag{22}$$

Both terms on the right of (22) are nonnegative—the first by the Cauchy–Schwarz inequality. We must only show they do not both vanish simultaneously. The first term vanishes only if $\mathbf{a}$ and $\mathbf{b}$ are proportional. This in turn implies that $\mathbf{x}$ and $\mathbf{q}_k$ are proportional, say $\mathbf{x} = \beta\mathbf{q}_k$. In that case, however,

$$\mathbf{p}_k'\mathbf{x} = \beta\mathbf{p}_k'\mathbf{q}_k = \beta\alpha_k\mathbf{g}_k'\mathbf{S}_k\mathbf{g}_k \neq 0$$

from (21). Thus $\mathbf{x}'\mathbf{S}_{k+1}\mathbf{x} > 0$ for all nonzero $\mathbf{x}$.

It is of interest to note that in the proof above the fact that α_k is chosen as the minimum point of the line search was used in (20), which led to the important conclusion $\mathbf{p}_k'\mathbf{q}_k > 0$. Actually any α_k, whether the minimum point or not, that gives $\mathbf{p}_k'\mathbf{q}_k > 0$ can be used in the algorithm, and $\mathbf{S}_{k+1}$ will be positive definite (see Exercises 7 and 8).

Finite Step Convergence

We assume now that f is quadratic with (constant) Hessian $\mathbf{F}$. We show in this case that the Davidon–Fletcher–Powell method produces direction vectors $\mathbf{p}_k$ that are $\mathbf{F}$-orthogonal and that if the method is carried n steps then $\mathbf{S}_n = \mathbf{F}^{-1}$.

Theorem. *If f is quadratic with positive definite Hessian $\mathbf{F}$, then for the Davidon–Fletcher–Powell method*

$$\mathbf{p}_i'\mathbf{F}\mathbf{p}_j = 0, \qquad 0 \leqslant i < j \leqslant k \tag{23}$$

$$\mathbf{S}_{k+1}\mathbf{F}\mathbf{p}_i = \mathbf{p}_i \quad \text{for} \quad 0 \leqslant i \leqslant k. \tag{24}$$

Proof. We note that for the quadratic case

$$\mathbf{q}_k = \mathbf{g}_{k+1} - \mathbf{g}_k = \mathbf{F}\mathbf{x}_{k+1} - \mathbf{F}\mathbf{x}_k = \mathbf{F}\mathbf{p}_k. \tag{25}$$

Also

$$\mathbf{S}_{k+1}\mathbf{F}\mathbf{p}_k = \mathbf{S}_{k+1}\mathbf{q}_k = \mathbf{p}_k \tag{26}$$

from (17).

We now prove (23) and (24) by induction. From (26) we see that they are true for $k = 0$. Assuming they are true for $k - 1$, we prove they are true for k. We have

$$\mathbf{g}_k = \mathbf{g}_{i+1} + \mathbf{F}(\mathbf{p}_{i+1} + \dots + \mathbf{p}_{k-1}).$$

Therefore from (23) and (20)

$$\mathbf{p}_i'\mathbf{g}_k = \mathbf{p}_i'\mathbf{g}_{i+1} = 0 \quad \text{for} \quad 0 \leqslant i < k. \tag{27}$$

Hence from (24)

$$\mathbf{p}_i'\mathbf{F}\mathbf{S}_k\mathbf{g}_k = 0. \tag{28}$$

Thus since $\mathbf{p}_k = -\alpha_k\mathbf{S}_k\mathbf{g}_k$ and since $\alpha_k \neq 0$, we obtain

$$\mathbf{p}_i'\mathbf{F}\mathbf{p}_k = 0 \quad \text{for} \quad i < k \tag{29}$$

which proves (23) for k.

Now since from (24) for $k - 1$, (25) and (29)

$$\mathbf{q}_k'\mathbf{S}_k\mathbf{F}\mathbf{p}_i = \mathbf{q}_k'\mathbf{p}_i = \mathbf{p}_k'\mathbf{F}\mathbf{p}_i = 0, \qquad 0 \leqslant i < k$$

we have

$$\mathbf{S}_{k+1}\mathbf{F}\mathbf{p}_i = \mathbf{S}_k\mathbf{F}\mathbf{p}_i = \mathbf{p}_i, \qquad 0 \leqslant i < k.$$

This together with (26) proves (24) for k. ∎

Since the $\mathbf{p}_k$'s are $\mathbf{F}$-orthogonal and since we minimize f successively in these directions, we see that the method is a conjugate direction method. Furthermore, if the initial approximation $\mathbf{S}_0$ is taken equal to the identity matrix, the method becomes the conjugate gradient method. In any case the process obtains the overall minimum point within n steps.

Finally, (24) shows that $\mathbf{p}_0, \mathbf{p}_1, \mathbf{p}_2, \ldots, \mathbf{p}_k$ are eigenvectors corresponding to unity eigenvalue for the matrix $\mathbf{S}_{k+1}\mathbf{F}$. These eigenvectors are linearly independent, since they are $\mathbf{F}$-orthogonal, and therefore $\mathbf{S}_n = \mathbf{F}^{-1}$.

General Case

For general nonquadratic objective functions the Davidon–Fletcher–Powell method seems to offer a combination of advantages that make it an extremely attractive general procedure. First, it requires only that first-order (that is, gradient) information be available. Second, the directions generated are always guaranteed to be directions of descent, since the matrix $\mathbf{S}_k$ is positive definite throughout the process. Third, since for a quadratic problem the matrices $\mathbf{S}_k$ converge to the inverse Hessian in at most n steps, it might be argued that in the general case $\mathbf{S}_k$ will converge to the inverse Hessian at the solution and hence convergence will be superlinear. Unfortunately, while the method is unquestionably an excellent one, in its pure form its convergence characteristics are not quite as outstanding as the above intuitive reasoning might suggest. We examine these properties more critically in the next section.

Partial Davidon Method

There is, of course, the possibility of restarting the Davidon–Fletcher–Powell method every $m + 1$ steps, where $m + 1 < n$. This would yield a *partial Davidon method* that, for small values of m, would have modest storage requirements, since the approximate inverse Hessian could be stored implicitly by storing only the vectors $\mathbf{p}_i$ and $\mathbf{q}_i$, $i \leqslant m + 1$. In the quadratic case this method exactly corresponds to the partial conjugate gradient method and hence it has similar convergence properties.

9.4 CONVERGENCE PROPERTIES

The various schemes for simultaneously generating and using an approximation to the inverse Hessian are difficult to analyze definitively. One must therefore, to some extent, resort to the use of analogy and approximate analyses to determine their effectiveness. Nevertheless, the machinery we developed earlier provides a basis for at least a preliminary analysis.

Global Convergence

In original form it was implicitly suggested that the Davidon–Fletcher–Powell method (and other quasi-Newton methods that update the inverse Hessian) should be executed in a continuing fashion, starting with an initial approximation and successively improving it throughout the iterative process. However, it has not been possible to prove that this procedure is globally convergent, and indeed some computational experiments suggest that the Davidon–Fletcher–Powell method operated this way can converge to a nonoptimal point. It is therefore advisable to restart these quasi-Newton methods every n or $n + 1$ steps by resetting the approximate inverse Hessian to its initial value, since then global convergence is guaranteed by the presence of the first descent step of each cycle (which acts as a spacer step).

Local Convergence

In view of the above remarks concerning global convergence, it is most worthwhile to consider the local convergence properties of these methods when executed with the restarting feature. Specifically, we consider the Davidon–Fletcher–Powell method and for simplicity we assume here that at the beginning of each cycle the approximate inverse Hessian is reset to the identity matrix. Each cycle, if at least n steps in duration, will then contain one complete cycle of an approximation to the conjugate gradient method. Asymptotically, in the tail of the generated sequence, this approximation becomes arbitrarily accurate, and hence we may conclude, as for any method that asymptotically approaches the conjugate gradient method, that the method converges superlinearly (at least if viewed at the end of each cycle). Although superlinear convergence is attractive, the fact that in this case it hinges on repeated cycles of n steps in duration can seriously detract from its practical significance for problems with large n, since we might hope to terminate the procedure before completing even a single full cycle of n steps.

To obtain insight into the defects of the method, let us consider a special situation. Suppose that f is quadratic and that the eigenvalues of the Hessian, $\mathbf{F}$, of f are close together but all very large. If, starting with the identity matrix, an approximation to the inverse Hessian is updated m times, the matrix $\mathbf{S}_m\mathbf{F}$ will have m eigenvalues equal to unity and the rest will still be large. Thus, the ratio of smallest to largest eigenvalue of $\mathbf{S}_m\mathbf{F}$, the condition number, will be worse than for $\mathbf{F}$ itself. Therefore, if the updating were discontinued and $\mathbf{S}_m$ were used as the approximation to $\mathbf{F}^{-1}$ in future iterations according to the procedure of Section 9.1, we see that convergence would be poorer than it would be for ordinary steepest descent. In other words, the approximations to $\mathbf{F}^{-1}$ generated by the updating formulas, although accurate over the subspace traveled, do not necessarily improve and, indeed, are likely to worsen, the eigenvalue structure of the iteration process.

In practice a poor eigenvalue structure arising in this manner will play a dominating role whenever there are factors that tend to weaken its approximation to the conjugate gradient method. Common factors of this type are round-off errors, inaccurate line searches, and nonquadratic terms in the objective function. Indeed, it has been frequently observed, empirically, that performance of the Davidon–Fletcher–Powell method is highly sensitive to the accuracy of the line search algorithm—to the point where superior step-wise convergence properties can only be obtained through excessive time expenditure in the line search phase.

Example. To illustrate some of these conclusions we consider the six-dimensional problem defined by

$$f(\mathbf{x}) = \tfrac{1}{2}\mathbf{x}'\mathbf{Q}\mathbf{x}$$

where

$$\mathbf{Q} = \begin{bmatrix} 40 & 0 & 0 & 0 & 0 & 0 \\ 0 & 38 & 0 & 0 & 0 & 0 \\ 0 & 0 & 36 & 0 & 0 & 0 \\ 0 & 0 & 0 & 34 & 0 & 0 \\ 0 & 0 & 0 & 0 & 32 & 0 \\ 0 & 0 & 0 & 0 & 0 & 30 \end{bmatrix}.$$

This function was minimized iteratively (the solution is obviously $\mathbf{x}^* = \mathbf{0}$) starting at $\mathbf{x}_0 = (10, 10, 10, 10, 10, 10)$, with $f(\mathbf{x}_0) = 10{,}500$, by using, alternatively, the method of steepest descent, the Davidon–Fletcher–Powell method, the Davidon–Fletcher–Powell method restarted every six steps, and the self-scaling method described in the next section. For this quadratic problem the appropriate step size to take at any stage can be calculated by a simple formula. On different computer runs of a given method, different levels of error were deliberately introduced into the step size in order to observe the effect of line search accuracy. This error took the form of a fixed percentage increase over the optimal value. The results are presented below:

Case 1. No error in step size α.

	Function value			
Iteration	Steepest descent	D–F–P	D–F–P (with restart)	Self-scaling
1	96.29630	96.29630	96.29630	96.29630
2	1.560669	6.900839×10^{-1}	6.900839×10^{-1}	6.900839×10^{-1}
3	2.932559×10^{-2}	3.988497×10^{-3}	3.988497×10^{-3}	3.988497×10^{-3}
4	5.787315×10^{-4}	1.683310×10^{-5}	1.683310×10^{-5}	1.683310×10^{-5}
5	1.164595×10^{-5}	3.878639×10^{-8}	3.878639×10^{-8}	3.878639×10^{-8}
6	2.359563×10^{-7}			

CASE 2. 0.1% error in step size α.

	Function value			
Iteration	Steepest descent	D–F–P	D–F–P (with restart)	Self-scaling
1	96.30669	96.30669	96.30669	96.30669
2	1.564971	6.994023×10^{-1}	6.994023×10^{-1}	6.902072×10^{-1}
3	2.939804×10^{-2}	1.225501×10^{-2}	1.225501×10^{-2}	3.989507×10^{-3}
4	5.810123×10^{-4}	7.301088×10^{-3}	7.301088×10^{-3}	1.684263×10^{-5}
5	1.169205×10^{-5}	2.636716×10^{-3}	2.636716×10^{-3}	3.881674×10^{-8}
6	2.372385×10^{-7}	1.031086×10^{-5}	1.031086×10^{-5}	
7		3.633330×10^{-9}	2.399278×10^{-8}	

CASE 3. 1% error in step size α.

	Function value			
Iteration	Steepest descent	D–F–P	D–F–P (with restart)	Self-scaling
1	97.33665	97.33665	97.33665	97.33665
2	1.586251	1.621908	1.621908	0.7024872
3	2.989875×10^{-2}	8.268893×10^{-1}	8.268893×10^{-1}	4.090350×10^{-3}
4	5.908101×10^{-4}	4.302943×10^{-1}	4.302943×10^{-1}	1.779424×10^{-5}
5	1.194144×10^{-5}	4.449852×10^{-3}	4.449852×10^{-3}	4.195668×10^{-8}
6	2.422985×10^{-7}	5.337835×10^{-5}	5.337835×10^{-5}	
7		3.767830×10^{-5}	4.493397×10^{-7}	
8		3.768097×10^{-9}		

CASE 4. 10% error in step size α.

	Function value			
Iteration	Steepest descent	D–F–P	D–F–P (with restart)	Self-scaling
1	200.333	200.333	200.333	200.333
2	2.732789	93.65457	93.65457	2.811061
3	3.836899×10^{-2}	56.92999	56.92999	3.562769×10^{-2}
4	6.376461×10^{-4}	1.620688	1.620688	4.200600×10^{-4}
5	1.219515×10^{-5}	5.251115×10^{-1}	5.251115×10^{-1}	4.726918×10^{-6}
6	2.457944×10^{-7}	3.323745×10^{-1}	3.323745×10^{-1}	
7		6.150890×10^{-3}	8.102700×10^{-3}	
8		3.025393×10^{-3}	2.973021×10^{-3}	
9		3.025476×10^{-5}	1.950152×10^{-3}	
10		3.025476×10^{-7}	2.769299×10^{-5}	
11			1.760320×10^{-5}	
12			1.123844×10^{-6}	

We note first that the error introduced is reported as a percentage of the step size itself. In terms of the change in function value, the quantity that is most often monitored to determine when to terminate a line search, the fractional error is the square of that in the step size. Thus, a one percent error in step size is equivalent to a 0.01% error in the change in function value.

Next we note that the method of steepest descent is not radically affected by an inaccurate line search while the Davidon–Fletcher–Powell methods are. Thus for this example while D-F-P is superior to steepest descent in the case of perfect accuracy, it becomes inferior at an error of only 0.1% in step size.

*9.5 SCALING

There is a general viewpoint about what makes up a desirable descent method that underlies much of our earlier discussions and which we now reiterate briefly in order to motivate the presentation of scaling. A method that converges to the exact solution after n steps when applied to a quadratic function on E^n has obvious appeal especially if, as is usually the case, it can be inferred that for nonquadratic problems repeated cycles of length n of the method will yield superlinear convergence. For problems having large n, however, a more sophisticated criterion of performance needs to be established, since for such problems one usually hopes to be able to terminate the descent process before completing even a single full cycle of length n. Thus, with these sorts of problems in mind, the finite-step convergence property serves at best only as a sign post indicating that the algorithm *might* make rapid progress in its early stages. It is essential to insure that in fact it *will* make rapid progress at every stage. Furthermore, the rapid convergence at each step must not be tied to an assumption on conjugate directions, a property easily destroyed by inaccurate line search and nonquadratic objective functions. With this viewpoint it is natural to look for a quasi-Newton method that simultaneously possesses favorable eigenvalue structure at each step (in the sense of Section 9.1) and reduces to the conjugate gradient method if the objective function happens to be quadratic. Such a method is developed in this section.

Improvement of Eigenvalue Ratio

Referring to the example presented in the last section where the Davidon–Fletcher–Powell method performed poorly, we can trace the difficulty to the simple observation that eigenvalues of $\mathbf{S}_0\mathbf{Q}$ are all much larger than unity. The Davidon–Fletcher–Powell algorithm essentially moves these eigenvalues, one at a time, to unity thereby producing an unfavorable eigenvalue ratio in each $\mathbf{S}_k\mathbf{Q}$ for $1 \leqslant k < n$. This phenomenon can be attributed to the fact that the method itself is sensitive to simple scale factors.

In particular if $\mathbf{S}_0$ were multiplied by a constant, the whole process would be different. In the example of the last section, if $\mathbf{S}_0$ were scaled by, for instance, multiplying it by 1/35, the eigenvalues of $\mathbf{S}_0\mathbf{Q}$ would be spread above and below unity, and in that case one might suspect that the poor performance would not show up.

Motivated by the above considerations, we in fact prove below that if $\mathbf{S}_k$ is any symmetric positive definite $n \times n$ matrix such that unity is contained in the interval spanned by the eigenvalues of $\mathbf{S}_k\mathbf{F}$, and $\mathbf{S}_{k+1}$ is determined by (17), then $\mathbf{S}_{k+1}\mathbf{F}$ will have at least as favorable an eigenvalue ratio as $\mathbf{S}_k\mathbf{F}$. This will then indicate how scaling can advantageously be introduced.

We use (but do not prove) the following matrix theoretic result due to Loewner.

Interlocking Eigenvalues Lemma. *Let the symmetric $n \times n$ matrix $\mathbf{A}$ have eigenvalues $\lambda_1 \leqslant \lambda_2 \leqslant \ldots \leqslant \lambda_n$. Let $\mathbf{a}$ be any vector in E^n and denote the eigenvalues of the matrix $\mathbf{A} + \mathbf{aa}'$ by $\mu_1 \leqslant \mu_2 \ldots \leqslant \mu_n$. Then $\lambda_1 \leqslant \mu_1 \leqslant \lambda_2 \leqslant \mu_2 \ldots \leqslant \lambda_n \leqslant \mu_n$.*

For convenience we make the following definitions

$$\mathbf{R}_k = \mathbf{F}^{1/2}\mathbf{S}_k\mathbf{F}^{1/2}$$

$$\mathbf{r}_k = \mathbf{F}^{1/2}\mathbf{p}_k.$$

Then using $\mathbf{q}_k = \mathbf{F}^{1/2}\mathbf{r}_k$ it can be readily verified that (17) is equivalent to

$$\mathbf{R}_{k+1} = \mathbf{R}_k - \frac{\mathbf{R}_k\mathbf{r}_k\mathbf{r}_k'\mathbf{R}_k}{\mathbf{r}_k'\mathbf{R}_k\mathbf{r}_k} + \frac{\mathbf{r}_k\mathbf{r}_k'}{\mathbf{r}_k'\mathbf{r}_k}. \tag{30}$$

Since $\mathbf{R}_k$ is similar to $\mathbf{S}_k\mathbf{F}$ (because $\mathbf{S}_k\mathbf{F} = \mathbf{F}^{-1/2}\mathbf{R}_k\mathbf{F}^{1/2}$) both have the same eigenvalues. It is most convenient, however, in view of (30) to study $\mathbf{R}_k$, obtaining the desired conclusions about $\mathbf{S}_k\mathbf{F}$ indirectly.

Suppose the eigenvalues of $\mathbf{R}_k$ are $\lambda_1, \lambda_2, \ldots, \lambda_n$ with $0 < \lambda_1 \leqslant \lambda_2 \leqslant \ldots \leqslant \lambda_n$. Suppose also that $1 \in [\lambda_1, \lambda_n]$. We will show that the eigenvalues of $\mathbf{R}_{k+1}$ are all contained in the interval $[\lambda_1, \lambda_n]$, which of course implies that $\mathbf{R}_{k+1}$ is no worse than $\mathbf{R}_k$ in terms of its condition number. Let us first consider the matrix

$$\mathbf{P} = \mathbf{R}_k - \frac{\mathbf{R}_k\mathbf{r}_k\mathbf{r}_k'\mathbf{R}_k}{\mathbf{r}_k'\mathbf{R}_k\mathbf{r}_k}.$$

We see that $\mathbf{P}\mathbf{r}_k = \mathbf{0}$ so one eigenvalue of $\mathbf{P}$ is zero. If we denote the eigenvalues of $\mathbf{P}$ by $\mu_1 \leqslant \mu_2 \leqslant \ldots \leqslant \mu_n$, we have from the above observation and the lemma on interlocking eigenvalues that

$$0 = \mu_1 \leqslant \lambda_1 \leqslant \mu_2 \leqslant \ldots \leqslant \mu_n \leqslant \lambda_n.$$

Next we consider

$$\mathbf{R}_{k+1} = \mathbf{P} + \frac{\mathbf{r}_k\mathbf{r}_k'}{\mathbf{r}_k'\mathbf{r}_k}.$$

Since $\mathbf{r}_k$ is an eigenvector of $\mathbf{P}$ and since all other eigenvectors of $\mathbf{P}$ are therefore orthogonal to $\mathbf{r}_k$, it follows that the only eigenvalue different in $\mathbf{R}_{k+1}$ from in $\mathbf{P}$ is the one corresponding to $\mathbf{r}_k$—it now being unity. Thus $\mathbf{R}_{k+1}$ has eigenvalues $\mu_2, \mu_3, \ldots, \mu_n$ and unity. These are all contained in the interval $[\lambda_1, \lambda_n]$. Thus updating does not worsen the eigenvalue ratio. It should be noted that this result in no way depends on α_k being selected to minimize f.

Scale Factors

In view of the result derived above, it is clearly advantageous to scale the matrix $\mathbf{S}_k$ so that the eigenvalues of $\mathbf{S}_k\mathbf{F}$ are spread both below and above unity. Of course in the ideal case of a quadratic problem with perfect line search this is strictly only necessary for $\mathbf{S}_0$, since unity is an eigenvalue of $\mathbf{S}_k\mathbf{F}$ for $k > 0$. But because of the inescapable deviations from the ideal, it is useful to consider the possibility of scaling every $\mathbf{S}_k$. Surprisingly, although this destroys the property that in the ideal case $\mathbf{S}_n = \mathbf{F}^{-1}$ it does not destroy the conjugate direction property.

A scale factor can be incorporated directly into the updating formula. Using $\gamma_0, \gamma_1, \ldots$ as arbitrary positive scale factors, we consider the algorithm: Start with any symmetric positive definite matrix $\mathbf{S}_0$ and any point $\mathbf{x}_0$, then starting with $k = 0$,

Step 1. Set $\mathbf{d}_k = -\mathbf{S}_k\mathbf{g}_k$.

Step 2. Minimize $f(\mathbf{x}_k + \alpha\mathbf{d}_k)$ with respect to $\alpha \geqslant 0$ to obtain

$\mathbf{x}_{k+1}$, $\mathbf{p}_k = \alpha_k\mathbf{d}_k$, and $\mathbf{g}_{k+1}$.

Step 3. Set $\mathbf{q}_k = \mathbf{g}_{k+1} - \mathbf{g}_k$ and

$$\mathbf{S}_{k+1} = \left(\mathbf{S}_k - \frac{\mathbf{S}_k\mathbf{q}_k\mathbf{q}_k'\mathbf{S}_k}{\mathbf{q}_k'\mathbf{S}_k\mathbf{q}_k}\right)\gamma_k + \frac{\mathbf{p}_k\mathbf{p}_k'}{\mathbf{p}_k'\mathbf{q}_k}. \tag{31}$$

We note that (31) is exactly what would be obtained by multiplying $\mathbf{S}_k$ by γ_k before updating by (17).

The following properties of this method can be proved as simple extensions of the results given in Section 9.3.

1) If $\mathbf{S}_k$ is positive definite and $\mathbf{p}_k'\mathbf{q}_k > 0$, (31) yields an $\mathbf{S}_{k+1}$ that is positive definite.
2) If f is quadratic with Hessian $\mathbf{F}$, then the vectors $\mathbf{p}_0, \mathbf{p}_1, \ldots \mathbf{p}_{n-1}$ are mutually $\mathbf{F}$-orthogonal, and, for each k, the vectors $\mathbf{p}_0, \mathbf{p}_1, \ldots, \mathbf{p}_k$ are eigenvectors of $\mathbf{S}_{k+1}\mathbf{F}$.

We can conclude that scale factors do not destroy the underlying conjugate behavior of the algorithm. Hence we can use scaling to ensure good single step convergence properties.

A Self-Scaling Quasi-Newton Algorithm

The question that arises next is how to select appropriate scale factors. If $\lambda_1 \leqslant \lambda_2 \leqslant \ldots \leqslant \lambda_n$ are the eigenvalues of $\mathbf{S}_k\mathbf{F}$, we want to multiply $\mathbf{S}_k$ by γ_k where $\lambda_1 \leqslant 1/\gamma_k \leqslant \gamma_n$. This will ensure that the new eigenvalues contain unity in the interval they span.

Note that in terms of our earlier notation

$$\frac{\mathbf{q}_k'\mathbf{S}_k\mathbf{q}_k}{\mathbf{p}_k'\mathbf{q}_k} = \frac{\mathbf{z}_k'\mathbf{R}_k\mathbf{z}_k}{\mathbf{z}_k'\mathbf{z}_k}.$$

Recalling that $\mathbf{R}_k$ has the same eigenvalues as $\mathbf{S}_k\mathbf{F}$ and noting that for any $\mathbf{z}_k$

$$\lambda_1 \leqslant \frac{\mathbf{z}_k'\mathbf{R}_k\mathbf{z}_k}{\mathbf{z}_k'\mathbf{z}_k} \leqslant \lambda_n$$

we see that

$$\gamma_k = \frac{\mathbf{p}_k'\mathbf{q}_k}{\mathbf{q}_k'\mathbf{S}_k\mathbf{q}_k}$$

serves as a suitable scale factor.

We state now a complete self-scaling, restarting, quasi-Newton method based on the ideas above. Start at any point $\mathbf{x}_0$, $k = 0$.

Step 1. Set $\mathbf{S}_k = \mathbf{I}$.

Step 2. Set $\mathbf{d}_k = -\mathbf{S}_k\mathbf{g}_k$.

Step 3. Minimize $f(\mathbf{x}_k + \alpha\mathbf{d}_k)$ with respect to $\alpha \geqslant 0$ to obtain α_k, $\mathbf{x}_{k+1}$, $\mathbf{p}_k = \alpha_k\mathbf{d}_k$, $\mathbf{g}_{k+1}$ and $\mathbf{q}_k = \mathbf{g}_{k+1} - \mathbf{g}_k$. (Select α_k accurately enough to ensure $\mathbf{p}_k'\mathbf{q}_k > 0$.)

Step 4. If k is not an integer multiple of n, set

$$\mathbf{S}_{k+1} = \left[\mathbf{S}_k - \frac{\mathbf{S}_k\mathbf{q}_k\mathbf{q}_k'\mathbf{S}_k}{\mathbf{q}_k'\mathbf{S}_k\mathbf{q}_k}\right]\frac{\mathbf{p}_k'\mathbf{q}_k}{\mathbf{q}_k'\mathbf{S}_k\mathbf{q}_k} + \frac{\mathbf{p}_k\mathbf{p}_k'}{\mathbf{p}_k'\mathbf{q}_k}.$$

Add one to k and return to step 2. If k is an integer multiple of n, return to step 1.

This algorithm was run, with various amounts of inaccuracy introduced in the line search, on the quadratic problem presented in Section 9.4. The results are presented in that section.

9.6 COMBINATION OF STEEPEST DESCENT AND NEWTON'S METHOD

In this section we digress from the study of quasi-Newton methods, and again expand our collection of basic principles. We present a combination of steepest descent and Newton's method which includes them both as special cases. The resulting combined method can be used as a foundation on which to build quasi-Newton methods, as illustrated in the next section, or for developing algorithms for problems having special structure, as illustrated in Chapter 12. This method and its analysis comprises a fundamental element of the modern theory of algorithms.

The method itself is quite simple. Suppose there is a subspace N of E^n on which the inverse Hessian of the objective function f is known (we shall make this statement more precise later). Then, in the quadratic case, the minimum of f over any linear variety parallel to N (that is, any translation of N) can be found in a single step. To minimize f over the whole space starting at any point $\mathbf{x}_k$, we could minimize f over the linear variety parallel to N and containing $\mathbf{x}_k$ to obtain $\mathbf{z}_k$; and then take a steepest descent step from there. This procedure is illustrated in Fig. 9.1. Since $\mathbf{z}_k$ is the minimum point of f over N, the gradient at $\mathbf{z}_k$ will be orthogonal to N, and hence the gradient step is orthogonal to N. If f is not quadratic we can, knowing the Hessian of f on N, approximate the minimum point of f over a linear variety parallel to N by one step of Newton's method. To implement this scheme, that we described in a geometric sense, it is necessary to agree on a method for defining the subspace N and to determine what information about the inverse Hessian is required so as to implement a Newton step over N. We now turn to these questions.

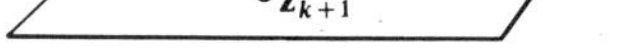

Fig. 9.1 Combined method

Often, the most convenient way to describe a subspace, and the one we follow in this development, is in terms of a set of vectors that generate it. Thus, if $\mathbf{B}$ is an $n \times m$ matrix consisting of m column vectors that generate N, we may write N as the set of all vectors of the form $\mathbf{Bu}$ where $\mathbf{u} \in E^m$. For simplicity we always assume that the columns of $\mathbf{B}$ are linearly independent.

To see what information about the inverse Hessian is required, imagine that we are at a point $\mathbf{x}_k$ and wish to find the approximate minimum point $\mathbf{z}_k$ of f with respect to movement in N. Thus, we seek $\mathbf{u}_k$ so that

$$\mathbf{z}_k = \mathbf{x}_k + \mathbf{B}\mathbf{u}_k$$

approximately minimizes f. By "approximately minimizes" we mean that $\mathbf{z}_k$ should be the Newton approximation to the minimum over this subspace. We write

$$f(\mathbf{z}_k) \cong f(\mathbf{x}_k) + \nabla f(\mathbf{x}_k)\mathbf{B}\mathbf{u}_k + \tfrac{1}{2}\mathbf{u}_k'\mathbf{B}'\mathbf{F}(\mathbf{x}_k)\mathbf{B}\mathbf{u}_k$$

and solve for $\mathbf{u}_k$ to obtain the Newton approximation. We find

$$\mathbf{u}_k = -\left(\mathbf{B}'\mathbf{F}(\mathbf{x}_k)\mathbf{B}\right)^{-1}\mathbf{B}'\nabla f(\mathbf{x}_k)'$$

$$\mathbf{z}_k = \mathbf{x}_k - \mathbf{B}\left(\mathbf{B}'\mathbf{F}(\mathbf{x}_k)\mathbf{B}\right)^{-1}\mathbf{B}'\nabla f(\mathbf{x}_k)'.$$

We see by analogy with the formula for Newton's method that the expression $\mathbf{B}(\mathbf{B}'\mathbf{F}(\mathbf{x}_k)\mathbf{B})^{-1}\mathbf{B}'$ can be interpreted as the inverse of $\mathbf{F}(\mathbf{x}_k)$ restricted to the subspace N.

Example. Suppose

$$\mathbf{B} = \begin{bmatrix} \mathbf{I} \\ \mathbf{0} \end{bmatrix}$$

where $\mathbf{I}$ is an $m \times m$ identity matrix. This corresponds to the case where N is the subspace generated by the first m unit basis elements of E^n. Let us partition $\mathbf{F} = \nabla^2 f(\mathbf{x}_k)$ as

$$\mathbf{F} = \begin{bmatrix} \mathbf{F}_{11} & \mathbf{F}_{12} \\ \mathbf{F}_{21} & \mathbf{F}_{22} \end{bmatrix}$$

where $\mathbf{F}_{11}$ is $m \times m$. Then, in this case

$$(\mathbf{B}'\mathbf{F}\mathbf{B})^{-1} = \mathbf{F}_{11}^{-1}$$

and

$$\mathbf{B}(\mathbf{B}'\mathbf{F}\mathbf{B})^{-1}\mathbf{B}' = \begin{bmatrix} \mathbf{F}_{11}^{-1} & \mathbf{0} \\ \mathbf{0} & \mathbf{0} \end{bmatrix}$$

which shows explicitly that it is the inverse of $\mathbf{F}$ on N that is required. The general case can be regarded as being obtained through partitioning in some skew coordinate system.

Now that the Newton approximation over N has been derived, it is possible to formalize the details of the algorithm suggested by Fig. 9.1. At a given point $\mathbf{x}_k$, the point $\mathbf{x}_{k+1}$ is determined through

$$
\begin{aligned}
&\text{a) set } \mathbf{d}_k = -\,\mathbf{B}(\mathbf{B}'\mathbf{F}(\mathbf{x}_k)\mathbf{B})^{-1}\,\mathbf{B}'\nabla f(\mathbf{x}_k)'.\\
&\text{b) } \mathbf{z}_k = \mathbf{x}_k + \beta_k \mathbf{d}_k\\
&\qquad\qquad \text{where } \beta_k \text{ minimizes } f(\mathbf{x}_k + \beta \mathbf{d}_k).\\
&\text{c) set } \mathbf{p}_k = -\,\nabla f(\mathbf{z}_k)'.\\
&\text{d) } \mathbf{x}_{k+1} = \mathbf{z}_k + \alpha_k \mathbf{p}_k\\
&\qquad\qquad \text{where } \alpha_k \text{ minimizes } f(\mathbf{z}_k + \alpha \mathbf{p}_k).
\end{aligned}
\tag{32}
$$

The scalar search parameter β_k is introduced in the Newton part of the algorithm simply to assure that the descent conditions required for global convergence are met. Normally β_k will be approximately equal to unity. (See Section 7.7.)

Analysis of Quadratic Case

Since the method is not a full Newton method, we can conclude that it possesses only linear convergence and that the dominating aspects of convergence will be revealed by an analysis of the method as applied to a quadratic function. Furthermore, as might be intuitively anticipated, the associated rate of convergence is governed by the steepest descent part of algorithm (32), and that rate is governed by a Kantorovich-like ratio defined over the subspace orthogonal to N.

Theorem (Combined method). *Let* $\mathbf{Q}$ *be an* $n \times n$ *symmetric positive definite matrix, and let* $\mathbf{x}^* \in E^n$. *Define the function*

$$E(\mathbf{x}) = \tfrac{1}{2}(\mathbf{x} - \mathbf{x}^*)'\,\mathbf{Q}(\mathbf{x} - \mathbf{x}^*)$$

and let $\mathbf{b} = \mathbf{Q}\mathbf{x}^*$. *Let* $\mathbf{B}$ *be an* $n \times m$ *matrix of rank* m. *Starting at an arbitrary point* $\mathbf{x}_0$, *define the iterative process*

$$
\begin{aligned}
&\text{a) } \mathbf{u}_k = -\,(\mathbf{B}'\mathbf{Q}\mathbf{B})^{-1}\mathbf{B}'\mathbf{g}_k \quad \textit{where} \quad \mathbf{g}_k = \mathbf{Q}x_k - \mathbf{b}.\\
&\text{b) } \mathbf{z}_k = \mathbf{x}_k + \mathbf{B}\mathbf{u}_k\\
&\text{c) } \mathbf{p}_k = \mathbf{b} - \mathbf{Q}\mathbf{z}_k\\
&\text{d) } \mathbf{x}_{k+1} = \mathbf{z}_k + \alpha_k \mathbf{p}_k \quad \textit{where} \quad \alpha_k = \frac{\mathbf{p}_k'\mathbf{p}_k}{\mathbf{p}_k'\mathbf{Q}\mathbf{p}_k}.
\end{aligned}
$$

This process converges to $\mathbf{x}^*$, *and satisfies*

$$E(\mathbf{x}_{k+1}) \leqslant (1 - \delta)E(\mathbf{x}_k) \tag{33}$$

where δ, $0 \leqslant \delta \leqslant 1$, *is the minimum of*

$$\frac{(\mathbf{p}'\mathbf{p})^2}{(\mathbf{p}'\mathbf{Q}\mathbf{p})(\mathbf{p}'\mathbf{Q}^{-1}\mathbf{p})}$$

over all vectors $\mathbf{p}$ *in the nullspace of* $\mathbf{B}'$.

Proof. The algorithm given in the theorem statement is exactly the general combined algorithm specialized to the quadratic situation. Next we note that

$$\begin{aligned}\mathbf{B}'\mathbf{p}_k &= \mathbf{B}'\mathbf{Q}(\mathbf{x}^* - \mathbf{z}_k) = \mathbf{B}'\mathbf{Q}(\mathbf{x}^* - \mathbf{x}_k) - \mathbf{B}'\mathbf{Q}\mathbf{B}\mathbf{u}_k \\ &= -\mathbf{B}'\mathbf{g}_k + \mathbf{B}\mathbf{Q}\mathbf{B}'(\mathbf{B}'\mathbf{Q}\mathbf{B})^{-1}\mathbf{B}'\mathbf{g}_k = \mathbf{0},\end{aligned} \tag{34}$$

which merely proves that the gradient at $\mathbf{z}_k$ is orthogonal to N. Next we calculate

$$\begin{aligned}2\{E(\mathbf{x}_k) - E(\mathbf{z}_k)\} &= (\mathbf{x}_k - \mathbf{x}^*)'\mathbf{Q}(\mathbf{x}_k - \mathbf{x}^*) - (\mathbf{z}_k - \mathbf{x}^*)'\mathbf{Q}(\mathbf{z}_k - \mathbf{x}^*) \\ &= -2\mathbf{u}_k'\mathbf{B}'\mathbf{Q}(\mathbf{x}_k - \mathbf{x}^*) - \mathbf{u}_k'\mathbf{B}'\mathbf{Q}\mathbf{B}\mathbf{u}_k \\ &= -2\mathbf{u}_k'\mathbf{B}'\mathbf{g}_k + \mathbf{u}_k'\mathbf{B}'\mathbf{Q}\mathbf{B}(\mathbf{B}'\mathbf{Q}\mathbf{B})^{-1}\mathbf{B}'\mathbf{g}_k \\ &= -\mathbf{u}_k'\mathbf{B}'\mathbf{g}_k = \mathbf{g}_k'\mathbf{B}(\mathbf{B}'\mathbf{Q}\mathbf{B})^{-1}\mathbf{B}'\mathbf{g}_k.\end{aligned} \tag{35}$$

Then we compute

$$\begin{aligned}2\{E(\mathbf{z}_k) - E(\mathbf{x}_{k+1})\} &= (\mathbf{z}_k - \mathbf{x}^*)'\mathbf{Q}(\mathbf{z}_k - \mathbf{x}^*) - (\mathbf{x}_{k+1} - \mathbf{x}^*)'\mathbf{Q}(\mathbf{x}_{k+1} - \mathbf{x}^*) \\ &= -2\alpha_k\mathbf{p}_k'\mathbf{Q}(\mathbf{z}_k - \mathbf{x}^*) - \alpha_k^2\mathbf{p}_k'\mathbf{Q}\mathbf{p}_k \\ &= 2\alpha_k\mathbf{p}_k'\mathbf{p}_k - \alpha_k^2\mathbf{p}_k'\mathbf{Q}\mathbf{p}_k \\ &= \alpha_k\mathbf{p}_k'\mathbf{p}_k = \frac{(\mathbf{p}_k'\mathbf{p}_k)^2}{\mathbf{p}_k'\mathbf{Q}\mathbf{p}_k}.\end{aligned} \tag{36}$$

Now using (34) and $\mathbf{p}_k = -\mathbf{g}_k - \mathbf{Q}\mathbf{B}\mathbf{u}_k$ we have

$$\begin{aligned}2E(\mathbf{x}_k) &= (\mathbf{x}_k - \mathbf{x}^*)'\mathbf{Q}(\mathbf{x}_k - \mathbf{x}^*) = \mathbf{g}_k'\mathbf{Q}^{-1}\mathbf{g}_k \\ &= (\mathbf{p}_k' + \mathbf{u}_k'\mathbf{B}'\mathbf{Q})\mathbf{Q}^{-1}(\mathbf{p}_k + \mathbf{Q}\mathbf{B}\mathbf{u}_k) \\ &= \mathbf{p}_k'\mathbf{Q}^{-1}\mathbf{p}_k + \mathbf{u}_k'\mathbf{B}'\mathbf{Q}\mathbf{B}\mathbf{u}_k \\ &= \mathbf{p}_k'\mathbf{Q}^{-1}\mathbf{p}_k + \mathbf{g}_k'\mathbf{B}(\mathbf{B}'\mathbf{Q}\mathbf{B})^{-1}\mathbf{B}'\mathbf{g}_k.\end{aligned} \tag{37}$$

Adding (35) and (36) and dividing by (37) there results

$$\begin{aligned}\frac{E(\mathbf{x}_k) - E(\mathbf{x}_{k+1})}{E(\mathbf{x}_k)} &= \frac{\mathbf{g}_k'\mathbf{B}(\mathbf{B}'\mathbf{Q}\mathbf{B})^{-1}\mathbf{B}'\mathbf{g}_k + (\mathbf{p}_k'\mathbf{p}_k)^2/\mathbf{p}_k'\mathbf{Q}\mathbf{p}_k}{\mathbf{p}_k'\mathbf{Q}^{-1}\mathbf{p}_k + \mathbf{g}_k'\mathbf{B}(\mathbf{B}'\mathbf{Q}\mathbf{B})^{-1}\mathbf{B}'\mathbf{g}_k} \\ &= \frac{q + (\mathbf{p}_k'\mathbf{p}_k)/(\mathbf{p}_k'\mathbf{Q}\mathbf{p}_k)}{q + (\mathbf{p}_k'\mathbf{Q}^{-1}\mathbf{p}_k)/(\mathbf{p}_k'\mathbf{p}_k)}\end{aligned}$$

where $q \geqslant 0$. This has the form $(q + a)/(q + b)$ with

$$a = \frac{\mathbf{p}_k'\mathbf{p}_k}{\mathbf{p}_k'\mathbf{Q}\mathbf{p}_k}, \qquad b = \frac{\mathbf{p}_k'\mathbf{Q}^{-1}\mathbf{p}_k}{\mathbf{p}_k'\mathbf{p}_k}.$$

But for any $\mathbf{p}_k$, it follows from the Kantorovich inequality that $a \leqslant b$. Hence

$$\frac{q + a}{q + b} \geqslant \frac{a}{b}$$

and thus

$$\frac{E(\mathbf{x}_k) - E(\mathbf{x}_{k+1})}{E(\mathbf{x}_k)} \geqslant \frac{(\mathbf{p}_k'\mathbf{p}_k)^2}{(\mathbf{p}_k'\mathbf{Q}\mathbf{p}_k)(\mathbf{p}_k'\mathbf{Q}^{-1}\mathbf{p}_k)}.$$

Finally

$$E(\mathbf{x}_{k+1}) \leqslant E(\mathbf{x}_k)\left[1 - \frac{(\mathbf{p}_k{}'\mathbf{p}_k)^2}{(\mathbf{p}_k{}'\mathbf{Q}\mathbf{p}_k)(\mathbf{p}_k{}'\mathbf{Q}^{-1}\mathbf{p}_k)}\right] \leqslant (1-\delta)E(\mathbf{x}_k)$$

since $\mathbf{B}'\mathbf{p}_k = \mathbf{0}$. ∎

The value δ associated with the above theorem is related to the eigenvalue structure of $\mathbf{Q}$. If $\mathbf{p}$ were allowed to vary over the whole space, then the Kantorovich inequality

$$\frac{(\mathbf{p}'\mathbf{p})^2}{(\mathbf{p}'\mathbf{Q}\mathbf{p})(\mathbf{p}'\mathbf{Q}^{-1}\mathbf{p})} \geqslant \frac{4aA}{(a+A)^2}, \tag{38}$$

where a and A are, respectively, the smallest and largest eigenvalues of $\mathbf{Q}$, gives explicitly

$$\delta = \frac{4aA}{(a+A)^2}.$$

When $\mathbf{p}$ is restricted to the nullspace of $\mathbf{B}'$, the corresponding value of δ is larger. In some special cases it is possible to obtain a fairly explicit estimate of δ. Suppose, for example, that the subspace N were the subspace spanned by m eigenvectors of $\mathbf{Q}$. Then the subspace in which $\mathbf{p}$ is allowed to vary is the space orthogonal to N and is thus, in this case, the space generated by the other $n - m$ eigenvectors of $\mathbf{Q}$. In this case since for $\mathbf{p}$ in $N^{\perp}$, both $\mathbf{Q}\mathbf{p}$ and $\mathbf{Q}^{-1}\mathbf{p}$ are also in $N^{\perp}$, the ratio δ satisfies

$$\delta = \frac{(\mathbf{p}'\mathbf{p})^2}{(\mathbf{p}'\mathbf{Q}\mathbf{p})(\mathbf{p}'\mathbf{Q}^{-1}\mathbf{p})} \geqslant \frac{4aA}{(a+A)^2}$$

where now a and A are, respectively, the smallest and largest of the $n - m$ eigenvalues of $\mathbf{Q}$ corresponding to $N^{\perp}$. Thus the convergence ratio (33) reduces to the familiar form

$$E(\mathbf{x}_{k+1}) \leqslant \left(\frac{A-a}{A+a}\right)^2 E(\mathbf{x}_k)$$

where a and A are these special eigenvalues. Thus if $\mathbf{B}$, or equivalently N, is chosen to include the eigenvectors corresponding to the most undesirable eigenvalues of $\mathbf{Q}$, the convergence rate of the combined method will be quite attractive.

Applications

The combination of steepest descent and Newton's method can be applied usefully in a number of important situations. Suppose, for example, we are faced with a problem of the form

$$\text{minimize} \quad f(\mathbf{x}, \mathbf{y})$$

where $\mathbf{x} \in E^n$, $\mathbf{y} \in E^m$, and where the second partial derivatives with respect to $\mathbf{x}$ are easily computible but those with respect to $\mathbf{y}$ are not. We may then employ Newton steps with respect to $\mathbf{x}$ and steepest descent with respect to $\mathbf{y}$.

Another instance where this idea can be greatly effective is when there are a few vital variables in a problem which, being assigned high costs, tend to dominate the value of the objective function; in other words, the partial second derivatives with respect to these variables are large. The poor conditioning induced by these variables can to some extent be reduced by proper scaling of variables, but more effectively, by carrying out Newton's method with respect to them and steepest descent with respect to the others.

9.7 UPDATING PROCEDURES

The combination of steepest descent and Newton's method can be used as a foundation for the development of quasi-Newton methods that update information about the inverse Hessian. In the terminology of the last section, the subspace N will, in these methods, be modified as the iterations progress. The number of ways in which this can be done effectively seems almost unbounded, and surely there is room for imaginative suggestions. The following is just one possibility.

We start the combined method with $N = \{\mathbf{0}\}$ and then as the iterations progress we enlarge N so as to include the gradient vector used in the steepest descent phase of the previous iteration. Thus, if N_k is represented by the corresponding matrix $\mathbf{B}_k$ and $\mathbf{p}_k = \mathbf{x}_{k+1} - \mathbf{z}_k$, we define N_{k+1} by its corresponding

$$\mathbf{B}_{k+1} = [\mathbf{B}_k, \mathbf{p}_k]. \tag{39}$$

The gradients at $\mathbf{x}_{k+1}$ and $\mathbf{z}_k$ yield new approximate information about $\mathbf{F}$ through the relation $\mathbf{q}_k = \mathbf{F}\mathbf{p}_k$ where $\mathbf{q}_k = \mathbf{g}_{k+1} - \mathbf{g}_k$. Thus if we have already calculated

$$\mathbf{D}_k = \mathbf{B}_k(\mathbf{B}_k{}'\mathbf{F}\mathbf{B}_k)^{-1}\mathbf{B}_k{}', \tag{40}$$

then it is easily verified that $\mathbf{D}_{k+1}$ can be found from

$$\mathbf{D}_{k+1} = \mathbf{D}_k + \frac{(\mathbf{p}_k - \mathbf{D}_k\mathbf{q}_k)(\mathbf{p}_k - \mathbf{D}_k\mathbf{q}_k)'}{(\mathbf{p}_k - \mathbf{D}_k\mathbf{q}_k)'\mathbf{q}_k} \tag{41}$$

which should be recognized as the rank one correction given in Section 9.2. As suggested in Section 9.2, the updating should be omitted if $(\mathbf{p}_k - \mathbf{D}_k\mathbf{q}_k)'\mathbf{q}_k \leqslant 0$. The difference between the method suggested here and that presented in Section 9.2 is that $\mathbf{D}_k$ is only positive semidefinite (while $\mathbf{S}_k$ was positive definite) and every iteration contains a steepest descent step. From the theory developed in this chapter, moreover, it can be concluded that this method has superior convergence characteristics, since it is guaranteed to be at least as fast per step as steepest descent and approaches Newton's method as the subspace N grows.

To implement the method it is, of course, not necessary to keep track of $\mathbf{B}_k$. The algorithm becomes: Given $\mathbf{x}_0$, $k = 0$,

Step 1. Put $\mathbf{D}_k = \mathbf{0}$

Step 2. Put $\mathbf{d}_k = -\mathbf{D}_k\nabla f(\mathbf{x}_k)'$

Step 3. Find $\mathbf{z}_k = \mathbf{x}_k + \beta_k\mathbf{d}_k$ where β_k minimizes $f(\mathbf{x}_k + \beta\mathbf{d}_k)$

Step 4. Find $\mathbf{x}_{k+1} = \mathbf{z}_k - \alpha_k\nabla f(\mathbf{z}_k)'$ where α_k minimizes $f(\mathbf{x}_k - \alpha\nabla f(\mathbf{z}_k)')$, and calculate

$$\mathbf{p}_k = \mathbf{x}_{k+1} - \mathbf{z}_k \quad \text{and} \quad \mathbf{q}_k = \nabla f(\mathbf{x}_{k+1})' - \nabla f(\mathbf{z}_k)'$$

Step 5. If $j = n$, set $j = 0$, increase k by one and return to step 1. Otherwise go to step 6.

Step 6. If $(\mathbf{p}_k - \mathbf{D}_k\mathbf{q}_k)'\mathbf{q}_k > 0$ set

$$\mathbf{D}_{k+1} = \mathbf{D}_k + \frac{(\mathbf{p}_k - \mathbf{D}_k\mathbf{q}_k)(\mathbf{p}_k - \mathbf{D}_k\mathbf{q}_k)'}{(\mathbf{p}_k - \mathbf{D}_k\mathbf{q}_k)'\mathbf{q}_k}$$

and increase j by one; Otherwise, set $\mathbf{D}_{k+1} = \mathbf{D}_k$. Increase k by one and return to step 2.

In this algorithm, after n updatings of N we will have $N = E^n$, and hence the corresponding $\mathbf{D}$ will be of full rank. In the case of a purely quadratic objective, it will also be true that $\mathbf{D} = \mathbf{F}^{-1}$. There are numerous variations possible depending on the storage and computation time per iteration available. For example, to limit storage the updating could be continued for only a few steps before restarting, or the process can be continued with $\mathbf{D}$ fixed and equal to a matrix of low rank.

9.8 SUMMARY

The basic motivation behind quasi-Newton methods is to try to obtain, at least on the average, the rapid convergence associated with Newton's method without explicitly evaluating the inverse Hessian at every step. This can be accomplished by constructing approximations to the inverse Hessian based on information gathered during the descent process, and results in methods which viewed in blocks of n steps (where n is the dimension of the problem) generally possess superlinear convergence.

Good, or even superlinear, convergence measured in terms of large blocks, however, is not always indicative of rapid convergence measured in terms of individual steps. It is important, therefore, to design quasi-Newton methods so that their single step convergence is rapid and relatively insensitive to line search inaccuracies. We discussed two general principles for examining these aspects of descent algorithms. The first of these is the modified Newton method in which the direction of descent is taken as the result of multiplication of the negative gradient by a positive definite matrix **S**. The single step convergence ratio of this method is determined by the usual steepest descent formula, but with the condition number of **SF** rather than just **F** used. This result was used to analyze some popular quasi-Newton methods and to develop the self-scaling method having good single step convergence properties.

The second principle method is the combined method in which Newton's method is executed over a subspace where the Hessian is known and steepest descent is executed elsewhere. This method converges at least as fast as steepest descent, and by incorporating the information gathered as the method progresses, the Newton portion can be executed over larger and larger subspaces.

At this point, it is perhaps valuable to summarize some of the main themes that have been developed throughout the four chapters comprising Part II. These chapters contain several important and popular algorithms that illustrate the range of possibilities available for minimizing a general nonlinear function. From a broad perspective, however, these individual algorithms can be considered simply as specific patterns on the analytical fabric that is woven through the chapters—the fabric that will support new algorithms and future developments.

One unifying element, that has reproved its value several times, is the Global Convergence Theorem. This result helped mold the final form of every algorithm presented in Part II and has effectively resolved the major questions concerning global convergence.

Another unifying element is the speed of convergence of an algorithm, which we have defined in terms of the asymptotic properties of the sequences an algorithm generates. Initially, it might have been argued that such measures, based on properties of the tail of the sequence, are perhaps not truly indicative of the actual time required to solve a problem—after all, a sequence generated in practice is a truncated version of the potentially infinite sequence, and asymptotic properties may not be representative of the finite version—a more complex measure of the speed of convergence may be required. It is fair to demand that the validity of the asymptotic measures we have proposed be judged in terms of how well they predict the performance of algorithms applied to specific examples. On this basis, as illustrated by the numerical examples presented in these chapters, and on others, the asymptotic rates are extremely reliable predictors of performance—provided that one carefully tempers his analysis with common sense (by, for example, not concluding that superlinear convergence is necessarily superior to linear convergence when the superlinear convergence is based on repeated cycles of length n). A major conclusion, therefore, of the previous chapters is the essential validity of the asymptotic approach to convergence analysis. This conclusion is a major strand in the analytical fabric of non-linear programming.

9.9 EXERCISES

1. Prove (4) directly for the modified Newton method by showing that each step of the modified Newton method is simply the ordinary method of steepest descent applied to a scaled version of the original problem.
2. Find the rate of convergence of the version of Newton's method defined by (44), (45) of Chapter 7. Show that convergence is only linear if δ is larger than the smallest eigenvalue of $\mathbf{F}(\mathbf{x}^*)$.
3. Consider the problem of minimizing a quadratic function
$$f(\mathbf{x}) = \tfrac{1}{2}\mathbf{x}'\mathbf{Q}\mathbf{x} - \mathbf{x}'\mathbf{b}$$
where $\mathbf{Q}$ is symmetric and sparse (that is, there are relatively few nonzero entries in $\mathbf{Q}$). The matrix $\mathbf{Q}$ has the form
$$\mathbf{Q} = \mathbf{I} + \mathbf{V}$$
where $\mathbf{I}$ is the identity and $\mathbf{V}$ is a matrix with eigenvalues bounded by $e < 1$ in magnitude.
   a) With the given information, what is the best bound you can give for the rate of convergence of steepest descent applied to this problem?
   b) In general it is difficult to invert $\mathbf{Q}$ but the inverse can be approximated by $\mathbf{I} - \mathbf{V}$, which is easy to calculate. (The approximation is very good for small e.) We are thus led to consider the iterative process
$$\mathbf{x}_{k+1} = \mathbf{x}_k - \alpha_k[\mathbf{I} - \mathbf{V}]\mathbf{g}_k$$
   where $\mathbf{g}_k = \mathbf{Q}\mathbf{x}_k - \mathbf{b}$ and α_k is chosen to minimize f in the usual way. With the information given, what is the best bound on the rate of convergence of this method?
   c) Show that for $e < \sqrt{2} - 1$ the method in part (b) is always superior to steepest descent.
4. This problem shows that the modified Newton's method is globally convergent under very weak assumptions.

   Let $a > 0$ and $b \geqslant a$ be given constants. Consider the collection $\mathscr{P}$ of all $n \times n$ symmetric positive definite matrices $\mathbf{P}$ having all eigenvalues greater than or equal to a and all elements bounded in absolute value by b. Define the point-to-set mapping $\mathbf{B}: E^n \to E^{n+n2}$ by $\mathbf{B}(\mathbf{x}) = \{(\mathbf{x}, \mathbf{P}) : \mathbf{P} \in \mathscr{P}\}$. Show that $\mathbf{B}$ is a closed mapping.

Now given an objective function $f \in C^1$, consider the iterative algorithm

$$\mathbf{x}_{k+1} = \mathbf{x}_k - \alpha_k \mathbf{P}_k \mathbf{g}_k$$

where $\mathbf{g}_k = \mathbf{g}(\mathbf{x}_k)$ is the gradient of f at $\mathbf{x}_k$, $\mathbf{P}_k$ is any matrix from $\mathscr{P}$ and α_k is chosen to minimize $f(\mathbf{x}_{k+1})$. This algorithm can be represented by $\mathbf{A}$ which can be decomposed as $\mathbf{A} = \mathbf{SCB}$ where $\mathbf{B}$ is defined above, $\mathbf{C}$ is defined by $\mathbf{C}(\mathbf{x}, \mathbf{P}) = (\mathbf{x}, -\mathbf{Pg}(\mathbf{x}))$, and $\mathbf{S}$ is the standard line search mapping. Show that if restricted to a compact set in E^n, the mapping $\mathbf{A}$ is closed.

Assuming that a sequence $\{\mathbf{x}_k\}$ generated by this algorithm is bounded, show that the limit $\mathbf{x}^*$ of any convergent subsequence satisfies $\mathbf{g}(\mathbf{x}^*) = \mathbf{0}$.

5. Suppose that the rank one correction method of Section 9.2 is applied to the quadratic problem (2) and suppose that the matrix $\mathbf{R}_0 = \mathbf{Q}^{1/2}\mathbf{S}_0\mathbf{Q}^{1/2}$ has $m < n$ eigenvalues less than unity and $n - m$ eigenvalues greater than unity. Show that the condition $\mathbf{q}_k'(\mathbf{p}_k - \mathbf{S}_k\mathbf{q}_k) > 0$ will be satisfied at most m times during the course of the method and hence, if updating is performed only when this condition holds, the sequence $\{\mathbf{S}_k\}$ will not converge to $\mathbf{Q}^{-1}$. Infer from this that, in using the rank one correction method, $\mathbf{S}_0$ should be taken very small; but that, despite such a precaution, on nonquadratic problems the method is subject to difficulty.

6. Show that if $\mathbf{S}_0 = \mathbf{I}$ the Davidon–Fletcher–Powell method is the conjugate gradient method. What similar statement can be made when $\mathbf{S}_0$ is an arbitrary symmetric positive definite matrix?

7. In the text it is shown that for the Davidon–Fletcher–Powell method $\mathbf{S}_{k+1}$ is positive definite if $\mathbf{S}_k$ is. The proof assumed that α_k is chosen to exactly minimize $f(\mathbf{x}_k + \alpha\mathbf{d}_k)$. Show that any $\alpha_k > 0$ which leads to $\mathbf{p}_k'\mathbf{q}_k > 0$ will guarantee the positive definiteness of $\mathbf{S}_{k+1}$. Show that for a quadratic problem any $\alpha_k \neq 0$ leads to a positive definite $\mathbf{S}_{k+1}$.

8. Suppose along the line $\mathbf{x}_k + \alpha\mathbf{d}_k$, $\alpha > 0$, the function $f(\mathbf{x}_k + \alpha\mathbf{d}_k)$ is unimodal and differentiable. Let $\bar{\alpha}_k$ be the minimizing value of α. Show that if any $\alpha_k > \bar{\alpha}_k$ is selected to define $\mathbf{x}_{k+1} = \mathbf{x}_k + \alpha_k\mathbf{d}_k$, then $\mathbf{p}_k\mathbf{q}_k > 0$. (Refer to Section 9.3).

9. Let $\{\mathbf{S}_k\}$, $k = 0, 1, 2 \ldots$ be the sequence of matrices generated by the Davidon–Fletcher–Powell method applied, without restarting, to a function f having continuous second partial derivatives. Assuming that there is $a > 0$, $A > 0$ such that for all k we have $\mathbf{S}_k - a\mathbf{I}$ and $A\mathbf{I} - \mathbf{S}_k$ positive definite and the corresponding sequence of $\mathbf{x}_k$'s is bounded, show that the method is globally convergent.

10. Explain how the partial Davidon method can be implemented for $m < n/2$, with less storage than required by the full method.

11. Prove statements (1) and (2) below Eq. (31) in Section 9.5.

12. Prove global convergence of the combination of steepest descent and Newton's method.

13. Formulate a rate of convergence theorem for the application of the combination of steepest and Newton's method to nonquadratic problems.

14. Prove that if $\mathbf{Q}$ is positive definite

$$\frac{(\mathbf{p}'\mathbf{p})}{\mathbf{p}'\mathbf{Q}\mathbf{p}} \leqslant \frac{\mathbf{p}'\mathbf{Q}^{-1}\mathbf{p}}{\mathbf{p}'\mathbf{p}}$$

for any vector $\mathbf{p}$.

15. It is possible to combine Newton's method and the partial conjugate gradient method. Given a subspace $N \subset E^n$, $\mathbf{x}_{k+1}$ is generated from $\mathbf{x}_k$ by first finding $\mathbf{z}_k$ by taking a Newton step in the linear variety through $\mathbf{x}_k$ parallel to N, and then taking m conjugate gradient steps from $\mathbf{z}_k$. What is a bound on the rate of convergence of this method?

16. Show that for quadratic problems the method of Section 9.7 is equivalent to the conjugate gradient method.

17. Show that the method of Section 9.7 is not dependent on an accurate line search routine as is the Davidon–Fletcher–Powell method.

REFERENCES

9.1. An early analysis of this method was given by Crockett and Chernoff [C6].

9.2—9.3. The variable metric method was developed originally by Davidon [D8] and its relation to the conjugate gradient method was descovered by Fletcher and Powell [F5]. For the rank one correction see Davidon [D9]. For general discussion of these kinds of quasi-Newton methods see Broyden [B4], Murtagh and Sargent [M2], Zeleznik [Z3], and Stewart [S4].

9.4. For some preliminary investigations into the convergence properties of these algorithms see McCormick [Mc2], McCormick and Pearson [Mc3], Goldfarb [G4], and Powell [P3]. The results of this section are new.

9.5. The lemma on interlocking eigenvalues is due to Loewner [L4]. An analysis similar to that contained in the first part of this section is contained in Fletcher [F4]. The scaling results, including the self-scaling algorithm, are due to Oren and Luenberger [O4] and were developed as the manuscript for this chapter evolved. See Oren [O3] for related methods.

9.6. The combined method appears in Luenberger [L8].

9.7. These methods and their analyses are new.

PART III
CONSTRAINED MINIMIZATION

Chapter 10 CONSTRAINED MINIMIZATION CONDITIONS

We turn now, in this final part of the book, to the study of minimization problems having constraints. We begin by studying in this chapter the necessary and sufficient conditions satisfied at solution points. These conditions, aside from their intrinsic value in characterizing solutions, define Lagrange multipliers and a certain Hessian matrix which, taken together, form the foundation for both the development and analysis of algorithms presented in subsequent chapters.

The general method used in this chapter to derive necessary and sufficient conditions is a straight-forward extension of that used in Chapter 6 for unconstrained problems. In the case of equality constraints, the feasible region is a curved surface embedded in E^n. Differential conditions satisfied at an optimal point are derived by considering the value of the objective function along curves on this surface passing through the optimal point. Thus, the arguments run almost identically to those for the unconstrained case; families of curves on the constraint surface replacing the earlier artifice of considering feasible directions.

10.1 CONSTRAINTS

We deal with general nonlinear programming problems of the form

$$\begin{array}{lll} \text{minimize} & f(\mathbf{x}) & \\ \text{subject to} & h_1(\mathbf{x}) = 0 & g_1(\mathbf{x}) \leqslant 0 \\ & h_2(\mathbf{x}) = 0 & g_2(\mathbf{x}) \leqslant 0 \\ & \vdots & \vdots \\ & h_m(\mathbf{x}) = 0 & g_p(\mathbf{x}) \leqslant 0 \\ & \mathbf{x} \in \Omega \subset E^n & \end{array}$$

where $m \leqslant n$ and the functions f, h_i, $i = 1, 2, \ldots, m$ and g_j, $j = 1, 2, \ldots, p$ are continuous, and usually assumed to possess continuous second partial derivatives. For notational simplicity, we introduce the vector-valued functions $\mathbf{h} = (h_1, h_2, \ldots, h_m)$ and $\mathbf{g} = (g_1, g_2, \ldots, g_p)$ and rewrite (1) as

$$\begin{aligned} &\text{minimize} \quad f(\mathbf{x}) \\ &\text{subject to} \quad \mathbf{h}(\mathbf{x}) = \mathbf{0}, \qquad \mathbf{g}(\mathbf{x}) \leqslant \mathbf{0} \\ &\qquad\qquad \mathbf{x} \in \Omega. \end{aligned} \tag{2}$$

The constraints $\mathbf{h}(\mathbf{x}) = \mathbf{0}$, $\mathbf{g}(\mathbf{x}) \leqslant \mathbf{0}$ are referred to as *functional constraints*, while the constraint $\mathbf{x} \in \Omega$ is a *set constraint*. As before we continue to de-emphasize the set constraint, assuming in most cases that either Ω is the whole space E^n or that the solution to (2) is in the interior of Ω. A point $\mathbf{x} \in E^n$ that satisfies all the functional constraints is said to be *feasible*.

A fundamental concept that provides a great deal of insight as well as simplifying the required theoretical development is that of an *active constraint*. An inequality constraint $g_i(\mathbf{x}) \leqslant 0$ is said to be *active* at a feasible point $\mathbf{x}$ if $g_i(\mathbf{x}) = 0$ and *inactive* at $\mathbf{x}$ if $g_i(\mathbf{x}) < 0$. By convention we refer to any equality constraint $h_i(\mathbf{x}) = 0$ as *active* at any feasible point. The constraints active at a feasible point $\mathbf{x}$ restrict the domain of feasibility in neighborhoods of $\mathbf{x}$, while the other, inactive constraints, have no influence in neighborhoods of $\mathbf{x}$. Therefore, in studying the properties of a local minimum point, it is clear that attention can be restricted to the active constraints. This is illustrated in Fig. 10.1 where local properties satisfied by the solution $\mathbf{x}^*$ obviously do not depend on the inactive constraints g_2 and g_3.

It is clear that, if it were known *a priori* which constraints were active at the solution to (1), the solution would be a local minimum point of the

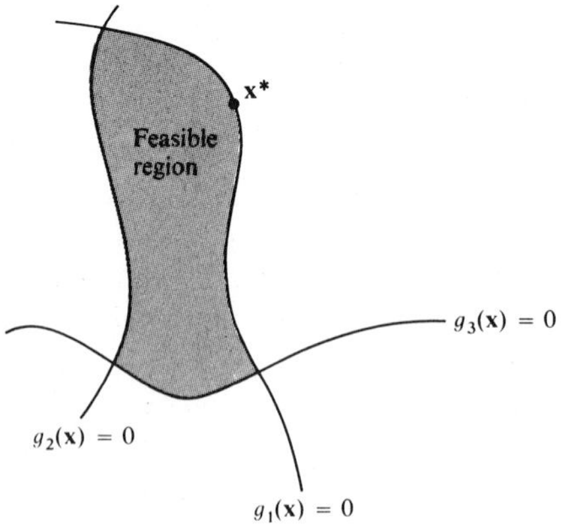

Fig. 10.1 Example of inactive constraints

problem defined by ignoring the inactive constraints and treating all active constraints as equality constraints. Hence, with respect to local (or relative) solutions, the problem could be regarded as having equality constraints only. This observation suggests that the majority of insight and theory applicable to (1) can be derived by consideration of equality constraints alone, later making additions to account for the selection of the active constraints. This is indeed so. Therefore in this chapter we first consider problems having only equality constraints, thereby both economizing on notation and isolating the primary ideas associated with constrained problems. We then extend these results to the more general situation.

Throughout the chapter our concern is with local solutions. A global theory is presented briefly in Exercise 3 and in Chapter 13.

10.2 TANGENT PLANE

A set of equality constraints on E^n

$$\begin{aligned} h_1(\mathbf{x}) &= 0 \\ h_2(\mathbf{x}) &= 0 \\ &\vdots \\ h_m(\mathbf{x}) &= 0 \end{aligned} \tag{3}$$

defines a subset of E^n which is best viewed as a hypersurface. If the constraints are independent, this hypersurface is of dimension $n - m$. If, as we assume in this section, the functions h_i, $i = 1, 2, \ldots, m$ belong to C^1, the surface defined by them is said to be *smooth*.

Associated with a point on a smooth surface is the *tangent plane* at that point, a term which in two or three dimensions has an obvious meaning. To formalize the general notion, we begin by defining curves on a surface. A *curve* on a surface S is a family of points $\mathbf{x}(t) \in S$ continuously parameterized by t for $a \leqslant t \leqslant b$. The curve is *differentiable* if $\dot{\mathbf{x}} \triangleq (\mathrm{d}/\mathrm{d}t)\,\mathbf{x}(t)$ exists, and is *twice differentiable* if $\ddot{\mathbf{x}}(t)$ exists. A curve $\mathbf{x}(t)$ is said to pass through the point $\mathbf{x}^*$ if $\mathbf{x}^* = \mathbf{x}(t^*)$ for some t^*, $a \leqslant t^* \leqslant b$. The derivative of the curve at $\mathbf{x}^*$ is, of course, defined as $\dot{\mathbf{x}}(t^*)$. It is itself a vector in E^n.

Now consider all differentiable curves on S passing through a point $\mathbf{x}^*$. The *tangent plane* at $\mathbf{x}^*$ is defined as the collection of the derivatives at $\mathbf{x}^*$ of all these differentiable curves. The tangent plane is a subspace of E^n.

For surfaces defined through a set of constraint relations such as (3), the problem of obtaining an explicit representation for the tangent plane is a fundamental problem that we now address. Ideally, we would like to express this tangent plane in terms of derivatives of functions h_i that define the surface. We introduce the subspace

$$M = \{\mathbf{y} : \nabla \mathbf{h}(\mathbf{x}^*)\mathbf{y} = \mathbf{0}\}$$

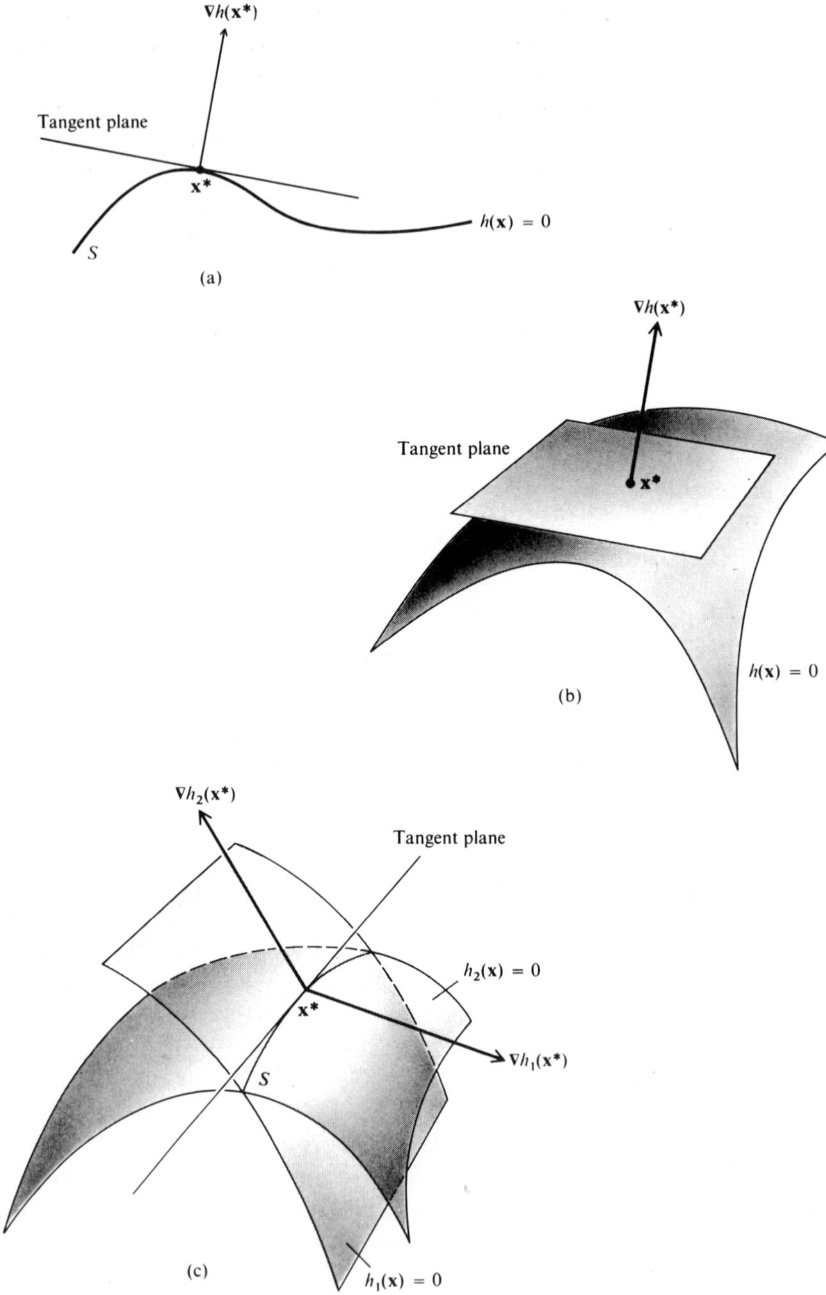

Fig. 10.2 Examples of tangent planes

and investigate under what conditions M is equal to the tangent plane at $\mathbf{x}^*$. See Fig. 10.2. The key concept for this purpose is that of a *regular point.*

Definition. A point $\mathbf{x}^*$ satisfying the constraint $\mathbf{h}(\mathbf{x}^*) = \mathbf{0}$ is said to be a *regular point* of the constraint if the gradient vectors $\nabla h_1(\mathbf{x}^*)$, $\nabla h_2(\mathbf{x}^*), \ldots, \nabla h_m(\mathbf{x}^*)$ are linearly independent.

At regular points it is possible to characterize the tangent plane in terms of the gradients of the constraint functions.

Theorem. *At a regular point* $\mathbf{x}^*$ *of the surface* S *defined by* $\mathbf{h}(\mathbf{x}) = \mathbf{0}$ *the tangent plane is equal to*

$$M = \{\mathbf{y} : \nabla\mathbf{h}(\mathbf{x}^*)\mathbf{y} = \mathbf{0}\}.$$

Proof. Let T be the tangent plane at $\mathbf{x}^*$. It is clear that $T \subset M$ whether $\mathbf{x}^*$ is regular or not, for any curve $\mathbf{x}(t)$ passing through $\mathbf{x}^*$ at $t = t^*$ having derivative $\dot{\mathbf{x}}(t^*)$ such that $\nabla\mathbf{h}(\mathbf{x}^*)\dot{\mathbf{x}}(t^*) \neq \mathbf{0}$ would not lie on S.

To prove that $M \subset T$ we must show that if $\mathbf{y} \in M$ then there is a curve on S passing through $\mathbf{x}^*$ with derivative $\mathbf{y}$. To construct such a curve we consider the equations

$$\mathbf{h}(\mathbf{x}^* + t\mathbf{y} + \nabla\mathbf{h}(\mathbf{x}^*)'\mathbf{u}(t)) = \mathbf{0} \tag{4}$$

where for fixed t we consider $\mathbf{u}(t) \in E^m$ to be the unknown. This is a nonlinear system of m equations and m unknowns, parameterized continuously by t. At $t = 0$ there is a solution $\mathbf{u}(0) = \mathbf{0}$. The Jacobian matrix of the system with respect to $\mathbf{u}$ at $t = 0$ is the $m \times m$ matrix

$$\nabla\mathbf{h}(\mathbf{x}^*)\nabla\mathbf{h}(\mathbf{x}^*)'$$

which is nonsingular, since $\nabla\mathbf{h}(\mathbf{x}^*)$ is of full rank if $\mathbf{x}^*$ is a regular point. Thus, by the Implicit Function Theorem (see Appendix A) there is a continuous solution $\mathbf{u}(t)$ in some region $-a \leqslant t \leqslant a$.

The curve $\mathbf{x}(t) = \mathbf{x}^* + t\mathbf{y} + \nabla\mathbf{h}(\mathbf{x}^*)'\mathbf{u}(t)$ is thus, by construction, a curve on S. By differentiating the system (4) with respect to t at $t = 0$ we obtain

$$\mathbf{0} = \frac{\mathrm{d}}{\mathrm{d}t}\mathbf{h}(\mathbf{x}(t))\bigg]_{t=0} = \nabla\mathbf{h}(\mathbf{x}^*)\mathbf{y} + \nabla\mathbf{h}(\mathbf{x}^*)\nabla\mathbf{h}(\mathbf{x}^*)'\dot{\mathbf{u}}(0).$$

By definition of $\mathbf{y}$ we have $\nabla\mathbf{h}(\mathbf{x}^*)\mathbf{y} = \mathbf{0}$ and thus, again since $\nabla\mathbf{h}(\mathbf{x}^*)\nabla\mathbf{h}(\mathbf{x}^*)'$ is nonsingular, we conclude that $\dot{\mathbf{u}}(0) = \mathbf{0}$. Therefore

$$\dot{\mathbf{x}}(0) = \mathbf{y} + \nabla\mathbf{h}(\mathbf{x}^*)'\dot{\mathbf{u}}(0) = \mathbf{y}$$

and the constructed curve has derivative $\mathbf{y}$ at $\mathbf{x}^*$. ∎

It is important to recognize that the condition of being a regular point is not a condition on the constraint surface itself but on its representation in terms of an **h**. The tangent plane is defined independently of the representation, while M is not.

Example. In E^2 let $h(x_1, x_2) = x_1$. Then $h(\mathbf{x}) = 0$ yields the x_2 axis, and every point on that axis is regular. If instead we put $h(x_1, x_2) = x_1{}^2$, again S is the x_2 axis but now no point on the axis is regular. Indeed in this case $M = E^2$, while the tangent plane is the x_2 axis.

10.3 NECESSARY AND SUFFICIENT CONDITIONS (EQUALITY CONSTRAINTS)

The derivation of necessary and sufficient conditions for a point to be a local minimum point subject to equality constraints is fairly simple now that the representation of the tangent plane is known. We begin by deriving the first-order necessary conditions.

Lemma. *Let* $\mathbf{x}^*$ *be a regular point of the constraints* $\mathbf{h}(\mathbf{x}) = \mathbf{0}$ *and a local extremum point (a minimum or maximum) of f subject to these constraints. Then all* $\mathbf{y} \in E^n$ *satisfying*

$$\nabla \mathbf{h}(\mathbf{x}^*)\mathbf{y} = \mathbf{0} \tag{5}$$

must also satisfy

$$\nabla f(\mathbf{x}^*)\mathbf{y} = 0. \tag{6}$$

Proof. Let $\mathbf{y}$ be any vector in the tangent plane at $\mathbf{x}^*$ and let $\mathbf{x}(t)$ be any smooth curve on the constraint surface passing through $\mathbf{x}^*$ with derivative $\mathbf{y}$ at $\mathbf{x}^*$; that is, $\mathbf{x}(0) = \mathbf{x}^*$, $\dot{\mathbf{x}}(0) = \mathbf{y}$, and $\mathbf{h}(\mathbf{x}(t)) = \mathbf{0}$ for $-a \leqslant t \leqslant a$ for some $a > 0$.

Since $\mathbf{x}^*$ is a regular point, the tangent plane is identical with the set of $\mathbf{y}$'s satisfying $\nabla \mathbf{h}(\mathbf{x}^*)\mathbf{y} = \mathbf{0}$. Then, since $\mathbf{x}^*$ is a constrained local extremum point of f, we have

$$\frac{d}{dt} f(\mathbf{x}(t))\Big]_{t=0} = 0,$$

or, equivalently,

$$\nabla f(\mathbf{x}^*)\mathbf{y} = 0. \quad \blacksquare$$

The above lemma says that $\nabla f(\mathbf{x}^*)$ is orthogonal to the tangent plane. Next we conclude that this implies that $\nabla f(\mathbf{x}^*)$ is a linear combination of the gradients of $\mathbf{h}$ at $\mathbf{x}^*$, a relation that leads to the introduction of Lagrange multipliers.

Theorem. *Let* $\mathbf{x}^*$ *be a local extremum point of f subject to the constraints* $\mathbf{h}(\mathbf{x}) = \mathbf{0}$. *Assume further that* $\mathbf{x}^*$ *is a regular point of these constraints. Then there is a* $\boldsymbol{\lambda} \in E_m$ *such that*

$$\nabla f(\mathbf{x}^*) + \boldsymbol{\lambda} \nabla \mathbf{h}(\mathbf{x}^*) = \mathbf{0}. \tag{7}$$

Proof. From the Lemma we may conclude that the value of the linear program

$$\begin{aligned} \text{maximize} \quad & \nabla f(\mathbf{x}^*)\mathbf{y} \\ \text{subject to} \quad & \nabla \mathbf{h}(\mathbf{x}^*)\mathbf{y} = \mathbf{0} \end{aligned}$$

is zero. Thus by the Duality Theorem of linear programming (Section 4.2) the dual problem is feasible. Specifically, there is $\boldsymbol{\lambda} \in E_m$ such that $\nabla f(\mathbf{x}^*) + \boldsymbol{\lambda}\nabla \mathbf{h}(\mathbf{x}^*) = \mathbf{0}$. ∎

It should be noted that the first-order necessary conditions

$$\nabla f(\mathbf{x}^*) + \boldsymbol{\lambda}\nabla \mathbf{h}(\mathbf{x}^*) = \mathbf{0}$$

together with the constraints

$$\mathbf{h}(\mathbf{x}^*) = \mathbf{0}$$

give a total of $n + m$ (generally nonlinear) equations in the $n + m$ variables comprising $\mathbf{x}^*$, $\boldsymbol{\lambda}$. Thus the necessary conditions are a complete set since, at least locally, they determine a unique solution.

It is convenient to introduce the *Lagrangian* associated with the constrained problem, defined as

$$l(\mathbf{x}, \boldsymbol{\lambda}) = f(\mathbf{x}) + \boldsymbol{\lambda}\mathbf{h}(\mathbf{x}). \tag{8}$$

The necessary conditions can then be expressed in the form

$$\nabla_{\mathbf{x}} l(\mathbf{x}, \boldsymbol{\lambda}) = \mathbf{0} \tag{9}$$

$$\nabla_{\boldsymbol{\lambda}} l(\mathbf{x}, \boldsymbol{\lambda}) = \mathbf{0} \tag{10}$$

the second of these being simply a restatement of the constraints.

Example. Consider the problem

$$\begin{aligned} \text{minimize} \quad & x_1x_2 + x_2x_3 + x_1x_3 \\ \text{subject to} \quad & x_1 + x_2 + x_3 = 3. \end{aligned}$$

The necessary conditions become

$$\begin{aligned} x_2 + x_3 + \lambda &= 0 \\ x_1 \qquad + x_3 + \lambda &= 0 \\ x_1 + x_2 \qquad + \lambda &= 0. \end{aligned}$$

These three equations together with the one constraint equation give four equations that can be solved for the four unknowns x_1, x_2, x_3, λ. Solution yields $x_1 = x_2 = x_3 = 1$, $\lambda = -2$.

Second-Order Condition

By an argument analogous to that used for the unconstrained case, we can also derive the corresponding second-order conditions for constrained problems. Throughout the remainder of this section it is assumed that $f, \mathbf{h} \in C^2$.

Second-Order Necessary Conditions. *Suppose that* $\mathbf{x}^*$ *is a local minimum of* f *subject to* $\mathbf{h}(\mathbf{x}) = \mathbf{0}$ *and that* $\mathbf{x}^*$ *is a regular point of these constraints. Then there is a* $\boldsymbol{\lambda} \in E_m$ *such that*

$$\nabla f(\mathbf{x}^*) + \boldsymbol{\lambda}\nabla \mathbf{h}(\mathbf{x}^*) = \mathbf{0}. \tag{11}$$

If we denote by M *the tangent plane* $M = \{\mathbf{y} : \nabla \mathbf{h}(\mathbf{x}^*)\mathbf{y} = \mathbf{0}\}$, *then the matrix*

$$\mathbf{L}(\mathbf{x}^*) = \mathbf{F}(\mathbf{x}^*) + \boldsymbol{\lambda}\mathbf{H}(\mathbf{x}^*) \tag{12}$$

is positive semidefinite on M, *that is,* $\mathbf{y}'\mathbf{L}(\mathbf{x}^*)\mathbf{y} \geqslant 0$ *for all* $\mathbf{y} \in M$.

Proof. From elementary calculus it is clear that for every twice differentiable curve on the constraint surface S through $\mathbf{x}^*$ (with $\mathbf{x}(0) = \mathbf{x}^*$) we have

$$\left.\frac{d^2}{dt^2} f(\mathbf{x}(t))\right]_{t=0} \geqslant 0 \tag{13}$$

By definition

$$\left.\frac{d^2}{dt^2} f(\mathbf{x}(t))\right]_{t=0} = \dot{\mathbf{x}}(0)\mathbf{F}(\mathbf{x}^*)\dot{\mathbf{x}}(0) + \nabla f(\mathbf{x}^*)\ddot{\mathbf{x}}(0). \tag{14}$$

Furthermore, differentiating the relation $\boldsymbol{\lambda}\mathbf{h}(\mathbf{x}(t)) = 0$ twice, we obtain

$$\dot{\mathbf{x}}(0)'\boldsymbol{\lambda}\mathbf{H}(\mathbf{x}^*)\dot{\mathbf{x}}(0) + \boldsymbol{\lambda}\nabla \mathbf{h}(\mathbf{x}^*)\ddot{\mathbf{x}}(0) = 0. \tag{15}$$

Adding (15) to (14), while taking account of (13), yields the result

$$\left.\frac{d^2}{dt^2} f(\mathbf{x}(t))\right]_{t=0} = \dot{\mathbf{x}}(0)'\mathbf{L}(\mathbf{x}^*)\dot{\mathbf{x}}(0) \geqslant 0.$$

Since $\dot{\mathbf{x}}(0)$ is arbitrary in M, we immediately have the stated conclusion. ▌

The above theorem is our first encounter with the matrix $\mathbf{L} = \mathbf{F} + \boldsymbol{\lambda}\mathbf{H}$ which is the matrix of second partial derivatives, with respect to $\mathbf{x}$, of the Lagrangian l. (See Appendix A, Section A.5, for a discussion of the notation $\boldsymbol{\lambda}\mathbf{H}$ used here.) This matrix is the backbone of the theory of algorithms for constrained problems, and it is encountered often in subsequent chapters.

We next state the corresponding set of sufficient conditions.

Second-Order Sufficiency Conditions. *Suppose there is a point* $\mathbf{x}^*$ *satisfying* $\mathbf{h}(\mathbf{x}^*) = \mathbf{0}$, *and a* $\boldsymbol{\lambda} \in E_m$ *such that*

$$\nabla f(\mathbf{x}^*) + \boldsymbol{\lambda}\nabla \mathbf{h}(\mathbf{x}^*) = \mathbf{0}. \tag{16}$$

Suppose also that the matrix $\mathbf{L}(\mathbf{x}^*) = \mathbf{F}(\mathbf{x}^*) + \boldsymbol{\lambda}\mathbf{H}(\mathbf{x}^*)$ *is positive definite on* $M = \{\mathbf{y} : \nabla \mathbf{h}(\mathbf{x}^*)\mathbf{y} = \mathbf{0}\}$, *that is, for* $\mathbf{y} \in M$, $\mathbf{y} \neq \mathbf{0}$ *there holds* $\mathbf{y}'\mathbf{L}(\mathbf{x}^*)\mathbf{y} > 0$. *Then* $\mathbf{x}^*$ *is a strict local minimum of* f *subject to* $\mathbf{h}(\mathbf{x}) = \mathbf{0}$.

Proof. If $\mathbf{x}^*$ is not a strict relative minimum point, there exists a sequence of feasible points $\{\mathbf{y}_k\}$ converging to $\mathbf{x}^*$ such that for each $k, f(\mathbf{y}_k) \leqslant f(\mathbf{x}^*)$. Write each $\mathbf{y}_k$ in the form $\mathbf{y}_k = \mathbf{x}^* + \delta_k \mathbf{s}_k$ where $\mathbf{s}_k \in E^n$, $|\mathbf{s}_k| = 1$, and $\delta_k > 0$ for each k. Clearly, $\delta_k \to 0$ and the sequence $\{\mathbf{s}_k\}$, being bounded, must have a convergent subsequence converging to some $\mathbf{s}^*$. For convenience of notation, we assume that the sequence $\{\mathbf{s}_k\}$ is itself convergent to $\mathbf{s}^*$. We also have $\mathbf{h}(\mathbf{y}_k) - \mathbf{h}(\mathbf{x}^*) = \mathbf{0}$, and dividing by δ_k and letting $k \to \infty$ we see that $\nabla \mathbf{h}(\mathbf{x}^*)\mathbf{s}^* = \mathbf{0}$.

Now by Taylor's theorem, we have for each j

$$0 = h_j(\mathbf{y}_k) = h_j(\mathbf{x}^*) + \delta_k \nabla h_j(\mathbf{x}^*)\mathbf{s}_k + \frac{\delta_k^{\,2}}{2} \mathbf{s}_k' \nabla^2 h_j(\boldsymbol{\eta}_j)\mathbf{s}_k \tag{17}$$

and

$$0 \geqslant f(\mathbf{y}_k) - f(\mathbf{x}^*) = \delta_k \nabla f(\mathbf{x}^*)\mathbf{s}_k + \frac{\delta_k^{\,2}}{2} \mathbf{s}_k' \nabla^2 f(\boldsymbol{\eta}_0)\mathbf{s}_k \tag{18}$$

where each $\boldsymbol{\eta}_j$ is a point on the line segment joining $\mathbf{x}^*$ and $\mathbf{s}_k$. Multiplying (17) by λ_j and adding these to (18) we obtain, on accounting for (16),

$$0 \geqslant \frac{\delta_k^{\,2}}{2} \mathbf{s}_k' \left\{ \nabla^2 f(\boldsymbol{\eta}_0) + \sum_{i=1}^{m} \lambda_i \nabla^2 h_i(\boldsymbol{\eta}_i) \right\} \mathbf{s}_k$$

which yields a contradiction as $k \to \infty$. ∎

Example. Consider the problem

$$\begin{aligned} &\text{maximize} && 2(x_1x_2 + x_2x_3 + x_1x_3) \\ &\text{subject to} && x_1 + x_2 + x_3 = 3. \end{aligned}$$

In Example 1 it was found that $x_1 = x_2 = x_3 = 1$, $\lambda = -2$ satisfy the first-order conditions. The matrix $\mathbf{F} + \boldsymbol{\lambda}\mathbf{H}$ becomes in this case

$$\mathbf{L} = \begin{bmatrix} 0 & 1 & 1 \\ 1 & 0 & 1 \\ 1 & 1 & 0 \end{bmatrix}$$

which itself is neither positive nor negative definite. On the subspace $M = \{\mathbf{y} : y_1 + y_2 + y_3 = 0\}$, however, we note that

$$\begin{aligned} \mathbf{y}'\mathbf{L}\mathbf{y} &= y_1(y_2 + y_3) + y_2(y_1 + y_3) + y_3(y_1 + y_2) \\ &= -(y_1^{\,2} + y_2^{\,2} + y_3^{\,2}) \end{aligned}$$

and thus $\mathbf{L}$ is negative definite on M. Therefore, the solution we found is at least a local maximum.

10.4 EIGENVALUES IN TANGENT SUBSPACE

In the last section it was shown that the matrix $\mathbf{L}$ restricted to the subspace M that is tangent to the constraint surface plays a role in second-order conditions entirely analogous to that of the Hessian of the objective function in the unconstrained case. It is perhaps not surprising, in view of this, that the structure of $\mathbf{L}$ restricted to M also determines rates of convergence of algorithms designed for constrained problems in the same way that the structure of the Hessian of the objective function does for unconstrained algorithms. Indeed, we shall see that the eigenvalues of $\mathbf{L}$ restricted to M determine the natural rates of convergence for algorithms designed for constrained problems. It is important, therefore, to understand what these restricted eigenvalues represent. We first determine geometrically what we mean by the restriction of $\mathbf{L}$ to M which we denote by $\mathbf{L}_M$. Next we define the eigenvalues of the operator $\mathbf{L}_M$. Finally we indicate how these various quantities can be computed.

Given any vector $\mathbf{y} \in M$, the vector $\mathbf{Ly}$ is in E^n but not necessarily in M. We project $\mathbf{Ly}$ orthogonally back onto M, as shown in Fig. 10.3, and the result is said to be the restriction of $\mathbf{L}$ to M operating on $\mathbf{y}$. In this way we obtain a linear transformation from M to M. The transformation is determined somewhat implicitly, however, since we do not have an explicit matrix representation.

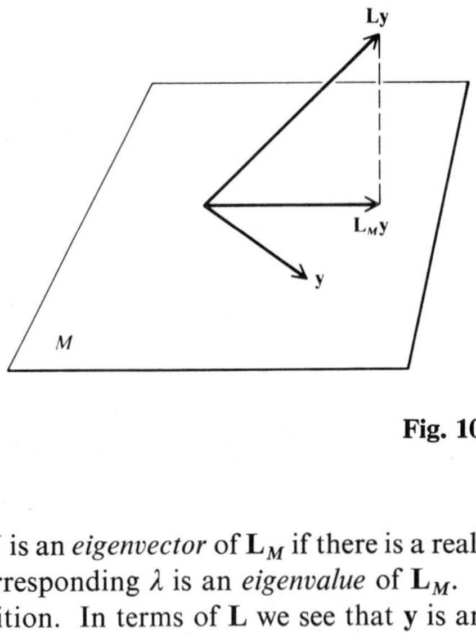

Fig. 10.3 Definition of $\mathbf{L}_M$

A vector $\mathbf{y} \in M$ is an *eigenvector* of $\mathbf{L}_M$ if there is a real number λ such that $\mathbf{L}_M\mathbf{y} = \lambda\mathbf{y}$; the corresponding λ is an *eigenvalue* of $\mathbf{L}_M$. This coincides with the standard definition. In terms of $\mathbf{L}$ we see that $\mathbf{y}$ is an eigenvector of $\mathbf{L}_M$ if $\mathbf{Ly}$ can be written as the sum of $\lambda\mathbf{y}$ and a vector orthogonal to M. See Fig. 10.4.

To obtain a matrix representation for $\mathbf{L}_M$ it is necessary to introduce a basis in the subspace M. For simplicity it is best to introduce an orthonormal basis, say $\mathbf{e}_1, \mathbf{e}_2, \ldots, \mathbf{e}_{n-m}$. Define the matrix $\mathbf{E}$ to be the $n \times (n - m)$ matrix whose columns consist of the vectors $\mathbf{e}_i$. Then any vector $\mathbf{y}$ in M can be written as $\mathbf{y} = \mathbf{Ez}$ for some $\mathbf{z} \in E^{n-m}$ and, of course,

LEz represents the action of **L** on such a vector. To project this result back into M and express the result in terms of the basis $\mathbf{e}_1, \mathbf{e}_2, \ldots, \mathbf{e}_{n-m}$, we merely multiply by $\mathbf{E}'$. Thus, $\mathbf{E}'\mathbf{L}\mathbf{E}\mathbf{z}$ is the vector whose components give the representation in terms of the basis; and, correspondingly, the $(n-m) \times (n-m)$ matrix $\mathbf{E}'\mathbf{L}\mathbf{E}$ is the matrix representation of **L** restricted to M.

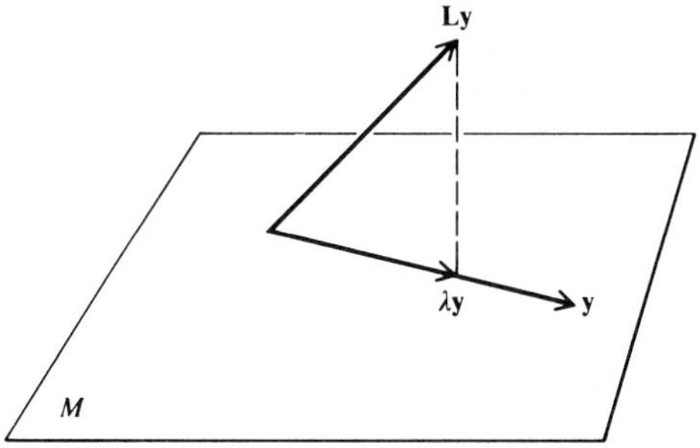

Fig. 10.4 Eigenvector of $\mathbf{L}_M$

The eigenvalues of **L** restricted to M can be found by determining the eigenvalues of $\mathbf{E}'\mathbf{L}\mathbf{E}$. These eigenvalues are independent of the particular orthonormal basis **E**.

Example. In the last section we considered

$$\mathbf{L} = \begin{bmatrix} 0 & 1 & 1 \\ 1 & 0 & 1 \\ 1 & 1 & 0 \end{bmatrix}$$

restricted to $M = \{\mathbf{y} : y_1 + y_2 + y_3 = 0\}$. To obtain an explicit matrix representation on M let us introduce the orthonormal basis.

$$\mathbf{e}_1 = \frac{1}{\sqrt{2}}(1, 0, -1)$$

$$\mathbf{e}_2 = \frac{1}{\sqrt{6}}(1, -2, 1)$$

This gives, upon expansion,

$$\mathbf{E}'\mathbf{L}\mathbf{E} = \begin{bmatrix} -1 & 0 \\ 0 & -1 \end{bmatrix}$$

and hence **L** restricted to M acts like the negative of the identity.

Example 2. Let us consider the problem

$$\begin{array}{ll} \text{extremize} & x_1 + x_2^2 + x_2x_3 + 2x_3^2 \\ \text{subject to} & x_1^2 + x_2^2 + x_3^2 = 1. \end{array}$$

The first order necessary conditions are

$$\begin{aligned} 1 + \qquad\quad 2\lambda x_1 &= 0 \\ 2x_2 + \ x_3 + 2\lambda x_2 &= 0 \\ x_2 + 4x_3 + 2\lambda x_3 &= 0. \end{aligned}$$

One solution to this set is easily seen to be $x_1 = 1$, $x_2 = 0$, $x_3 = 0$, $\lambda = -\frac{1}{2}$. Let us examine the second-order conditions at this solution point. The Lagrangian matrix there is

$$\mathbf{L} = \begin{bmatrix} -1 & 0 & 0 \\ 0 & 1 & 1 \\ 0 & 1 & 3 \end{bmatrix}$$

and the corresponding subspace M is

$$M = \{\mathbf{y} : y_1 = 0\}.$$

In this case M is the subspace spanned by the second two basis vectors in $\mathbf{E}^3$ and hence the restriction of $\mathbf{L}$ to M can be found by taking the corresponding submatrix of $\mathbf{L}$. Thus, in this case,

$$\mathbf{E}'\mathbf{L}\mathbf{E} = \begin{bmatrix} 1 & 1 \\ 1 & 3 \end{bmatrix}.$$

Since this matrix is positive definite, we conclude that the point found is a relative minimum point. This example illustrates that, in general, the restriction of $\mathbf{L}$ to M can be thought of as a submatrix of $\mathbf{L}$, although it can be read directly from the original matrix only if the subspace M is spanned by a subset of the original basis vectors.

10.5 SENSITIVITY

The Lagrange multipliers associated with a constrained minimization problem have an interpretation as prices, similar to the prices associated with constraints in linear programming. In the nonlinear case the multipliers are associated with the particular solution point and correspond to incremental or marginal prices, that is, prices associated with small variations in the constraint requirements.

Suppose the problem

$$\begin{array}{ll} \text{minimize} & f(\mathbf{x}) \\ \text{subject to} & \mathbf{h}(\mathbf{x}) = \mathbf{0} \end{array} \tag{19}$$

has a solution at the point $\mathbf{x}^*$ which is a regular point of the constraints. Let $\boldsymbol{\lambda}$ be the corresponding Lagrange multiplier vector. Now consider the family of problems

$$\begin{aligned} &\text{minimize} \quad f(\mathbf{x}) \\ &\text{subject to} \quad \mathbf{h}(\mathbf{x}) = \mathbf{c} \end{aligned} \tag{20}$$

where $\mathbf{c} \in E^m$. For a sufficiently small range of $\mathbf{c}$ near the zero vector, the problem will have a solution point $\mathbf{x}(\mathbf{c})$ near $\mathbf{x}(\mathbf{0}) \triangleq \mathbf{x}^*$. For each of these solutions there is a corresponding value $f(\mathbf{x}(\mathbf{c}))$, and this value can be regarded as a function of $\mathbf{c}$, the right-hand side of the constraints. The components of the gradient of this function can be interpreted as the incremental rate of change in value per unit change in the constraint requirements. Thus, they are the incremental prices of the constraint requirements measured in units of the objective. We show below how these prices are related to the Lagrange multipliers of the problem having $\mathbf{c} = \mathbf{0}$.

Sensitivity Theorem. *Let $f, \mathbf{h} \in C^2$ and consider the family of problems*

$$\begin{aligned} &\text{minimize} \quad f(\mathbf{x}) \\ &\text{subject to} \quad \mathbf{h}(\mathbf{x}) = \mathbf{c}. \end{aligned} \tag{20}$$

Suppose for $\mathbf{c} = \mathbf{0}$ there is a local solution $\mathbf{x}^*$ that is a regular point and that, together with its associated Lagrange multiplier vector $\boldsymbol{\lambda}$, satisfies the second order sufficiency conditions for a strict local minimum. Then for every $\mathbf{c} \in E^m$ in a region containing $\mathbf{0}$ there is an $\mathbf{x}(\mathbf{c})$, depending continuously on $\mathbf{c}$, such that $\mathbf{x}(\mathbf{0}) = \mathbf{x}^*$ and such that $\mathbf{x}(\mathbf{c})$ is a local minimum of (20). *Furthermore,*

$$\nabla_c f(\mathbf{x}(\mathbf{c}))\Big]_{\mathbf{c}=\mathbf{0}} = -\boldsymbol{\lambda}.$$

Proof. Consider the system of equations

$$\nabla f(\mathbf{x}) + \boldsymbol{\lambda} \nabla \mathbf{h}(\mathbf{x}) = \mathbf{0} \tag{21}$$

$$\mathbf{h}(\mathbf{x}) = \mathbf{c}. \tag{22}$$

By hypothesis, there is a solution $\mathbf{x}^*$, $\boldsymbol{\lambda}$ to this system when $\mathbf{c} = \mathbf{0}$. The Jacobian matrix of the system at this solution is

$$\begin{bmatrix} \mathbf{L}(\mathbf{x}^*) & \nabla \mathbf{h}(\mathbf{x}^*)' \\ \nabla \mathbf{h}(\mathbf{x}^*) & \mathbf{0} \end{bmatrix}.$$

Because by assumption $\mathbf{x}^*$ is a regular point and $\mathbf{L}(\mathbf{x}^*)$ is positive definite on M, it follows that this matrix is nonsingular (see Exercise 9). Thus, by the Implicit Function Theorem, there is a solution $\mathbf{x}(\mathbf{c})$, $\boldsymbol{\lambda}(\mathbf{c})$ to the system which is in fact twice continuously differentiable.

By the chain rule we have

$$\nabla_c f(\mathbf{x}(\mathbf{c}))\Big]_{\mathbf{c}=0} = \nabla_x f(\mathbf{x}^*)\nabla_c \mathbf{x}(\mathbf{0})$$

and

$$\nabla_c \mathbf{h}(\mathbf{x}(\mathbf{c}))\Big]_{\mathbf{c}=0} = \nabla_x \mathbf{h}(\mathbf{x}^*)\nabla_c \mathbf{x}(\mathbf{0}).$$

In view of (22), the second of these is equal to the identity $\mathbf{I}$ on E^m, while this, in view of (21), implies that the first can be written

$$\nabla_c f(\mathbf{x}(\mathbf{c}))\Big]_{\mathbf{c}=0} = -\boldsymbol{\lambda}. \quad \blacksquare$$

10.6 INEQUALITY CONSTRAINTS

We consider now problems of the form

$$\begin{aligned} \text{minimize} \quad & f(\mathbf{x}) \\ \text{subject to} \quad & \mathbf{h}(\mathbf{x}) = \mathbf{0}, \\ & \mathbf{g}(\mathbf{x}) \leqslant \mathbf{0}. \end{aligned} \tag{23}$$

We assume that f and $\mathbf{h}$ are as before and that $\mathbf{g}$ is a p-dimensional function. Initially, we assume $f, \mathbf{h}, \mathbf{g} \in C^1$.

There are a number of distinct theories concerning this problem, based on various regularity conditions or constraint qualifications, which are directed toward obtaining definitive general statements of necessary and sufficient conditions. One can by no means pretend that all such results can be obtained as minor extensions of the theory for problems having equality constraints only. To date, however, these alternative results concerning necessary conditions have been of isolated theoretical interest only—for they have not had an influence on the development of algorithms, and have not contributed to the theory of algorithms. Their use has been limited to small-scale programming problems of two or three variables. We therefore choose to emphasize the simplicity of incorporating inequalities rather than the possible complexities, not only for ease of presentation and insight, but also because it is this viewpoint that forms the basis for work beyond that of obtaining necessary conditions.

First-Order Necessary Conditions

With the following generalization of our previous definition it is possible to parallel the development of necessary conditions for equality constraints.

Definition. Let $\mathbf{x}^*$ be a point satisfying the constraints

$$\mathbf{h}(\mathbf{x}^*) = \mathbf{0}, \qquad \mathbf{g}(\mathbf{x}^*) \leqslant \mathbf{0} \tag{24}$$

and let J be the set of indices j for which $g_j(\mathbf{x}^*) = 0$. Then $\mathbf{x}^*$ is said to be a *regular point* of the constraints (24) if the gradient vectors $\nabla h_i(\mathbf{x}^*)$, $\nabla g_j(\mathbf{x}^*)$, $1 \leqslant i \leqslant m$, $j \in J$ are linearly independent.

We note that, following the definition of active constraints given in Section 10.1, a point $\mathbf{x}^*$ is a regular point if the gradients of the active constraints are linearly independent. Or, equivalently, $\mathbf{x}^*$ is regular for the constraints if it is regular in the sense of the earlier definition for equality constraints applied to the active constraints.

Kuhn–Tucker Conditions. *Let* $\mathbf{x}^*$ *be a relative minimum point for the problem*

$$\begin{aligned} &\text{minimize} \quad f(\mathbf{x}) \\ &\text{subject to} \quad \mathbf{h}(\mathbf{x}) = \mathbf{0}, \qquad \mathbf{g}(\mathbf{x}) \leqslant \mathbf{0} \end{aligned} \tag{25}$$

and suppose $\mathbf{x}^*$ *is a regular point for the constraints. Then there is a vector* $\boldsymbol{\lambda} \in E_m$ *and a vector* $\boldsymbol{\mu} \in E_p$ *with* $\boldsymbol{\mu} \geqslant \mathbf{0}$ *such that*

$$\nabla f(\mathbf{x}^*) + \boldsymbol{\lambda}\nabla\mathbf{h}(\mathbf{x}^*) + \boldsymbol{\mu}\nabla\mathbf{g}(\mathbf{x}^*) = \mathbf{0} \tag{26}$$

$$\boldsymbol{\mu}\mathbf{g}(\mathbf{x}^*) = 0. \tag{27}$$

Proof. We note first, since $\boldsymbol{\mu} \geqslant \mathbf{0}$ and $\mathbf{g}(\mathbf{x}^*) \leqslant \mathbf{0}$, (27) is equivalent to the statement that a component of $\boldsymbol{\mu}$ is nonzero only if the corresponding constraint is active.

Since $\mathbf{x}^*$ is a relative minimum point over the constraint set, it is also a relative minimum over the subset of that set defined by setting the active constraints to zero. Thus for the resulting equality constrained problem defined in a neighborhood of $\mathbf{x}^*$, there are Lagrange multipliers. Therefore, we conclude that (26) holds with $\mu_j = 0$ if $g_j(\mathbf{x}^*) \neq 0$ (and hence (27) also holds).

It remains to be shown that $\boldsymbol{\mu} \geqslant \mathbf{0}$. Suppose $\mu_k < 0$ for some $k \in J$. Let S and M be the surface and tangent plane, respectively, defined by all other active constraints at $\mathbf{x}^*$. By the regularity assumption, there is a $\mathbf{y}$ such that $\mathbf{y} \in M$ and $\nabla g_k(\mathbf{x}^*)\mathbf{y} < 0$. Let $\mathbf{x}(t)$ be a curve on S passing through $\mathbf{x}^*$ (at $t = 0$) with $\dot{\mathbf{x}}(0) = \mathbf{y}$. Then for small $t \geqslant 0$, $\mathbf{x}(t)$ is feasible, and

$$\left.\frac{df}{dt}(\mathbf{x}(t))\right]_{t=0} = \nabla f(\mathbf{x}^*)\mathbf{y} < 0$$

by (26), which contradicts the minimality of $\mathbf{x}^*$. ∎

Example. Consider the problem

$$\begin{aligned} &\text{minimize} \quad 2x_1^2 + 2x_1x_2 + x_2^2 - 10x_1 - 10x_2 \\ &\text{subject to} \quad x_1^2 + x_2^2 \leqslant 5 \\ &\qquad\qquad\quad 3x_1 + x_2 \leqslant 6. \end{aligned}$$

The first-order necessary conditions, in addition to the constraints, are

$$
\begin{aligned}
4x_1 + 2x_2 - 10 + 2\mu_1 x_1 + 3\mu_2 &= 0\\
2x_1 + 2x_2 - 10 + 2\mu_1 x_2 + \mu_2 &= 0\\
\mu_1 \geqslant 0, \qquad \mu_2 &\geqslant 0\\
\mu_1({x_1}^2 + {x_2}^2 - 5) &= 0\\
\mu_2(3x_1 + x_2 - 6) &= 0.
\end{aligned}
$$

To find a solution we define various combinations of active constraints and check the signs of the resulting Lagrange multipliers. In this problem we can try setting none, one, or two constraints active. Assuming the first constraint is active and the second is inactive yields the equations

$$
\begin{aligned}
4x_1 + 2x_2 - 10 + 2\mu_1 x_1 &= 0\\
2x_1 + 2x_2 - 10 + 2\mu_1 x_2 &= 0\\
{x_1}^2 + {x_2}^2 &= 5
\end{aligned}
$$

which has the solution

$$x_1 = 1, \quad x_2 = 2, \quad \mu_1 = 1.$$

This yields $3x_1 + x_2 = 5$ and hence the second constraint is satisfied. Thus, since $\mu_1 > 0$, we conclude that this solution satisfies the first-order necessary conditions.

Second-Order Conditions

The second-order conditions, both necessary and sufficient, for problems with inequality constraints, are derived essentially by consideration only of the equality constrained problem that is implied by the active constraints. The appropriate tangent plane for these problems is the plane tangent to the active constraints.

Second-Order Necessary Conditions. *Suppose the functions f, $\mathbf{g}$, $\mathbf{h} \in C^2$ and that $\mathbf{x}^*$ is a regular point of the constraints (24). If $\mathbf{x}^*$ is a relative minimum point for problem (23), then there is a $\boldsymbol{\lambda} \in E_m$, $\boldsymbol{\mu} \in E_p$, $\boldsymbol{\mu} \geqslant \mathbf{0}$ such that (26) and (27) hold and such that*

$$\mathbf{L}(\mathbf{x}^*) = \mathbf{F}(\mathbf{x}^*) + \boldsymbol{\lambda}\mathbf{H}(\mathbf{x}^*) + \boldsymbol{\mu}\mathbf{G}(\mathbf{x}^*) \tag{28}$$

is positive semidefinite on the tangent subspace of the active constraints at $\mathbf{x}^*$.

Proof. If $\mathbf{x}^*$ is a relative minimum point over the constraints (24), it is also a relative minimum point for the problem with the active constraints taken as equality constraints. ∎

Just as in the theory of unconstrained minimization, it is possible to formulate a converse to the Second-Order Necessary Condition Theorem and thereby obtain a Second-Order Sufficiency Condition Theorem. By analogy with the unconstrained situation, one can guess that the required hypothesis is that $\mathbf{L}(\mathbf{x}^*)$ be positive definite on the tangent plane M. This is indeed sufficient in most situations. However, if there are *degenerate inequality constraints* (that is, active inequality constraints having zero as associated Lagrange multiplier), we must require $\mathbf{L}(\mathbf{x}^*)$ to be positive definite on a subspace that is larger than M.

Second-Order Sufficiency Conditions. *Let f, $\mathbf{g}$, $\mathbf{h} \in C^2$. Sufficient conditions that a point $\mathbf{x}^*$ satisfying (24) be a strict relative minimum point of problem (23) is that there exist $\boldsymbol{\lambda} \in E_m$, $\boldsymbol{\mu} \in E_p$, such that*

$$\boldsymbol{\mu} \geqslant \mathbf{0} \tag{29}$$

$$\boldsymbol{\mu}\mathbf{g}(\mathbf{x}^*) = 0 \tag{30}$$

$$\nabla f(\mathbf{x}^*) + \boldsymbol{\lambda}\nabla\mathbf{h}(\mathbf{x}^*) + \boldsymbol{\mu}\nabla\mathbf{g}(\mathbf{x}^*) = \mathbf{0} \tag{31}$$

and the Hessian matrix

$$\mathbf{L}(\mathbf{x}^*) = \mathbf{F}(\mathbf{x}^*) + \boldsymbol{\lambda}\mathbf{H}(\mathbf{x}^*) + \boldsymbol{\mu}\mathbf{G}(\mathbf{x}^*) \tag{32}$$

is positive definite on the subspace

$$M' = \{\mathbf{y} : \nabla\mathbf{h}(\mathbf{x}^*)\mathbf{y} = \mathbf{0},\ \nabla g_j(\mathbf{x}^*)\mathbf{y} = 0 \text{ for all } j \in J\}$$

where

$$J = \{j : g_j(\mathbf{x}^*) = 0,\ \mu_j > 0\}.$$

Proof. As in the proof of the corresponding theorem for equality constraints in Section 10.3, assume that $\mathbf{x}^*$ is not a strict relative minimum point; let $\{\mathbf{y}_k\}$ be a sequence of feasible points converging to $\mathbf{x}^*$ such that $f(\mathbf{y}_k) \leqslant f(\mathbf{x}^*)$, and write each $\mathbf{y}_k$ in the form $\mathbf{y}_k = \mathbf{x}^* + \delta_k\mathbf{s}_k$ with $|\mathbf{s}_k| = 1$, $\delta_k > 0$. We may assume that $\delta_k \to 0$ and $\mathbf{s}_k \to \mathbf{s}^*$. We have $0 \geqslant \nabla f(\mathbf{x}^*)\mathbf{s}^*$, and for each $i = 1, \ldots, m$ we have

$$\nabla h_i(\mathbf{x}^*)\mathbf{s}^* = 0.$$

Also for each active constraint g_j we have $g_j(\mathbf{y}_k) - g_j(\mathbf{x}^*) \leqslant 0$, and hence

$$\nabla g_j(\mathbf{x}^*)\mathbf{s}^* \leqslant 0.$$

If $\nabla g_j(\mathbf{x}^*)\mathbf{s}^* = 0$ for all $j \in J$, then the proof goes through just as in Section 10.3. If $\nabla g_j(\mathbf{x}^*)\mathbf{s}^* < 0$ for at least one $j \in J$, then

$$0 \geqslant \nabla f(\mathbf{x}^*)\mathbf{s}^* = -\boldsymbol{\lambda}\nabla\mathbf{h}(\mathbf{x}^*)\mathbf{s}^* - \boldsymbol{\mu}\nabla\mathbf{g}(\mathbf{x}^*)\mathbf{s}^* > 0$$

which is a contradiction. ∎

We note in particular that if all active inequality constraints have strictly positive corresponding Lagrange multipliers (no degenerate inequalities), then the set J includes all of the active inequalities. In this case the sufficient condition is that the Lagrangian be positive definite on the tangent plane M.

Sensitivity

The sensitivity result for problems with inequalities is a simple restatement of the result for equalities. In this case, a nondegeneracy assumption is introduced so that the small variations produced in Lagrange multipliers when the constraints are varied will not violate the positivity requirement.

Sensitivity Theorem. *Let* $f, \mathbf{g}, \mathbf{h} \in C^2$ *and consider the family of problems*

$$\begin{aligned} &\text{minimize} && f(\mathbf{x}) \\ &\text{subject to} && \mathbf{h}(\mathbf{x}) = \mathbf{c} \\ & && \mathbf{g}(\mathbf{x}) \leqslant \mathbf{d}. \end{aligned} \tag{33}$$

Suppose that for $\mathbf{c} = \mathbf{0}$, $\mathbf{d} = \mathbf{0}$, *there is a local solution* $\mathbf{x}^*$ *that is a regular point and that, together with the associated Lagrange multipliers,* $\boldsymbol{\lambda}$, $\boldsymbol{\mu} \geqslant \mathbf{0}$, *satisfies the second-order sufficiency conditions for a strict local minimum. Assume further that no active inequality constraint is degenerate. Then for every* $(\mathbf{c}, \mathbf{d}) \in E_{m+p}$ *in a region containing* $(\mathbf{0}, \mathbf{0})$ *there is a solution* $\mathbf{x}(\mathbf{c}, \mathbf{d})$, *depending continuously on* $(\mathbf{c}, \mathbf{d})$, *such that* $\mathbf{x}(\mathbf{0}, \mathbf{0}) = \mathbf{x}^*$, *and such that* $\mathbf{x}(\mathbf{c}, \mathbf{d})$ *is a relative minimum point of* (33). *Furthermore,*

$$\nabla_c f(\mathbf{x}(\mathbf{c}, \mathbf{d}))\Big]_{0,0} = -\boldsymbol{\lambda} \tag{34}$$

$$\nabla_d f(\mathbf{x}(\mathbf{c}, \mathbf{d}))\Big]_{0,0} = -\boldsymbol{\mu}. \tag{35}$$

10.7 SUMMARY

Given a minimization problem subject to equality constraints in which all functions are smooth, a necessary condition satisfied at a minimum point is that the gradient of the objective function is orthogonal to the tangent plane of the constraint surface. If the point is regular, then the tangent plane has a simple representation in terms of the gradients of the constraint functions, and the above condition can be expressed in terms of Lagrange multipliers.

If the functions have continuous second partial derivatives and Lagrange multipliers exist, then the Hessian of the Lagrangian restricted to the tangent plane plays a role in second-order conditions analogous to that played by the Hessian of the objective function in unconstrained problems. Specifically, the restricted Hessian must be positive semidefinite at a relative minimum point and, conversely, if it is positive definite at a point satisfying the first-order conditions, that point is a strict local minimum point.

Inequalities are treated by determining which of them are active at a solution. An active inequality then acts just like an equality, except that its associated Lagrange multiplier can never be negative because of the sensitivity interpretation of the multipliers.

10.8 EXERCISES

1. In E^2 consider the constraints

$$\begin{aligned} x_1 &\geqslant 0, \\ x_2 &\geqslant 0, \\ x_2 - (x_1 - 1)^2 &\leqslant 0. \end{aligned}$$

Show that the point $x_1 = 1, x_2 = 0$ is feasible but is not a regular point.

2. Find the rectangle of given perimeter that has greatest area by solving the first-order necessary conditions. Verify that the second-order sufficiency conditions are satisfied.

3. Consider the problem

$$\begin{aligned} &\text{minimize } f(\mathbf{x}) \\ &\text{subject to } g_1(\mathbf{x}) \leqslant 0 \\ &\qquad\qquad g_2(\mathbf{x}) \leqslant 0 \\ &\qquad\qquad\quad \vdots \\ &\qquad\qquad g_p(\mathbf{x}) \leqslant 0 \end{aligned}$$

where the functions f and $g_i, i = 1, , \ldots p$ are convex and have continuous first partial derivatives. Suppose that $\mathbf{x}^*$ is a regular point of the constraints. Show that a necessary and sufficient condition for $\mathbf{x}^*$ to be a (global) solution to this problem is that there exist $\mu_1 \geqslant 0, \mu_2 \geqslant 0, \ldots, \mu_p \geqslant 0$ such that

$$f(\mathbf{x}^*) = \min \{f(\mathbf{x}) + \mu_1 g_1(\mathbf{x}) + \cdots + \mu_p g_p(\mathbf{x})\}$$

$$\mu_i g_i(\mathbf{x}^*) = 0 \qquad \text{for} \qquad i = 1, 2, \ldots, p.$$

4. Find the eigenvalues of

$$\mathbf{F} = \begin{bmatrix} 4 & 3 & 2 \\ 3 & 1 & 1 \\ 2 & 1 & 1 \end{bmatrix}$$

restricted to the subspace tangent to the surface defined as those points $\mathbf{x} = (x_1, x_2, x_3)$ satisfying $x_1 + x_2 = 1$.

5. After a heavy military campaign a certain army requires many new shoes. The quartermaster can order three sizes of shoes. Although he does not know precisely how many of each size are required, he feels that the demand for the three sizes are independent and the demand for each size is uniformly distributed between zero and three thousand pairs. He wishes to allocate his shoe budget of four thousand dollars among the three sizes so as to maximize the expected number of men properly shod. Small shoes cost one dollar per pair, medium shoes cost two dollars per pair, and large shoes cost four dollars per pair. How many pairs of each size should he order?

6. *Optimal Control.* A one-dimensional dynamic process is governed by a difference equation

$$x(k+1) = \phi\big(x(k), u(k), k\big)$$

with initial condition $x(0) = x_0$. In this equation the value $x(k)$ is called the *state* at step k and $u(k)$ is the *control* at step k. Associated with this system there is an *objective function* of the form

$$J = \sum_{k=0}^{N} \psi\big(x(k), u(k), k\big).$$

In addition, there is a *terminal constraint* of the form

$$g\big(x(N+1)\big) = 0.$$

The problem is to find the sequence of controls $u(0), u(1), u(2), \ldots, u(N)$ and corresponding state values to minimize the objective function while satisfying the terminal constraint. Assuming all functions have continuous first partial derivatives and that the regularity condition is satisfied, show that associated with an optimal solution there is a sequence $\lambda(k)$, $k = 0, 1, \ldots N$ and a μ such that

$$\lambda(k-1) = \lambda(k)\,\phi_x\big(x(k), u(k), k\big) + \psi_x\big(x(k), u(k), k\big), \qquad k = 1, 2, \ldots, N$$

$$\lambda(N) = \mu g_x\big(x(N+1)\big)$$

$$\psi_u\big(x(k), u(k), k\big) - \lambda(k)\,\phi_u\big(x(k), u(k), k\big) = 0, \qquad k = 0, 1, 2, \ldots, N.$$

7. Generalize Exercise 6 to include the case where the state $\mathbf{x}(k)$ is an n-dimensional vector and the control $\mathbf{u}(k)$ is an m-dimensional vector at each stage k.

8. An egocentric young man has just inherited a fortune F and is now planning how to spend it so as to maximize his total lifetime enjoyment. He deduces that if $x(k)$ denotes his capital at the beginning of year k, his holdings will be approximately governed by the difference equation

$$x(k+1) = \alpha x(k) - u(k)$$
$$x(0) = F$$

where $\alpha \geqslant 1$ (with $\alpha - 1$ as the interest rate of investment) and where $u(k)$ is the amount spent in year k. He decides that the enjoyment achieved in year k can be expressed as $\psi(u(k))$ where ψ, his utility function, is a smooth function, and that his total lifetime enjoyment is

$$J = \sum_{k=0}^{N} \psi(u(k))\beta^k$$

where the term β^k $(0 < \beta < 1)$ reflects the notion that future enjoyment is counted less today. The young man wishes to determine the sequence of expenditures that will maximize his total enjoyment subject to the condition $x(N + 1) = 0$.

a) Find the general solution to this problem.

b) Find the solution for the special case $\psi(u) = u^{1/2}$.

9. Let $\mathbf{A}$ be an $m \times n$ matrix of rank m and let $\mathbf{L}$ be an $n \times n$ matrix that is symmetric and positive definite on the subspace $M = \{\mathbf{x} : \mathbf{Ax} = \mathbf{0}\}$. Show that the $(n + m) \times (n + m)$ matrix

$$\begin{bmatrix} \mathbf{L} & \mathbf{A}' \\ \mathbf{A} & \mathbf{0} \end{bmatrix}$$

is nonsingular.

10. Consider the quadratic program

$$\begin{aligned} &\text{minimize} \quad \tfrac{1}{2}\mathbf{x}'\mathbf{Qx} - \mathbf{b}'\mathbf{x} \\ &\text{subject to} \quad \mathbf{Ax} = \mathbf{c}. \end{aligned}$$

Prove that $\mathbf{x}^*$ is a local minimum point if and only if it is a global minimum point. (No convexity is assumed.)

11. Maximize $14x - x^2 + 6y - y^2 + 7$ subject to $x + y \leqslant 2, x + 2y \leqslant 3$.

REFERENCES

10.1—10.5 For general discussions of Lagrange multipliers and their interpretations see Hancock [H3], Hadley [H2], Fiacco and McCormick [F3] or Luenberger [L6].

10.6 The systematic treatment of inequality constraints was initiated by Kuhn and Tucker [K4].

Chapter 11 PRIMAL METHODS

In this chapter we initiate the presentation, analysis, and comparison of algorithms designed to solve constrained minimization problems. From our study there soon emerges the theme that the rate of convergence of such algorithms is determined by the structure of the Hessian of the Lagrangian restricted to the subspace tangent to the constraint surface. Thus, various algorithms described in this chapter and the next, although differing substantially in their motivation and detail, are ultimately found to have nearly identical convergence properties.

11.1 ADVANTAGE OF PRIMAL METHODS

We consider the question of solving the general nonlinear programming problem

$$\begin{aligned} \text{minimize} \quad & f(\mathbf{x}) \\ \text{subject to} \quad & \mathbf{g}(\mathbf{x}) \leqslant \mathbf{0} \\ & \mathbf{h}(\mathbf{x}) = \mathbf{0} \end{aligned} \tag{1}$$

where $\mathbf{x}$ is of dimension n, while f, $\mathbf{g}$, and $\mathbf{h}$ have dimensions 1, p, and m, respectively. It is assumed throughout the chapter that all of the functions have continuous partial derivatives of order three. Geometrically, we regard the problem as that of minimizing f over the region in E^n defined by the constraints.

By a *primal method* of solution we mean a search method that works on the original problem directly by searching through the feasible region for the optimal solution. Each point in the process is feasible and the value of the objective function constantly decreases. This is in contrast to the methods presented in the following two chapters where, rather than attacking the original problem, either an approximating problem or the dual problem is solved.

Primal methods possess three significant advantages that recommend their use as general procedures applicable to almost all nonlinear programming problems. First, since each point generated in the search procedure is feasible, if the process is terminated before reaching the solution (as practicality almost always dictates for nonlinear problems), the terminating point is feasible. Thus this final point is a feasible and probably nearly optimal solution to the original problem and therefore may represent an acceptable solution to the practical problem that motivated the nonlinear program. A second attractive feature of primal methods is that, often, it can be guaranteed that if they generate a convergent sequence, the limit point of that sequence must be at least a local constrained minimum. In other words, the global convergence characteristics of these methods is often satisfactory. Finally, a major advantage is that most primal methods do not rely on special problem structure, such as convexity, and hence these methods are applicable to general nonlinear programming problems.

Primal methods are not, however, without major disadvantages. They require a phase I procedure (see Section 3.5) to obtain an initial feasible point, and they are all plagued, particularly for problems with nonlinear constraints, with computational difficulties arising from the necessity to remain within the feasible region as the method progresses. Thus, the defining feature of these methods that is responsible for their attractiveness also leads to disadvantage.

The convergence rates of primal methods are competitive with those of other methods, and particularly for linear constraints, they are often among the most efficient. On balance their general applicability, simplicity, and good convergence properties place these methods in a role of central importance among nonlinear programming algorithms.

11.2 FEASIBLE DIRECTION METHODS

The idea of feasible direction methods is to take steps through the feasible region of the form

$$\mathbf{x}_{k+1} = \mathbf{x}_k + \alpha_k \mathbf{d}_k \tag{2}$$

where $\mathbf{d}_k$ is a direction vector and α_k is a nonnegative scalar. The scalar is chosen to minimize the objective function f with the restriction that the point $\mathbf{x}_{k+1}$ and the line segment joining $\mathbf{x}_k$ and $\mathbf{x}_{k+1}$ be feasible. Thus, in order that the process of minimizing with respect to α be nontrivial, an initial segment of the ray $\mathbf{x}_k + \alpha\mathbf{d}_k$, $\alpha > 0$ must be contained in the feasible region. This motivates the use of *feasible directions* for the directions of search. We recall from Section 6.1 that a vector $\mathbf{d}_k$ is a *feasible direction* (at $\mathbf{x}_k$) if there is an $\bar{\alpha} > 0$ such that $\mathbf{x}_k + \alpha\mathbf{d}_k$ is feasible for all α, $0 \leqslant \alpha \leqslant \bar{\alpha}$. A feasible direction method can be considered as a natural extension of our unconstrained descent methods. Each step is the composition of selecting a feasible direction, and a constrained line search.

Example 1 (Zoutendijk's method). One of the earliest proposals for a feasible direction method uses a linear programming subproblem. Consider the problem with linear inequality constraints

$$\begin{aligned} \text{minimize} \quad & f(\mathbf{x}) \\ \text{subject to} \quad & \mathbf{a}_1{}' \mathbf{x} \leqslant b_1 \\ & \quad \cdot \\ & \quad \cdot \\ & \quad \cdot \\ & \mathbf{a}_m{}' \mathbf{x} \leqslant b_m. \end{aligned} \tag{3}$$

Given a feasible point, $\mathbf{x}_k$, let I be the set of indices representing active constraints, that is, $\mathbf{a}_i{}' \mathbf{x}_k = b_i$ for $i \in I$. The direction vector $\mathbf{d}_k$ is then chosen as the solution to the linear program

$$\begin{aligned} \text{minimize} \quad & \nabla f(\mathbf{x}_k)\mathbf{d} \\ \text{subject to} \quad & \mathbf{a}_i{}' \mathbf{d} \leqslant 0, \qquad i \in I \\ & \sum_{i=1}^{n} |d_i| = 1 \end{aligned} \tag{4}$$

where $\mathbf{d} = (d_1, d_2, \ldots, d_n)$. The last equation is a normalizing equation that ensures a bounded solution. (Even though it is written in terms of absolute values, the problem can be converted to a linear program; see Exercise 1.) The other constraints assure that vectors of the form $\mathbf{x}_k + \alpha \mathbf{d}_k$ will be feasible for sufficiently small $\alpha > 0$, and subject to these conditions, $\mathbf{d}$ is chosen to line up as closely as possible with the negative gradient of f. In some sense this will result in the locally best direction in which to proceed. The overall procedure progresses by generating feasible directions in this manner, and moving along them to decrease the objective. We examine this algorithm further in the next section.

There are two major shortcomings of feasible direction methods that require that they be modified in most cases. The first shortcoming is that for general problems there may not exist any feasible directions. If, for example, a problem had nonlinear equality constraints, we might find ourselves in the situation depicted by Fig. 11.1 where no straight line from $\mathbf{x}_k$ has a feasible segment. For such problems it is necessary either to relax our requirement of feasibility by allowing points to deviate slightly from the constraint surface or to introduce the concept of moving along curves rather than straight lines.

Fig. 11.1 No feasible direction

A second shortcoming is that in simplest form most feasible direction methods are not globally convergent. They are subject to *jamming* (sometimes referred to as *zigzagging*) where the sequence of points generated by the process converges to a point that is not even a constrained local minimum point. This phenomenon can be explained by the fact that the algorithmic map is not closed.

11.3 GLOBAL CONVERGENCE

The algorithm associated with a method of feasible directions can generally be written as the composition of two maps $\mathbf{A} = \mathbf{MD}$, where $\mathbf{D}$ is a map that selects a direction and $\mathbf{M}$ is the map corresponding to constrained minimization in the given direction. (We use the new notation $\mathbf{M}$ rather than $\mathbf{S}$, since now the line search is constrained to the feasible region.) Unfortunately, it is quite often the case in feasible direction methods that $\mathbf{M}$ and $\mathbf{D}$ are not both closed.

Example 1 ($\mathbf{M}$ not closed). Consider the region shown in Fig. 11.2 together with the sequence of feasible points $\{\mathbf{x}_k\}$ and feasible directions $\{\mathbf{d}_k\}$. We have $\mathbf{x}_k \to \mathbf{x}^*$ and $\mathbf{d}_k \to \mathbf{d}^*$. Also from the diagram and the direction of ∇f it is clear that

$$\mathbf{M}(\mathbf{x}_k, \mathbf{d}_k) = \mathbf{x}_{k+1} \to \mathbf{x}^*, \qquad \mathbf{M}(\mathbf{x}^*, \mathbf{d}^*) = \mathbf{y} \neq \mathbf{x}^*.$$

Thus $\mathbf{M}$ is not closed at $(\mathbf{x}^*, \mathbf{d}^*)$.

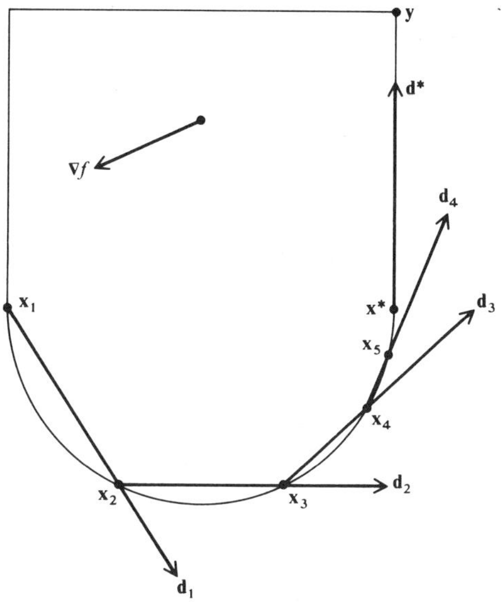

Fig. 11.2 Example of M not closed

Example 2 (**D** not closed). In Zoutendijk's method presented in Example 1 of the last section, the feasible direction selection map **D** is not closed. This can be seen from Fig. 11.3 where the directions are shown for a convergent sequence of points, and the limiting direction is not equal to the direction at the limiting point. Basically, nonclosedness is caused in this case by the fact that the method used for generating the feasible direction changes suddenly when an additional constraint becomes active.

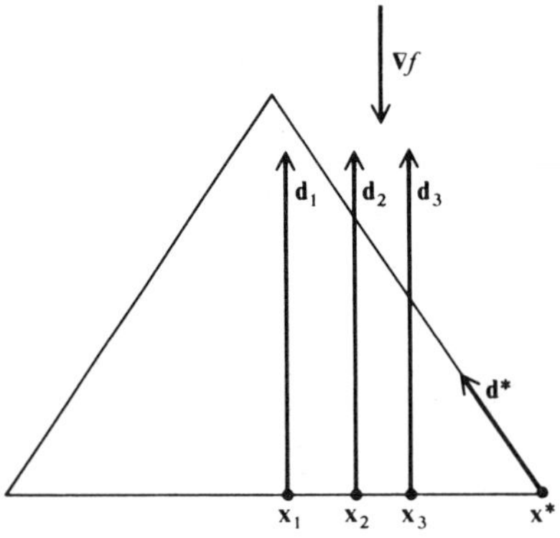

Fig. 11.3 Example of D not closed

It is possible to develop feasible direction algorithms that are closed and hence not subject to jamming. First, we show that by requiring that all directions be *uniformly feasible* it is possible to define a constrained line search map that is closed.

Definition. Let $\Omega \subset E^n$ be a given feasible region. A set $\Gamma \subset E^{2n}$ consisting of pairs $(\mathbf{x}, \mathbf{d})$, with $\mathbf{x} \in \Omega$ and $\mathbf{d}$ a feasible direction at $\mathbf{x}$, is said to be a set of *uniformly feasible direction vectors* if there is a $\delta > 0$ such that $(\mathbf{x}, \mathbf{d}) \in \Gamma$ implies that $\mathbf{x} + \alpha\mathbf{d}$ is feasible for all α, $0 \leqslant \alpha \leqslant \delta$. The number δ is referred to as the *feasibility constant* of the set Γ.

Definition. Let $\Gamma \subset E^{2n}$ be a set of uniformly feasible direction vectors for Ω with feasibility constant δ. The map $\mathbf{M}_\delta$ is defined on Γ by the relation

$$\mathbf{M}_\delta(\mathbf{x}, \mathbf{d}) = \{\mathbf{y}: f(\mathbf{y}) \leqslant f(\mathbf{x} + \tau\mathbf{d}), \quad \text{for all} \quad \tau, \ 0 \leqslant \tau \leqslant \delta;$$

$$\mathbf{y} = \mathbf{x} + \alpha\mathbf{d} \quad \text{for some} \quad \alpha, \ 0 \leqslant \alpha < \infty, \ \mathbf{y} \in \Omega\}.$$

In words, the map $\mathbf{M}_\delta$ yields the set consisting of those points that have value lower than any point on the interval $\mathbf{x} + \tau\mathbf{d}$, $0 \leqslant \tau \leqslant \delta$. This set thus includes the minimum point over this interval as well as other feasible points on the whole line that have value at least as small.

Theorem. *If* $\mathbf{d} \neq \mathbf{0}$, *the map* $\mathbf{M}_\delta$ *is closed at* $(\mathbf{x}, \mathbf{d})$.

Proof. Suppose that $\{(\mathbf{x}_k, \mathbf{d}_k)\}$ is a sequence with $(\mathbf{x}_k, \mathbf{d}_k) \to (\mathbf{x}, \mathbf{d})$, $\mathbf{d} \neq \mathbf{0}$. Suppose also that $\mathbf{y}_k \in \mathbf{M}_\delta(\mathbf{x}_k, \mathbf{d}_k)$ and that $\mathbf{y}_k \to \mathbf{y}$. We must show that $\mathbf{y} \in \mathbf{M}_\delta(\mathbf{x}, \mathbf{d})$.

For each k we have $\mathbf{y}_k = \mathbf{x}_k + \alpha_k \mathbf{d}_k$, and hence $\alpha_k \to \alpha = \|\mathbf{y} - \mathbf{x}\|/\|\mathbf{d}\|$. It follows that $\mathbf{y} = \mathbf{x} + \alpha\mathbf{d}$.

For each k and each α, $0 \leqslant \alpha \leqslant \delta$,

$$f(\mathbf{y}_k) \leqslant f(\mathbf{x}_k + \alpha\mathbf{d}_k).$$

Taking the limit and using the continuity of f

$$f(\mathbf{y}) \leqslant f(\mathbf{x} + \alpha\mathbf{d}), \qquad 0 \leqslant \alpha \leqslant \delta,$$

and hence

$$\mathbf{y} \in \mathbf{M}_\delta(\mathbf{x}, \mathbf{d}). \quad \blacksquare$$

It should be noted that for any $(\mathbf{x}, \mathbf{d}) \in \Gamma$ it is not necessary to restrict the line search to the interval $0 \leqslant \alpha \leqslant \delta$ to obtain a point in $\mathbf{M}_\delta(\mathbf{x}, \mathbf{d})$. Indeed, in practice we can apply the map $\mathbf{M}$ and search the whole line in the feasible region, for although this map itself is not closed, the set $\mathbf{M}(\mathbf{x}, \mathbf{d})$ is a subset of $\mathbf{M}_\delta(\mathbf{x}, \mathbf{d})$, and hence the closedness of $\mathbf{M}_\delta$ will ensure convergence. Thus we may continue to decompose algorithms in the form $\mathbf{A} = \mathbf{MD}$, and as long as $\mathbf{D}$ is closed and generates uniformly feasible directions, we need not check the closedness of $\mathbf{A}$.

Next, to develop a complete globally convergent algorithm it is necessary to propose a direction generating algorithm that is closed and which generates uniformly feasible directions of descent. We present below one way in which Zoutendijk's method can be modified to guarantee global convergence. There are other methods, as discussed in the exercises and in other sections of the chapter.

Example 3. We generalize Example 1 of Section 11.2 to the case where the constraints are nonlinear. Thus, we consider the problem

$$\begin{aligned} &\text{minimize} && f(\mathbf{x}) \\ &\text{subject to} && \mathbf{g}(\mathbf{x}) \leqslant \mathbf{0} \end{aligned} \tag{5}$$

where it is assumed that the set $\{\mathbf{x} : \mathbf{g}(\mathbf{x}) \leqslant \mathbf{0}\}$ is convex, compact, and contains a point satisfying $\mathbf{g}(\mathbf{x}) < \mathbf{0}$. We also assume that each feasible point is a regular point of the constraints.

Given a feasible $\mathbf{x}$ we solve the linear program

$$\begin{aligned} &\text{minimize} \quad \eta \\ &\text{subject to} \qquad \nabla f(\mathbf{x})\mathbf{y} - \eta \leqslant 0 \\ &\qquad \mathbf{g}(\mathbf{x}) + \nabla \mathbf{g}(\mathbf{x})\mathbf{y} - \eta \mathbf{u} \leqslant \mathbf{0} \\ &\qquad \sum_i |\mathbf{y}_i| \leqslant 1 \end{aligned} \tag{6}$$

where $\mathbf{u} = (1, 1, \ldots, 1)$ and $\mathbf{y} = (y_1, y_2, \ldots, y_n)$. Then put $\mathbf{d} = \alpha\mathbf{y}$, where α is the largest α, $0 < \alpha \leqslant 1$ for which the segment $[\mathbf{x}, \mathbf{x} + \alpha\mathbf{y}]$ is feasible.

Intuitively, the difference between this method of determining the direction $\mathbf{d}$ and that given by (4) is that inactive as well as active constraints play a role; in this way the determination of the direction does not change suddenly when approaching a constraint boundary. The vector $\mathbf{y}$ is chosen by (6) as an optimal compromise between lining-up with the negative gradient while avoiding the boundary of the feasible region.

It follows from the continuity of the functions involved that this mapping is closed. It generates uniformly feasible directions with $\delta = 1$. The set Γ of pairs $(\mathbf{x}, \mathbf{d})$ is compact. Thus, to prove global convergence of the algorithm $\mathbf{A} = \mathbf{MD}$ it only remains to be shown that it is a descent algorithm. This will be established if it is shown that for a non-solution point $\mathbf{x}$ we have $\nabla f(\mathbf{x})\mathbf{d} < 0$, since then it is possible to descend along $\mathbf{d}$. But this, in turn, will be impossible only if the value of the linear program (with $\mathbf{x}$ fixed)

$$\begin{aligned} &\text{minimize} \quad \eta \\ &\text{subject to} \quad \nabla f(\mathbf{x})\mathbf{y} - \eta \leqslant 0 \\ &\qquad \mathbf{g}(\mathbf{x}) + \nabla \mathbf{g}(\mathbf{x})\mathbf{y} - \eta \mathbf{u} \leqslant \mathbf{0} \end{aligned} \tag{7}$$

is zero. If the value is zero, there is a solution $[\lambda_0, \lambda_1, \ldots, \lambda_m]$ to the corresponding dual linear program also having value zero. Let $\boldsymbol{\lambda} = [\lambda_1, \lambda_2, \ldots, \lambda_m]$. The dual to (7) is

$$\text{maximize} \quad \boldsymbol{\lambda}\mathbf{g}(\mathbf{x}) \tag{8a}$$

$$\text{subject to} \quad \sum_{i=0}^{m} \lambda_i = 1 \tag{8b}$$

$$\lambda_0 \nabla f(\mathbf{x}) + \boldsymbol{\lambda} \nabla \mathbf{g}(\mathbf{x}) = \mathbf{0} \tag{8c}$$

$$\lambda_i \geqslant 0, \qquad i = 0, 1, 2, \ldots, m. \tag{8d}$$

An optimal solution of this problem must have $\boldsymbol{\lambda}\mathbf{g}(\mathbf{x}) = 0$ and hence only λ_i's corresponding to active constraints are nonzero. It then follows that $\lambda_0 \neq 0$. Otherwise (8c) would yield $\boldsymbol{\lambda}\nabla\mathbf{g}(\mathbf{x}) = 0$ which, since $\mathbf{x}$ is a regular point, would imply $\boldsymbol{\lambda} = \mathbf{0}$, violating (8b). Thus $\lambda_0 > 0$. Dividing $\boldsymbol{\lambda}$ by λ_0, we can then conclude that there holds

$$\begin{aligned} \nabla f(\mathbf{x}) + \boldsymbol{\lambda} \nabla \mathbf{g}(\mathbf{x}) &= 0 \\ \boldsymbol{\lambda} \mathbf{g}(\mathbf{x}) &= \mathbf{0} \\ \boldsymbol{\lambda} &\geqslant \mathbf{0} \end{aligned} \tag{9}$$

which are precisely the Kuhn–Tucker conditions of the original problem (5). Thus, the algorithm terminates only at points satisfying the Kuhn–Tucker conditions.

By this modification we have developed a version of the linear programming feasible direction method that is globally convergent to a point satisfying the Kuhn–Tucker conditions. The modification is fairly expensive, however, since at each step a linear program as large as the original problem must be solved, whereas in the original method a smaller linear program is solved at each step. Another approach which accounts only for constraints that are active or nearly active is discussed in Exercises 6 and 7.

11.4 THE GRADIENT PROJECTION METHOD

The gradient projection method is motivated by a desire to implement the feasible direction philosophy while not requiring the solution of a linear program at each step. Instead of attempting to find the direction that leads to the best local improvement while maintaining feasibility, one seeks a direction that is perhaps not quite as good but much easier to compute.

Linear Constraints

Consider first problems of the form

$$\begin{aligned} &\text{minimize} \quad f(\mathbf{x}) \\ &\text{subject to} \quad \mathbf{a}_i'\,\mathbf{x} \leqslant b_i \qquad i \in I_1 \\ &\qquad\qquad\quad \mathbf{a}_i'\,\mathbf{x} = b_i \qquad i \in I_2 \end{aligned} \tag{10}$$

having linear equalities and inequalities.

A feasible solution to the constraints, if one exists, can be found by application of the phase I procedure of linear programming; so we shall always assume that our descent process is initiated at such a feasible point. At a given feasible point $\mathbf{x}$ there will be a certain number q of active constraints satisfying $\mathbf{a}_i'\,\mathbf{x} = b_i$ and some inactive constraints $\mathbf{a}_i'\,\mathbf{x} < b_i$. We denote the set of indices corresponding to the active constraints by $I(\mathbf{x})$. Clearly $I(\mathbf{x}) \supset I_2$.

At the feasible point $\mathbf{x}$ we seek a feasible direction vector $\mathbf{d}$ satisfying $\nabla f(\mathbf{x})\mathbf{d} < 0$, so that movement in the direction $\mathbf{d}$ will cause a decrease in the function f. Initially, we consider directions satisfying $\mathbf{a}_i'\,\mathbf{d} = 0$, $i \in I(\mathbf{x})$ so that all active constraints remain active. This requirement amounts to requiring that the direction vector **d** lie in the tangent subspace **M** defined by the active constraints. The particular direction vector that we shall use is the projection of the negative gradient onto this subspace.

To compute this projection let $\mathbf{A}_q$ be defined as composed of the rows of active constraints. Assuming regularity of the constraints, as we shall always assume, $\mathbf{A}_q$ will be a $q \times n$ matrix of rank $q < n$. The tangent subspace M in which **d** must lie is the subspace of vectors satisfying $\mathbf{A}_q\mathbf{d} = \mathbf{0}$. This means that the subspace N consisting of the vectors making up the rows of $\mathbf{A}_q$ (that is, all vectors of the form $\mathbf{A}_q'\,\boldsymbol{\beta}$ for $\boldsymbol{\beta} \in E^q$) is orthogonal to M. Indeed, any vector can be written as the sum of vectors from each of these two complementary subspaces. In particular, the negative gradient vector $-\mathbf{g}_k$ can be written

$$-\mathbf{g}_k = \mathbf{d}_k + \mathbf{A}_q'\,\boldsymbol{\beta}_k \tag{11}$$

where $\mathbf{d}_k \in M$ and $\boldsymbol{\beta}_k \in E^q$. We may solve for $\boldsymbol{\beta}_k$ through the requirement that $\mathbf{A}_q\mathbf{d}_k = \mathbf{0}$. Thus

$$\mathbf{A}_q\mathbf{d}_k = -\mathbf{A}_q\mathbf{g}_k - (\mathbf{A}_q\mathbf{A}_q')\boldsymbol{\beta}_k = \mathbf{0} \tag{12}$$

which leads to

$$\boldsymbol{\beta}_k = -(\mathbf{A}_q\mathbf{A}_q')^{-1}\mathbf{A}_q\mathbf{g}_k \tag{13}$$

and

$$\mathbf{d}_k = -[\mathbf{I} - \mathbf{A}_q'(\mathbf{A}_q\mathbf{A}_q')^{-1}\mathbf{A}_q]\mathbf{g}_k = -\mathbf{P}_k\mathbf{g}_k. \tag{14}$$

The matrix

$$\mathbf{P}_k = [\mathbf{I} - \mathbf{A}_q'(\mathbf{A}_q\mathbf{A}_q')^{-1}\mathbf{A}_q] \tag{15}$$

is called the projection matrix corresponding to the subspace M. Action by it on any vector yields the projection of that vector onto M. See Exercises 8 and 9 for other derivations of this result.

We easily check that if $\mathbf{d}_k \neq \mathbf{0}$, then it is a direction of descent. Since $\mathbf{g}_k + \mathbf{d}_k$ is orthogonal to $\mathbf{d}_k$, we have

$$\mathbf{g}_k'\,\mathbf{d}_k = (\mathbf{g}_k' + \mathbf{d}_k' - \mathbf{d}_k')\mathbf{d}_k = -|\mathbf{d}_k|^2.$$

Thus if $\mathbf{d}_k$ as computed from (14) turns out to be nonzero, it is a feasible direction of descent.

We next consider selection of the step size. As α is increased from zero, the point $\mathbf{x} + \alpha\mathbf{d}$ will initially remain feasible and the corresponding value of f will decrease. We find the length of the feasible segment of the line emanating from **x** and then minimize f over this segment.

Next, consider the possibility that the projected negative gradient is zero. We have in that case

$$\nabla f(\mathbf{x}_k) + \boldsymbol{\beta}_k'\,\mathbf{A}_q = \mathbf{0}. \tag{16}$$

We recall that $\mathbf{A}_q$ is made up of rows corresponding to the active constraints at $\mathbf{x}_k$. If the components of $\boldsymbol{\beta}_k$ corresponding to the active inequalities are all nonnegative, then this fact together with (16) implies that the Kuhn–Tucker conditions are satisfied at $\mathbf{x}_k$ and the process terminates. If, however, at least one of those components of $\boldsymbol{\beta}_k$ is negative, it is possible, by relaxing the corresponding inequality, to move in a new direction to an improved point. Suppose that β_{jk}, the jth component of $\boldsymbol{\beta}_k$, is negative and the indexing is arranged so that the corresponding constraint is the inequality $\mathbf{a}_j'\mathbf{x} \leqslant b_j$. We determine the new direction vector by relaxing the jth constraint and projecting the negative gradient onto the subspace determined by the remaining $q-1$ active constraints. Let $\mathbf{A}_{\bar{q}}$ denote the matrix $\mathbf{A}_q$ with row $\mathbf{a}_j$ deleted. We have

$$-\mathbf{g}_k = \mathbf{A}_q'\,\boldsymbol{\beta}_k \tag{17}$$

$$-\mathbf{g}_k = \bar{\mathbf{d}}_k + \mathbf{A}_{\bar{q}}'\,\bar{\boldsymbol{\beta}}_k \tag{18}$$

where $\bar{\mathbf{d}}_k$ is the projection of $-\mathbf{g}_k$ using $\mathbf{A}_{\bar{q}}$. It is immediately clear that $\bar{\mathbf{d}}_k \neq \mathbf{0}$, since otherwise (18) would be a special case of (17) with $\beta_{jk} = 0$ which is impossible, since the rows of $\mathbf{A}_q$ are linearly independent. From our previous work we know that $\mathbf{g}_k'\,\bar{\mathbf{d}}_k < 0$. Multiplying the transpose of (17) by $\bar{\mathbf{d}}_k$ and using $\mathbf{A}_{\bar{q}}\bar{\mathbf{d}}_k = \mathbf{0}$ we obtain

$$0 > \mathbf{g}_k'\,\bar{\mathbf{d}}_k = -\beta_{jk}\mathbf{a}_j'\,\bar{\mathbf{d}}_k. \tag{19}$$

Since $\beta_{jk} < 0$ we conclude that $\mathbf{a}_j'\,\bar{\mathbf{d}}_k < 0$. Thus the vector $\bar{\mathbf{d}}_k$ is not only a direction of descent, but it is a feasible direction, since $\mathbf{a}_i'\,\bar{\mathbf{d}}_k = 0$, $i \in I(\mathbf{x}_k)$, $i \neq j$, and $\mathbf{a}_j'\,\bar{\mathbf{d}}_k < 0$.

In summary, one step of the algorithm is as follows: Given a feasible point $\mathbf{x}$

1. Find the subspace of active constraints M, and form $\mathbf{A}_q$.
2. Calculate $\mathbf{P} = \mathbf{I} - \mathbf{A}_q'(\mathbf{A}_q\mathbf{A}_q')^{-1}\mathbf{A}_q$ and $\mathbf{d} = -\mathbf{P}\nabla f(\mathbf{x})'$.
3. If $\mathbf{d} \neq \mathbf{0}$, find α_1 and α_2 achieving, respectively,

$$\max\,\{\alpha : \mathbf{x} + \alpha\mathbf{d} \text{ is feasible}\}$$
$$\min\,\{f(\mathbf{x} + \alpha\mathbf{d}) : 0 \leqslant \alpha \leqslant \alpha_1\}.$$

   Set $\mathbf{x}$ to $\mathbf{x} + \alpha_2\mathbf{d}$ and return to (1).

4. If $\mathbf{d} = \mathbf{0}$, find $\boldsymbol{\beta} = -(\mathbf{A}_q\mathbf{A}_q')^{-1}\mathbf{A}_q\nabla f(\mathbf{x})'$.
   a) If $\beta_j \geqslant 0$, for all j corresponding to active inequalities, stop, $\mathbf{x}$ satisfies the Kuhn–Tucker conditions.
   b) Otherwise, delete the row from $\mathbf{A}_q$ corresponding to the inequality with the most negative component of $\boldsymbol{\beta}$ and return to (2).

The projection matrix need not be recomputed in its entirety at each new point. Since the set of active constraints changes by at most one constraint at a time, it is possible to calculate one required projection matrix from the previous one by an updating procedure. (See Exercise 11.) This is an important feature of the gradient projection method and greatly reduces the computation required at each step.

Example. Consider the problem

$$\begin{aligned} &\text{minimize} \quad x_1^2 + x_2^2 + x_3^2 + x_4^2 - 2x_1 - 3x_4 \\ &\text{subject to} \quad 2x_1 + x_2 + x_3 + 4x_4 = 7 \\ &\qquad\qquad x_1 + x_2 + 2x_3 + x_4 = 6 \\ &\qquad\qquad x_i \geqslant 0 \qquad i = 1, 2, 3, 4. \end{aligned} \tag{20}$$

Suppose that given the feasible point $\mathbf{x} = (2, 2, 1, 0)$ we wish to find the direction of the projected negative gradient. The active constraints are the two equalities and the inequality $x_4 \geqslant 0$. Thus

$$\mathbf{A}_q = \begin{bmatrix} 2 & 1 & 1 & 4 \\ 1 & 1 & 2 & 1 \\ 0 & 0 & 0 & 1 \end{bmatrix} \tag{21}$$

and hence

$$\mathbf{A}_q\mathbf{A}_q' = \begin{bmatrix} 22 & 9 & 4 \\ 9 & 7 & 1 \\ 4 & 1 & 1 \end{bmatrix}.$$

After considerable calculation we then find

$$(\mathbf{A}_q\mathbf{A}_q')^{-1} = \frac{1}{11}\begin{bmatrix} 6 & -5 & -19 \\ -5 & 6 & 14 \\ -19 & 14 & 73 \end{bmatrix}$$

and finally

$$\mathbf{P} = \frac{1}{11}\begin{bmatrix} 1 & -3 & 1 & 0 \\ -3 & 9 & -3 & 0 \\ 1 & -3 & 1 & 0 \\ 0 & 0 & 0 & 0 \end{bmatrix}. \tag{22}$$

The gradient at the point $(2, 2, 1, 0)$ is $\mathbf{g} = (2, 4, 2, -3)$ and hence we find

$$\mathbf{d} = -\mathbf{P}\mathbf{g} = \tfrac{1}{11}(-8, 24, -8, 0)$$

or normalizing by $8/11$

$$\mathbf{d} = (-1, 3, -1, 0). \tag{23}$$

It can be easily verified that movement in this direction does not violate the constraints.

Global Convergence Properties

Although gradient projection provides a simple solution to the problem of determining feasible directions of descent, without requiring the solution of a linear program, the resulting algorithm is not closed and therefore is subject to the possibility of jamming. The algorithm is not closed essentially because the method for selecting the direction of descent changes abruptly when reaching either a point where the projected gradient vanishes or a point where a new constraint becomes active. The first of these sources of discontinuity can be easily patched up by a simple modification of the algorithm (see Exercise 14) but the second source of discontinuity, introduced by changes in the set of active constraints, apparently cannot be eliminated without destroying the basic simplicity of the method. It remains, therefore, an open question as to whether the gradient projection method can experience jamming, for although no convergence proof is available, neither is there a counterexample. Indeed, the method has been used successfully for numerous nonlinear programming problems and jamming has never been detected. Thus it seems fair to conclude that the great simplicity of the method, over the linear programming approach to direction finding, more than offsets the small risk of jamming.

Nonlinear Constraints

In extending the gradient projection method to problems of the form

$$\begin{aligned} &\text{minimize} && f(\mathbf{x}) \\ &\text{subject to} && \mathbf{h}(\mathbf{x}) = \mathbf{0} \\ & && \mathbf{g}(\mathbf{x}) \leqslant \mathbf{0} \end{aligned} \tag{24}$$

the basic idea is that at a feasible point $\mathbf{x}_k$ one determines the active constraints and projects the negative gradient onto the subspace tangent to the surface determined by these constraints. This vector, if it is nonzero, determines the direction for the next step. The vector itself, however, is not in general a feasible direction, since the surface may be curved as illustrated in Fig. 11.4. It is therefore not always possible to move along this projected negative gradient to obtain the next point.

What is typically done in the face of this difficulty is to make an initial move along the projected negative gradient to a point $\mathbf{y}$ that decreases (or perhaps minimizes) the objective, and then move to a nearby feasible point. The move to the nearby feasible point is made in the direction perpendicular to the tangent plane at the original point, as illustrated in Fig. 11.4. This procedure of tentatively moving away from the feasible region and then coming back introduces a number of additional difficulties that require a series of interpolations, line searches, and nonlinear equation solutions for their resolution. A satisfactory general routine implementing the gradient projection philosophy is therefore of necessity quite complex. It is not our purpose here to elaborate on these details but simply to point out the general nature of the difficulties and the basic devices for surmounting them.

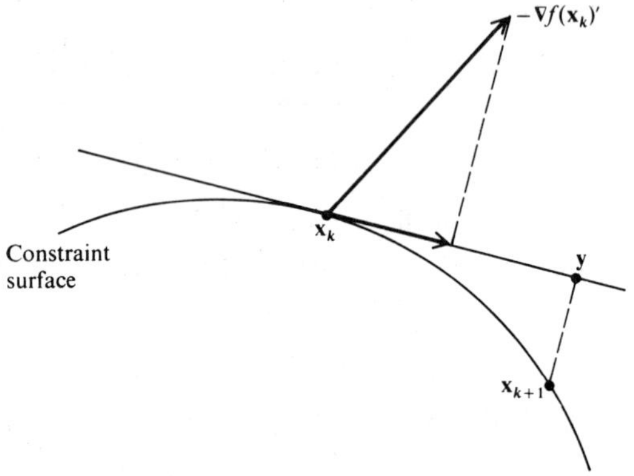

Fig. 11.4 Gradient projection method

One difficulty is illustrated in Fig. 11.5. If, after moving along the projected negative gradient to a point $\mathbf{y}$, one attempts to return to a point that satisfies the old active constraints, some inequalities that were originally satisfied may then be violated. One must in this circumstance use an interpolation scheme to find a new point $\bar{y}$ along the negative gradient so that when returning to the active constraints no originally nonactive constraint is violated. Finding an appropriate $\bar{y}$ is to some extent a trial and error process.

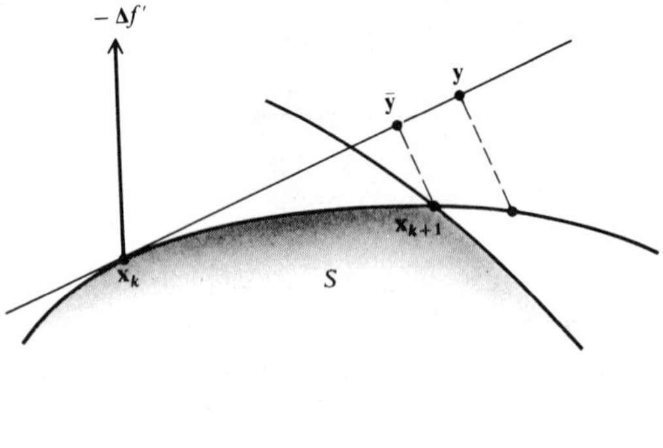

Fig. 11.5 Interpolation to obtain feasible point

A second difficulty is that when moving from $\mathbf{y}$ to the constraint boundary, the value of the objective may increase so that, overall, the move from the starting point $\mathbf{x}_k$ does not lead to a decreased value. This can be remedied only by selecting a new $\mathbf{y}$ closer to the starting point $\mathbf{x}_k$. Finally, the job of returning to the active constraints is itself a nonlinear problem which must be solved with an iterative technique. Such a technique is described below, but within a finite number of iterations, it cannot exactly reach the surface. Thus, typically an error tolerance δ is introduced, and throughout the procedure the constraints are satisfied only to within δ.

Computation of the projections is also more difficult in the nonlinear case. Lumping, for notational convenience, the active inequalities together with the equalities into $\mathbf{h}(\mathbf{x}_k)$, the projection matrix at $\mathbf{x}_k$ is

$$\mathbf{P}_k = \mathbf{I} - \nabla\mathbf{h}(\mathbf{x}_k)' [\nabla\mathbf{h}(\mathbf{x}_k)\nabla\mathbf{h}(\mathbf{x}_k)']^{-1}\nabla\mathbf{h}(\mathbf{x}_k). \tag{25}$$

At the point $\mathbf{x}_k$ this matrix can be updated to account for one more or one less constraint, just as in the linear case. When moving from $\mathbf{x}_k$ to $\mathbf{x}_{k+1}$, however, $\nabla\mathbf{h}$ will change and the new projection matrix cannot be found from the old, and hence, this matrix must be recomputed at each step.

The most important new feature of the method is the problem of returning to the feasible region from points outside this region. The type of iterative technique employed is a common one in nonlinear programming and we describe it here. The idea is, from any point near $\mathbf{x}_k$, to move back to the constraint surface in a direction orthogonal to the tangent plane at $\mathbf{x}_k$. Thus from a point $\mathbf{y}$ we seek a point of the form $\mathbf{y} + \nabla\mathbf{h}(\mathbf{x}_k)'\boldsymbol{\alpha} = \mathbf{y}^*$ such that $\mathbf{h}(\mathbf{y}^*) = \mathbf{0}$. As shown in Fig. 11.6 such a solution may not always exist, but it does for $\mathbf{y}$ sufficiently close to $\mathbf{x}_k$.

Fig. 11.6 Case in which it is impossible to return to surface

To find a suitable first approximation to $\boldsymbol{\alpha}$, and hence to $\mathbf{y}^*$, we linearize the equation at $\mathbf{x}_k$ obtaining

$$\mathbf{h}(\mathbf{y} + \nabla\mathbf{h}(\mathbf{x}_k)'\boldsymbol{\alpha}) \simeq \mathbf{h}(\mathbf{y}) + \nabla\mathbf{h}(\mathbf{x}_k)\nabla\mathbf{h}(\mathbf{x}_k)'\boldsymbol{\alpha} \tag{26}$$

the approximation being accurate for $|\boldsymbol{\alpha}|$ and $|\mathbf{y} - \mathbf{x}|$ small. This motivates the first approximation

$$\boldsymbol{\alpha}_1 = -[\nabla \mathbf{h}(\mathbf{x}_k)\nabla \mathbf{h}(\mathbf{x}_k)']^{-1}\mathbf{h}(\mathbf{y}) \tag{27}$$

$$\mathbf{y}_1 = \mathbf{y} - \nabla \mathbf{h}(\mathbf{x}_k)'[\nabla \mathbf{h}(\mathbf{x}_k)\nabla \mathbf{h}(\mathbf{x}_k)']^{-1}\mathbf{h}(\mathbf{y}). \tag{28}$$

Substituting $\mathbf{y}_1$ for $\mathbf{y}$ and successively repeating the process yields the sequence $\{\mathbf{y}_j\}$ generated by

$$\mathbf{y}_{j+1} = \mathbf{y}_j - \nabla \mathbf{h}(\mathbf{x}_k)'[\nabla \mathbf{h}(\mathbf{x}_k)\nabla \mathbf{h}(\mathbf{x}_k)']^{-1}\mathbf{h}(\mathbf{y}_j) \tag{29}$$

which, started close enough to $\mathbf{x}_k$ and the constraint surface, will converge to a solution $\mathbf{y}^*$. We note that this process requires the same matrices as the projection operation.

The gradient projection method has been successfully implemented and has been found to be effective in solving general nonlinear programming problems. Successful implementation resolving the several difficulties introduced by the requirement of staying in the feasible region requires, as one would expect, some degree of skill. The true value of the method, however, can be determined only through an analysis of its rate of convergence.

11.5 CONVERGENCE RATE OF THE GRADIENT PROJECTION METHOD

An analysis that directly attacked the nonlinear version of the gradient projection method, with all of its iterative and interpolative devices, would quickly become monsterous. To obtain the asymptotic rate of convergence, however, it is not necessary to analyze this complex algorithm directly—instead it is sufficient to analyze an alternate simplified algorithm that asymptotically duplicates the gradient projection method near the solution. Through the introduction of this idealized algorithm we show that the rate of convergence of the gradient projection method is governed by the eigenvalue structure of the Hessian of the Lagrangian restricted to the constraint tangent subspace.

Geodesic Descent

For simplicity we consider first the problem having only equality constraints

$$\begin{aligned} &\text{minimize} \quad f(\mathbf{x}) \\ &\text{subject to} \quad \mathbf{h}(\mathbf{x}) = \mathbf{0}. \end{aligned} \tag{30}$$

The constraints define a continuous surface Ω in E^n.

In considering our own difficulties with this problem, owing to the fact that the surface is nonlinear thereby making directions of descent difficult to define, it is well to also consider the problem as it would be viewed by a small bug confined to the constraint surface who imagines it to be his total universe. To him the problem seems to be a simple one. It is unconstrained, with respect to his universe, and is only $n - m$ dimensional. He would characterize a solution point as a point where the gradient of f (as measured on the surface) vanishes and where the appropriate $n - m$ dimensional Hessian of f is positive semidefinite. If asked to develop a computational procedure for this problem, he would undoubtedly suggest, since he views the problem as unconstrained, the method of steepest descent. He would compute the gradient, as measured on his surface, and would move along what would appear to him to be straight lines.

Exactly what the bug would compute as the gradient and exactly what he would consider as straight lines would depend basically on how distance between two points on his surface were measured. If, as is most natural, we assume that he inherits his notion of distance from the one which we are using in E^n, then the path $\mathbf{x}(t)$ between two points $\mathbf{x}_1$ and $\mathbf{x}_2$ on his surface that minimizes $\int_{x_1}^{x_2} |\dot{\mathbf{x}}(t)|dt$ would be considered a straight line by him. Such a curve, having minimum arc length between two given points, is called a *geodesic*.

Returning to our own view of the problem, we note, as we have previously, that if we project the negative gradient onto the tangent plane of the constraint surface at a point $\mathbf{x}_k$, we cannot move along this projection itself and remain feasible. We might, however, consider moving along a curve which had the same initial heading as the projected negative gradient but which remained on the surface. Exactly which such curve to move along is somewhat arbitrary, but a natural choice, inspired perhaps by the considerations of the bug, is a geodesic. Specifically, at a given point on the surface, we would determine the geodesic curve passing through that point that had an initial heading identical to that of the projected negative gradient. We would then move along this geodesic to a new point on the surface having a lesser value of f.

The idealized procedure then, which the bug would use without a second thought, and which we would use if it were computationally feasible (which it definitely is not), would at a given feasible point $\mathbf{x}_k$ (see Fig. 11.7):

1. Calculate the projection $\mathbf{p}$ of $-\nabla f(\mathbf{x}_k)'$ onto the tangent plane at $\mathbf{x}_k$.
2. Find the geodesic, $\mathbf{x}(t)$, $t \geqslant 0$, of the constraint surface having $\mathbf{x}(0) = \mathbf{x}_k$, $\dot{\mathbf{x}}(0) = \mathbf{p}$.
3. Minimize $f(\mathbf{x}(t))$ with respect to $t \geqslant 0$, obtaining t_k and $\mathbf{x}_{k+1} = \mathbf{x}(t_k)$.

At this point we emphasize that this technique (which we refer to as geodesic descent) is proposed essentially for theoretical purposes only. It does, however, capture the main philosophy of the gradient projection method. Furthermore, as the step size of the methods go to zero, as it does near the solution point, the distance between the point that would be determined by the gradient projection method and the point found by the idealized method goes to zero even faster. Thus, the asymptotic rates of convergence for the two methods will be equal, and it is, therefore, appropriate to concentrate on the idealized method only.

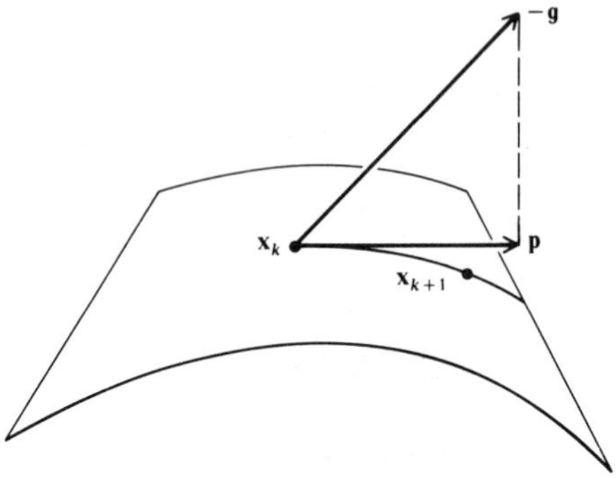

Fig. 11.7 Geodesic descent

Our bug confined to the surface would have no hesitation in estimating the rate of convergence of this method. He would simply express it in terms of the smallest and largest eigenvalues of the Hessian of f as measured on his surface. It should not be surprising, then, that we show that the asymptotic convergence ratio is

$$\left(\frac{A - a}{A - a}\right)^2 \tag{31}$$

where a and A are, respectively, the smallest and largest eigenvalues of $\mathbf{L}$, the Hessian of the Lagrangian, restricted to the tangent subspace M. This result parallels the convergence rate of the method of steepest descent, but with the eigenvalues determined from the same restricted Hessian matrix that is important in the general theory of necessary and sufficient conditions for constrained problems. This rate, which almost invariably arises when studying algorithms designed for constrained problems, will be referred to as the *canonical rate*.

We emphasize again that, since this convergence ratio governs the convergence of a large family of algorithms, it is the formula itself rather than its numerical value that is important. For any given problem we do not suggest that this ratio be evaluated, since this would be extremely difficult. Instead, the potency of the result derives from the fact that fairly comprehensive comparisons among algorithms can be made, on the basis of this formula, that apply to general classes of problems rather than simply to particular problems.

The remainder of this section is devoted to the analysis that is required to establish the convergence rate. Since this analysis is somewhat involved and not crucial for an understanding of remaining material, some readers may wish to simply read the theorem statement and proceed to the next section.

Geodesics

Given the surface $\Omega = \{\mathbf{x}: \mathbf{h}(\mathbf{x}) = \mathbf{0}\} \subset E^n$, a smooth curve, $\mathbf{x}(t) \in \Omega$, $0 \leqslant t \leqslant T$ starting at $\mathbf{x}(0)$ and terminating at $\mathbf{x}(T)$ that minimizes the total arc length

$$\int_0^T |\dot{\mathbf{x}}(t)|dt$$

with respect to all other such curves on Ω is said to be a *geodesic* connecting $\mathbf{x}(0)$ and $\mathbf{x}(T)$.

It is common to parameterize a geodesic $\mathbf{x}(t)$, $0 \leqslant t \leqslant T$ so that $|\dot{\mathbf{x}}(t)| = 1$. The parameter t is then itself the arc length. If the parameter t is also regarded as time, then this parameterization corresponds to moving along the geodesic curve with unit velocity. Parameterized in this way, the geodesic is said to be *normalized*. On any linear subspace of E^n geodesics are straight lines. On a three-dimensional sphere, the geodesics are arcs of great circles.

It can be shown, using the calculus of variations, that any normalized geodesic on Ω satisfies the condition

$$\ddot{\mathbf{x}}(t) = \nabla \mathbf{h}'(\mathbf{x}(t))\boldsymbol{\omega}(t) \tag{32}$$

for some function $\boldsymbol{\omega}$ taking values in E^m. Geometrically, this condition says that if one moves along the geodesic curve with unit velocity, the acceleration at every point will be orthogonal to the surface. Indeed, this property can be regarded as the fundamental defining characteristic of a geodesic. To stay on the surface Ω, the geodesic must also satisfy the equation

$$\nabla \mathbf{h}(\mathbf{x}(t))\dot{\mathbf{x}}(t) = \mathbf{0} \tag{33}$$

since the velocity vector at every point is tangent to Ω. At a regular point $\mathbf{x}_0$ these two differential equations, together with the initial conditions $\mathbf{x}(0) = \mathbf{x}_0$, $\dot{\mathbf{x}}(0)$ specified, and $|\dot{\mathbf{x}}(0)| = 1$, uniquely specify a curve $\mathbf{x}(t)$, $t \geqslant 0$ that can be continued as long as points on the curve are regular. Furthermore, $|\dot{\mathbf{x}}(t)| = 1$ for $t \geqslant 0$. Hence, geodesic curves emanate in every direction from a regular point. Thus, for example, at any point on a sphere there is a unique great circle passing through the point in a given direction.

Lagrangian and Geodesics

Corresponding to any regular point $\mathbf{x} \in \Omega$ we may define a corresponding Lagrange multiplier $\boldsymbol{\lambda}(\mathbf{x})$ by calculating the projection of the gradient of f onto the tangent subspace at $\mathbf{x}$, denoted $M(\mathbf{x})$. The matrix that, when operating on a vector, projects it onto $M(\mathbf{x})$ is

$$\mathbf{P}(\mathbf{x}) = \mathbf{I} - \nabla\mathbf{h}(\mathbf{x})'[\nabla\mathbf{h}(\mathbf{x})\nabla\mathbf{h}(\mathbf{x})']^{-1}\nabla\mathbf{h}(\mathbf{x})$$

and it follows immediately that the projection of $\nabla\mathbf{f}(\mathbf{x})'$ onto $M(\mathbf{x})$ has the form

$$\mathbf{y}(\mathbf{x}) = [\nabla f(\mathbf{x}) + \boldsymbol{\lambda}(\mathbf{x})\nabla\mathbf{h}(\mathbf{x})]' \tag{34}$$

where $\boldsymbol{\lambda}(\mathbf{x})$ is given explicitly as

$$\boldsymbol{\lambda}(\mathbf{x}) = -\nabla f(\mathbf{x})\nabla\mathbf{h}(\mathbf{x})'[\nabla\mathbf{h}(\mathbf{x})\nabla\mathbf{h}(\mathbf{x})']^{-1}. \tag{35}$$

Thus in terms of the Lagrangian function $l(\mathbf{x}, \boldsymbol{\lambda}) = f(\mathbf{x}) + \boldsymbol{\lambda}\mathbf{h}(\mathbf{x})$, the projected gradient is

$$\mathbf{y}(\mathbf{x}) = l_{\mathbf{x}}(\mathbf{x}, \boldsymbol{\lambda}(\mathbf{x}))'. \tag{36}$$

If a local solution to the original problem occurs at a regular point $\mathbf{x}^* \in \Omega$, then as we know

$$l_{\mathbf{x}}(\mathbf{x}^*, \boldsymbol{\lambda}(\mathbf{x}^*)) = \mathbf{0} \tag{37}$$

which states that the projected gradient must vanish at $\mathbf{x}^*$. Defining $\mathbf{L}(\mathbf{x}) = l_{\mathbf{xx}}(\mathbf{x}, \boldsymbol{\lambda}(\mathbf{x})) = \mathbf{F}(\mathbf{x}) + \boldsymbol{\lambda}(\mathbf{x})\mathbf{H}(\mathbf{x})$ we also know that at $\mathbf{x}^*$ we have the second-order necessary condition that $\mathbf{L}(\mathbf{x}^*)$ is positive semidefinite on $M(\mathbf{x}^*)$; that is, $\mathbf{z}'\mathbf{L}(\mathbf{x}^*)\mathbf{z} \geqslant 0$ for all $\mathbf{z} \in M(\mathbf{x}^*)$. Equivalently, letting

$$\overline{\mathbf{L}}(\mathbf{x}) = \mathbf{P}(\mathbf{x})\mathbf{L}(\mathbf{x})\mathbf{P}(\mathbf{x}); \tag{38}$$

it follows that $\overline{\mathbf{L}}(\mathbf{x}^*)$ is positive semidefinite.

We then have the following fundamental and simple result, valid along a geodesic.

Proposition 1. *Let* $\mathbf{x}(t)$, $0 \leqslant t \leqslant T$, *be a geodesic on* Ω. *Then*

$$\frac{d}{dt} f(\mathbf{x}(t)) = l_{\mathbf{x}}(\mathbf{x}, \boldsymbol{\lambda}(\mathbf{x}))\dot{\mathbf{x}}(t) \tag{39}$$

$$\frac{d^2}{dt^2} f(\mathbf{x}(t)) = \dot{\mathbf{x}}(t)'\mathbf{L}(\mathbf{x}(t))\dot{\mathbf{x}}(t). \tag{40}$$

Proof. We have

$$\frac{d}{dt} f(\mathbf{x}(t)) = \nabla f(\mathbf{x}(t))\dot{\mathbf{x}}(t) = l_{\mathbf{x}}(\mathbf{x}, \boldsymbol{\lambda}(\mathbf{x}))\dot{\mathbf{x}}(t)$$

the second equality following from the fact that $\dot{\mathbf{x}}(t) \in M(\mathbf{x})$. Next,

$$\frac{d^2}{dt^2} f(\mathbf{x}(t)) = \mathbf{x}(t)'\,\mathbf{F}(\mathbf{x}(t))\dot{\mathbf{x}}(t) + \nabla f(\mathbf{x}(t))\ddot{\mathbf{x}}(t). \tag{41}$$

But differentiating the relation $\boldsymbol{\lambda}\mathbf{h}(\mathbf{x}(t)) = 0$ twice, for fixed $\boldsymbol{\lambda}$, yields

$$\dot{\mathbf{x}}(t)'\,\boldsymbol{\lambda}\mathbf{H}(\mathbf{x}(t))\dot{\mathbf{x}}(t) + \boldsymbol{\lambda}\nabla\mathbf{h}(\mathbf{x}(t))\ddot{\mathbf{x}}(t) = 0.$$

Adding this to (41) we have

$$\frac{d^2}{dt^2}f(\mathbf{x}(t)) = \dot{\mathbf{x}}(t)'(\mathbf{F} + \boldsymbol{\lambda}\mathbf{H})\dot{\mathbf{x}}(t) + (\nabla f(\mathbf{x}) + \boldsymbol{\lambda}\nabla\mathbf{h}(\mathbf{x}))\ddot{\mathbf{x}}(t)$$

which is true for any fixed $\boldsymbol{\lambda}$. Setting $\boldsymbol{\lambda} = \boldsymbol{\lambda}(\mathbf{x})$ determined as above, $(\nabla\mathbf{f} + \boldsymbol{\lambda}\nabla\mathbf{h})'$ is then in $M(\mathbf{x})$ and hence orthogonal to $\ddot{\mathbf{x}}(t)$, since $\mathbf{x}(t)$ is a normalized geodesic. This gives (40). ∎

It should be noted that we proved a simplified version of this result in Chapter 10. There the result was given only for the optimal point $\mathbf{x}^*$, although it was valid for any curve. Here, we have shown that essentially the same result is valid at any point provided that we move along a geodesic.

Rate of Convergence

We now prove the main theorem regarding the rate of convergence. We assume that all functions are three times continuously differentiable and that every point in a region near the solution $\mathbf{x}^*$ is regular. This theorem only establishes the rate of convergence and not convergence itself so for that reason the stated hypotheses assume that the method of geodesic descent generates a sequence $\{\mathbf{x}_k\}$ converging to $\mathbf{x}^*$.

> ***Theorem.*** *Let $\mathbf{x}^*$ be a local solution to the problem* (30) *and suppose that A and $a > 0$ are, respectively, the largest and smallest eigenvalues of $\mathbf{L}(\mathbf{x}^*)$ restricted to the tangent subspace $M(\mathbf{x}^*)$. If $\{\mathbf{x}_k\}$ is a sequence generated by the method of geodesic descent that converges to $\mathbf{x}^*$, then the sequence of objective values $\{f(\mathbf{x}_k)\}$ converges to $f(\mathbf{x}^*)$ linearly with a ratio no greater than $[(A - a)/(A + a)]^2$.*

Proof. Without loss of generality we may assume $f(\mathbf{x}^*) = 0$. Given a point $\mathbf{x}_k$ it will be convenient to define its distance from the solution point $\mathbf{x}^*$ as the arc length of the geodesic connecting $\mathbf{x}^*$ and $\mathbf{x}_k$. Thus if $\mathbf{x}(t)$ is a parameterized version of the geodesic with $\mathbf{x}(0) = \mathbf{x}^*$, $|\dot{\mathbf{x}}(t)| = 1$, $\mathbf{x}(T) = \mathbf{x}_k$, then T is the distance of $\mathbf{x}_k$ from $\mathbf{x}^*$. Associated with such a geodesic we also have the family $\mathbf{y}(t)$, $0 \leqslant t \leqslant T$, of corresponding projected gradients $\mathbf{y}(t) = l_{\mathbf{x}}(\mathbf{x}, \boldsymbol{\lambda}(\mathbf{x}))'$, and Hessians $\mathbf{L}(t) = \mathbf{L}(\mathbf{x}(t))$. We write $\mathbf{y}_k = \mathbf{y}(\mathbf{x}_k)$, $\mathbf{L}_k = \mathbf{L}(\mathbf{x}_k)$.

We now derive an estimate for $f(\mathbf{x}_k)$. Using the geodesic discussed above we can write (setting $\dot{\mathbf{x}}_k = \dot{\mathbf{x}}(T)$)

$$f(\mathbf{x}^*) - f(\mathbf{x}_k) = -f(\mathbf{x}_k) = -\mathbf{y}_k{}'\,\dot{\mathbf{x}}_k T + \tfrac{1}{2}T^2\dot{\mathbf{x}}_k{}'\,\mathbf{L}_k\dot{\mathbf{x}}_k + o(T^2) \qquad (42)$$

which follows from Proposition 1. We also have

$$\mathbf{y}_k = -\mathbf{y}(\mathbf{x}^*) + \mathbf{y}(\mathbf{x}_k) = \dot{\mathbf{y}}_k T + o(T). \tag{43}$$

But differentiating (34) we obtain

$$\dot{\mathbf{y}}_k = \mathbf{L}_k \dot{\mathbf{x}}_k + \nabla \mathbf{h}(\mathbf{x}_k)' \dot{\boldsymbol{\lambda}}_k' \tag{44}$$

and hence if $\mathbf{P}_k$ is the projection matrix onto $M(\mathbf{x}_k) = M_k$, we have

$$\mathbf{P}_k \dot{\mathbf{y}}_k = \mathbf{P}_k \mathbf{L}_k \dot{\mathbf{x}}_k. \tag{45}$$

Multiplying (43) by $\mathbf{P}_k$ and accounting for $\mathbf{P}_k \mathbf{y}_k = \mathbf{y}_k$ we have

$$\mathbf{P}_k \dot{\mathbf{y}}_k T = \mathbf{y}_k + o(T). \tag{46}$$

Substituting (45) into this we obtain

$$\mathbf{P}_k \mathbf{L}_k \dot{\mathbf{x}}_k T = \mathbf{y}_k + o(T).$$

Since $\mathbf{P}_k \dot{\mathbf{x}}_k = \dot{\mathbf{x}}_k$ we have, defining $\bar{\mathbf{L}}_k = \mathbf{P}_k \mathbf{L}_k \mathbf{P}_k$,

$$\bar{\mathbf{L}}_k \dot{\mathbf{x}}_k T = \mathbf{y}_k + o(T). \tag{47}$$

The matrix $\bar{\mathbf{L}}_k$ is related to $\mathbf{L}_{M_k}$, the restriction of $\mathbf{L}_k$ to M_k, the only difference being that while $\mathbf{L}_{M_k}$ is defined only on M_k, the matrix $\bar{\mathbf{L}}_k$ is defined on all of E^n but in such a way that it agrees with $\mathbf{L}_{M_k}$ on M_k and is zero on $M_k^{\perp}$. The matrix $\bar{\mathbf{L}}_k$ is not invertible, but for $\mathbf{y}_k \in M_k$ there is a unique solution $\mathbf{z} \in M_k$ to the equation $\bar{\mathbf{L}}_k \mathbf{z} = \mathbf{y}_k$ which we denote* $\bar{\mathbf{L}}_k^{-1} \mathbf{y}_k$. With this notation we obtain from (47)

$$\dot{\mathbf{x}}_k T = \bar{\mathbf{L}}_k^{-1} \mathbf{y}_k + o(T). \tag{48}$$

Substituting this last result into (42) and accounting for $|\mathbf{y}_k| = O(T)$ (see (43)) we have

$$f(\mathbf{x}_k) = \tfrac{1}{2} \mathbf{y}_k' \bar{\mathbf{L}}_k^{-1} \mathbf{y}_k + o(T^2) \tag{49}$$

which expresses the objective value at $\mathbf{x}_k$ in terms of the projected gradient.

Since $|\dot{\mathbf{x}}_k| = 1$ and since $\bar{\mathbf{L}}_k \to \bar{\mathbf{L}}^*$ as $\mathbf{x}_k \to \mathbf{x}^*$, we see from (47) that

$$o(T) + aT \leqslant |\mathbf{y}_k| \leqslant AT + o(T) \tag{50}$$

which means that not only do we have $|\mathbf{y}_k| = O(T)$, which was known before, but also $|\mathbf{y}_k| \neq o(T)$. We may therefore write our estimate (49) in the alternate form

$$f(\mathbf{x}_k) = \tfrac{1}{2} \mathbf{y}_k' \bar{\mathbf{L}}_k^{-1} \mathbf{y}_k \left(1 + \frac{o(T^2)}{\mathbf{y}_k' \bar{\mathbf{L}}_k^{-1} \mathbf{y}_k} \right) \tag{51}$$

---

* Actually a more standard procedure is to define the pseudoinverse $\bar{\mathbf{L}}_k^{\dagger}$, and then $\mathbf{z} = \bar{\mathbf{L}}_k^{\dagger} \mathbf{y}_k$.

and since $o(T^2) \neq \mathbf{y}_k'\bar{\mathbf{L}}_k^{-1}\mathbf{y}_k = O(T^2)$, we have

$$f(\mathbf{x}_k) = \tfrac{1}{2}\mathbf{y}_k'\,\bar{\mathbf{L}}_k^{-1}\mathbf{y}_k(1 + O(T)) \tag{52}$$

which is the desired estimate.

Next, we estimate $f(\mathbf{x}_{k+1})$ in terms of $f(\mathbf{x}_k)$. Given $\mathbf{x}_k$ now let $\mathbf{x}(t)$, $t \geqslant 0$, be the normalized geodesic emanating from $\mathbf{x}_k \triangleq \mathbf{x}(0)$ in the direction of the negative projected gradient, that is,

$$\dot{\mathbf{x}}(0) \triangleq \dot{\mathbf{x}}_k = -\mathbf{y}_k/|\mathbf{y}_k|.$$

Then

$$f(\mathbf{x}(t)) = f(\mathbf{x}_k) + t\mathbf{y}_k'\,\dot{\mathbf{x}}_k + \frac{t^2}{2}\dot{\mathbf{x}}_k'\,\mathbf{L}_k\dot{\mathbf{x}}_k + o(t^2). \tag{53}$$

This is minimized at

$$t_k = -\frac{\mathbf{y}_k'\,\dot{\mathbf{x}}_k}{\dot{\mathbf{x}}_k'\,\mathbf{L}_k\dot{\mathbf{x}}_k} + o(t_k). \tag{54}$$

In view of (50) this implies that $t_k = O(T)$, $t_k \neq o(T)$. Thus t_k goes to zero at essentially the same rate as T. Thus, we have

$$f(\mathbf{x}_{k+1}) = f(\mathbf{x}_k) - \frac{1}{2}\frac{(\mathbf{y}_k'\,\dot{\mathbf{x}}_k)^2}{\dot{\mathbf{x}}_k'\,\mathbf{L}_k\dot{\mathbf{x}}_k} + o(T^2). \tag{55}$$

Using the same argument as before we can express this as

$$f(\mathbf{x}_k) - f(\mathbf{x}_{k+1}) = \frac{1}{2}\frac{(\mathbf{y}_k'\,\mathbf{y}_k)^2}{\mathbf{y}_k'\,\mathbf{L}_k\mathbf{y}_k}(1 + O(T)), \tag{56}$$

which is the other required estimate.

Finally, dividing (56) by (52) we find

$$\frac{f(\mathbf{x}_k) - f(\mathbf{x}_{k+1})}{f(\mathbf{x}_k)} = \frac{(\mathbf{y}_k'\,\mathbf{y}_k)^2(1 + O(T))}{(\mathbf{y}_k'\,\mathbf{L}_k\mathbf{y}_k)(\mathbf{y}_k'\,\bar{\mathbf{L}}^{-1}\mathbf{y}_k)} \tag{57}$$

and thus

$$f(\mathbf{x}_{k+1}) = \left[1 - \frac{(\mathbf{y}_k'\,\mathbf{y}_k)^2(1 + O(T))}{(\mathbf{y}_k'\,\mathbf{L}_k\mathbf{y}_k)(\mathbf{y}_k'\,\bar{\mathbf{L}}^{-1}\mathbf{y}_k)}\right] f(\mathbf{x}_k). \tag{58}$$

Using the fact that $\mathbf{L}_k \to \mathbf{L}^*$ and applying the Kantorovich inequality leads to

$$f(\mathbf{x}_{k+1}) \leqslant \left[\left(\frac{A - a}{A + a}\right)^2 + O(T)\right] f(\mathbf{x}_k). \;\blacksquare \tag{59}$$

Problems with Inequalities

The idealized version of gradient projection could easily be extended to problems having nonlinear inequalities as well as equalities by following the pattern of Section 11.4. Such an extension, however, has no real value, since the idealized scheme cannot be implemented. The idealized procedure was devised only as a technique for analyzing the asymptotic rate of convergence of the analytically more complex, but more practical, gradient projection method.

The analysis of the idealized version of gradient projection given above, nevertheless, does apply to problems having inequality as well as equality constraints. If a computationally feasible procedure is employed that avoids jamming and does not bounce on and off constraint boundaries an infinite number of times, then near the solution the active constraints will remain fixed. This means that near the solution the method acts just as if it were solving a problem having the active constraints as equality constraints. Thus, the asymptotic rate of convergence of the gradient projection method applied to a problem with inequalities is also given by (59) but with $\mathbf{L}(\mathbf{x}^*)$ and $M(\mathbf{x}^*)$ (and hence a and A) determined by the active constraints at the solution point $\mathbf{x}^*$. In every case, therefore, the rate of convergence is determined by the eigenvalues of the same restricted Hessian that arises in the necessary conditions.

11.6 THE REDUCED GRADIENT METHOD

From a computational viewpoint, the reduced gradient method, discussed in this section and the next, is closely related to the simplex method of linear programming in that the problem variables are partitioned into basic and nonbasic groups. From a theoretical viewpoint, the method can be shown to behave very much like the gradient projection method.

Linear Constraints

Consider the problem

$$\begin{aligned} &\text{minimize} \quad f(\mathbf{x}) \\ &\text{subject to} \quad \mathbf{A}\mathbf{x} = \mathbf{b}, \qquad \mathbf{x} \geqslant \mathbf{0} \end{aligned} \tag{60}$$

where $\mathbf{x} \in E^n$, $\mathbf{b} \in E^m$, $\mathbf{A}$ is $m \times n$, and f is a function in C^2. The constraints are expressed in the format of the standard form of linear programming. For simplicity of notation it is assumed that each variable is required to be nonnegative—if some variables were free, the procedure (but not the notation) would be somewhat simplified.

We invoke the *nondegeneracy assumptions* that every collection of m columns from $\mathbf{A}$ is linearly independent and every basic solution to the constraints has m strictly positive variables. With these assumptions any feasible solution will have at most $n - m$ variables taking the value zero. Given a vector $\mathbf{x}$ satisfying the constraints, we partition the variables into two groups: $\mathbf{x} = (\mathbf{y}, \mathbf{z})$ where $\mathbf{y}$ has dimension m and $\mathbf{z}$ has dimension $n - m$. This partition is formed in such a way that all variables in $\mathbf{y}$ are strictly positive (for simplicity of notation we indicate the basic variables as being the first m components of $\mathbf{x}$ but, of course, in general this will not be so). With respect to the partition, the original problem can be expressed as

$$\text{minimize} \quad f(\mathbf{y}, \mathbf{z}) \tag{61a}$$

$$\text{subject to} \quad \mathbf{By} + \mathbf{Cz} = \mathbf{b} \tag{61b}$$

$$\mathbf{y} \geqslant \mathbf{0}, \qquad \mathbf{z} \geqslant \mathbf{0} \tag{61c}$$

where, of course, $\mathbf{A} = [\mathbf{B}, \mathbf{C}]$. We can regard $\mathbf{z}$ as consisting of the independent variables and $\mathbf{y}$ the dependent variables, since if $\mathbf{z}$ is specified, (61b) can be uniquely solved for $\mathbf{y}$. Furthermore, a small change $\Delta\mathbf{z}$ from the original value that leaves $\mathbf{z} + \Delta\mathbf{z}$ nonnegative will, upon solution of (61b), yield another feasible solution, since $\mathbf{y}$ was originally taken to be strictly positive and thus $\mathbf{y} + \Delta\mathbf{y}$ will also be positive for small $\Delta\mathbf{y}$. We may therefore move from one feasible solution to another by selecting a $\Delta\mathbf{z}$ and moving $\mathbf{z}$ on the line $\mathbf{z} + \alpha\Delta\mathbf{z}$, $\alpha \geqslant 0$. Accordingly, $\mathbf{y}$ will move along a corresponding line $\mathbf{y} + \alpha\Delta\mathbf{y}$. If in moving this way some independent variable becomes zero, a new direction $\Delta\mathbf{z}$ must be chosen. If in moving this way a dependent (basic) variable becomes zero, the partition must be modified. The zero-valued basic variable is declared independent and one of the strictly positive independent variables is made dependent. Operationally, this interchange will be associated with a pivot operation.

The idea of the reduced gradient method is to consider, at each stage, the problem only in terms of the independent variables. Since the vector of dependent variables $\mathbf{y}$ is determined through the constraints (61b) from the vector of independent variables $\mathbf{z}$, the objective function can be considered to be a function of $\mathbf{z}$ only. From this point of view, the only constraints are nonnegativity constraints on the independent variables, and hence a simple modification of steepest descent, accounting for these constraints, can be executed. The gradient with respect to the independent variables $\mathbf{z}$ (the *reduced gradient*) is found by evaluating the gradient of $f(\mathbf{B}^{-1}\mathbf{b} - \mathbf{B}^{-1}\mathbf{Cz}, \mathbf{z})$. It is equal to

$$\mathbf{r} = \nabla_{\mathbf{z}} f(\mathbf{y}, \mathbf{z}) - \nabla_{\mathbf{y}} f(\mathbf{y}, \mathbf{z})\mathbf{B}^{-1}\mathbf{C}. \tag{62}$$

The elementary form of the reduced gradient method progresses by moving the vector $\mathbf{z}$ in the direction of the negative reduced gradient, except that zero-valued components of $\mathbf{z}$ that would thereby become negative are held at zero. One step of the procedure is as follows:

1. Let $\Delta\mathbf{z} = \begin{cases} r_i & \text{if } r_i > 0 \text{ or } z_i > 0 \\ 0 & \text{otherwise} \end{cases}$
2. If $\Delta\mathbf{z}$ is zero, stop; the current point is a solution. Otherwise, find $\Delta\mathbf{y} = \mathbf{B}^{-1}\mathbf{C}\Delta\mathbf{z}$.

3. Find α_1, α_2, α_3 achieving, respectively,
   max $\{\alpha: \mathbf{y} + \alpha\Delta\mathbf{y} \geqslant \mathbf{0}\}$
   max $\{\alpha: \mathbf{z} + \alpha\Delta\mathbf{z} \geqslant \mathbf{0}\}$
   min $\{f(\mathbf{x} + \alpha\Delta\mathbf{x}): 0 \leqslant \alpha \leqslant \alpha_1, 0 \leqslant \alpha \leqslant \alpha_2\}$
   Let $\bar{\mathbf{x}} = \mathbf{x} + \alpha_3\Delta\mathbf{x}$.

4. If $\alpha_3 < \alpha_1$, return to (1). Otherwise, declare the vanishing variable in the dependent set independent and declare a strictly positive variable in the independent set dependent. Update **B** and **C**.

Example. We consider the example presented in Section 11.4 where the projected negative gradient was computed:

$$\begin{aligned} \text{minimize} \quad & x_1^2 + x_2^2 + x_3^2 + x_4^2 - 2x_1 - x_4 \\ \text{subject to} \quad & 2x_1 + x_2 + x_3 + 4x_4 = 7 \\ & x_1 + x_2 + 2x_3 + x_4 = 6 \\ & x_i \geqslant 0, \qquad i = 1, 2, 3, 4. \end{aligned}$$

We are given the feasible point $\mathbf{x} = (2, 2, 1, 0)$. We may select any two of the strictly positive variables to be the basic variables. Suppose $\mathbf{y} = (x_1, x_2)$ is selected. In standard form the constraints are then

$$\begin{aligned} x_1 + 0 - x_3 + 3x_4 &= 1 \\ 0 + x_2 + 3x_3 - 2x_4 &= 5 \\ x_i \geqslant 0, \qquad i &= 1, 2, 3, 4. \end{aligned}$$

The gradient at the current point is $\mathbf{g} = (2, 4, 2, -3)$. The corresponding reduced gradient (with respect to $\mathbf{z} = (x_3, x_4)$) is then found by *pricing-out* in the usual manner. The situation at the current point can then be summarized by the tableau

Variable	x_1	x_2	x_3	x_4	
Constraints	1	0	−1	3	1
	0	1	3	−2	5
r	0	0	−8	−1	
Current value	2	2	1	0	

Tableau for example

In this solution x_3 and x_4 would be increased together in a ratio of eight to one. As they increase, x_1 and x_2 would follow in such a way as to keep the constraints satisfied. Overall, in E^4, the implied direction of movement is thus

$$\mathbf{d} = (5, 22, 8, 1).$$

If the reader carefully supplies the computational details not shown in the presentation of the example as worked here and in Section 11.4, he will undoubtedly develop a considerable appreciation for the relative simplicity of the reduced gradient method.

It should be clear that the reduced gradient method can, as illustrated in the example above, be executed with the aid of a tableau. At each step the tableau of constraints is arranged so that an identity matrix appears over the m dependent variables, and thus the dependent variables can be easily calculated from the values of the independent variables. The reduced gradient at any step is calculated by evaluating the n-dimensional gradient and "pricing out" the dependent variables just as the reduced cost vector is calculated in linear programming. And when the partition of basic and nonbasic variables must be changed, a simple pivot operation is all that is required.

Global Convergence

The perceptive reader will note the direction finding algorithm that results from this primary form of the reduced gradient method is not closed, since slight movement away from the boundary of an inequality constraint can cause a sudden change in the direction of search. Thus one might suspect, and correctly so, that this method is subject to jamming. However, a trivial modification will yield a closed mapping; and hence global convergence. This is discussed in Exercise 18.

Nonlinear Constraints

The *generalized reduced gradient method* solves nonlinear programming problems in the *standard form*

$$\begin{aligned} &\text{minimize} \quad f(\mathbf{x}) \\ &\text{subject to} \quad \mathbf{h}(\mathbf{x}) = \mathbf{0}, \qquad \mathbf{a} \leqslant \mathbf{x} \leqslant \mathbf{b} \end{aligned}$$

where $\mathbf{h}(\mathbf{x})$ is of dimension m. A general nonlinear programming problem can always be expressed in this form by the introduction of slack variables, if required, and by allowing some components of $\mathbf{a}$ and $\mathbf{b}$ to take on the values $+\infty$ or $-\infty$, if necessary.

In a manner quite analogous to that of the case of linear constraints, we introduce a *nondegeneracy* assumption that, at each point $\mathbf{x}$, hypothesizes the existence of a partition of $\mathbf{x}$ into $\mathbf{x} = (\mathbf{y}, \mathbf{z})$ having the following properties:

i) $\mathbf{y}$ is of dimension m, and $\mathbf{z}$ is of dimension $n - m$.

ii) If $\mathbf{a} = (\mathbf{a_y}, \mathbf{a_z})$ is the corresponding partition of $\mathbf{a}$, $\mathbf{b}$; then $\mathbf{a_y} < \mathbf{y} < \mathbf{b_y}$.

iii) The $m \times m$ matrix $\nabla_y \mathbf{h}(\mathbf{y}, \mathbf{z})$ is nonsingular at $\mathbf{x} = (\mathbf{y}, \mathbf{z})$.

Again $\mathbf{y}$ and $\mathbf{z}$ are referred to as the vectors of *dependent* and *independent variables*, respectively.

The reduced gradient (with respect to $\mathbf{z}$) is in this case:

$$\mathbf{r} = \Delta_z f(\mathbf{y}, \mathbf{z}) + \boldsymbol{\lambda} \nabla_z \mathbf{h}(\mathbf{y}, \mathbf{z})$$

where $\boldsymbol{\lambda}$ satisfies

$$\nabla_y f(\mathbf{y}, \mathbf{z}) + \boldsymbol{\lambda} \nabla_y \mathbf{h}(\mathbf{y}, \mathbf{z}) = \mathbf{0}.$$

Equivalently, we have

$$\mathbf{r} = \nabla_z f(\mathbf{y}, \mathbf{z}) - \nabla_y f(\mathbf{y}, \mathbf{z})[\nabla_y \mathbf{h}(\mathbf{y}, \mathbf{z})]^{-1} \nabla_z \mathbf{h}(\mathbf{y}, \mathbf{z}). \tag{63}$$

The actual procedure is roughly the same as for linear constraints in that moves are taken by changing $\mathbf{z}$ in the direction of the negative reduced gradient (with components of $\mathbf{z}$ on their boundary held fixed if the movement would violate the bound). The difference here is that although $\mathbf{z}$ moves along a straight line as before, the vector of dependent variables $\mathbf{y}$ must move nonlinearly to continuously satisfy the equality constraints. Computationally, this is accomplished by first moving linearly along the tangent to the surface defined by $\mathbf{z} \to \mathbf{z} + \Delta\mathbf{z}$, $\mathbf{y} \to \mathbf{y} + \Delta\mathbf{y}$ with $\Delta\mathbf{y} = -[\nabla_y \mathbf{h}]^{-1} \nabla_z \mathbf{h} \Delta\mathbf{z}$. Then a correction procedure, much like that employed in the gradient projection method, is used to return to the constraint surface and the magnitude bounds on the dependent variables are checked for feasibility. As with the gradient projection method, a feasibility tolerance must be introduced to acknowledge the impossibility of returning exactly to the constraint surface. An example corresponding to $n = 3$, $m = 1$, $a = 0$, $b = +\infty$ is shown in Fig. 11.8.

To return to the surface once a tentative move along the tangent is made, an iterative scheme is employed. If the point $\mathbf{x}_k$ was the point at the previous step, then from any point $\mathbf{x} = (\mathbf{v}, \mathbf{w})$ near $\mathbf{x}_k$ one gets back to the constraint surface by solving the nonlinear equation

$$\mathbf{h}(\mathbf{y}, \mathbf{w}) = \mathbf{0} \tag{64}$$

for $\mathbf{y}$ (with $\mathbf{w}$ fixed). This is accomplished through the iterative process

$$\mathbf{y}_{j+1} = \mathbf{y}_j - [\nabla_y \mathbf{h}(\mathbf{x}_k)]^{-1} \mathbf{h}(\mathbf{y}_j, \mathbf{w}) \tag{65}$$

which, if started close enough to $\mathbf{x}_k$, will produce $\{\mathbf{y}_j\}$ with $\mathbf{y}_j \to \mathbf{y}$, solving (64).

The reduced gradient method suffers from the same basic difficulties as the gradient projection method, but as with the latter method, these difficulties can all be more or less successfully resolved. Computation is somewhat less complex in the case of the reduced gradient method, because rather than compute with $[\nabla \mathbf{h}(\mathbf{x})\nabla \mathbf{h}(\mathbf{x})']^{-1}$ at each step, the matrix $[\nabla_{\mathbf{y}}\mathbf{h}(\mathbf{y}, \mathbf{z})]^{-1}$ is used.

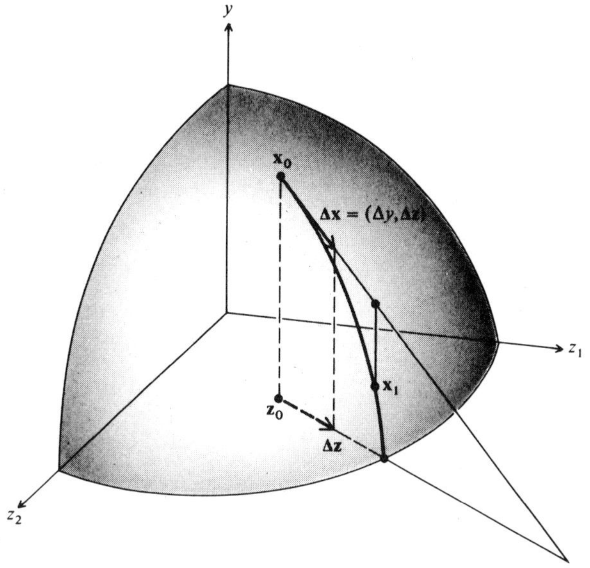

Fig. 11.8 Reduced gradient method

11.7 CONVERGENCE RATE OF THE REDUCED GRADIENT METHOD

As argued before, for purposes of analyzing the rate of convergence, it is sufficient to consider the problem having only equality constraints

$$\begin{aligned} &\text{minimize} \quad f(\mathbf{x}) \\ &\text{subject to} \quad \mathbf{h}(\mathbf{x}) = \mathbf{0}. \end{aligned} \tag{66}$$

We then regard the problem as being defined over a surface Ω of dimension $n - m$. At this point it is again timely to consider the view of our bug, who lives on this constraint surface. Invariably, he continues to regard the problem as extremely elementary, and indeed would have little appreciation for the complexity that seems to face us. To him the problem is an unconstrained problem in $n - m$ dimensions not, as we see it, a constrained problem in n dimensions. The bug will tenaciously hold to the method of steepest descent. We can emulate him provided that we know how he measures distance on his surface and thus how he calculates gradients and what he considers to be straight lines.

Rather than imagine that the measure of distance on his surface is the one that would be inherited from us in n dimensions, as we did when studying the gradient projection method, we, in this instance, follow the construction shown in Fig. 11.9. In our n-dimensional space, $n - m$ coordinates are selected as independent variables in such a way that, given their values, the values of the remaining (dependent) variables are determined by the surface. There is already a coordinate system in the space of independent variables, and it can be used on the surface by projecting it parallel to the space of the remaining dependent variables. Thus, an arc on the surface is considered to be straight if its projection onto the space of independent variables is a segment of a straight line. With this method for inducing a geometry on the surface, the bug's notion of steepest descent exactly coincides with an idealized version of the reduced gradient method.

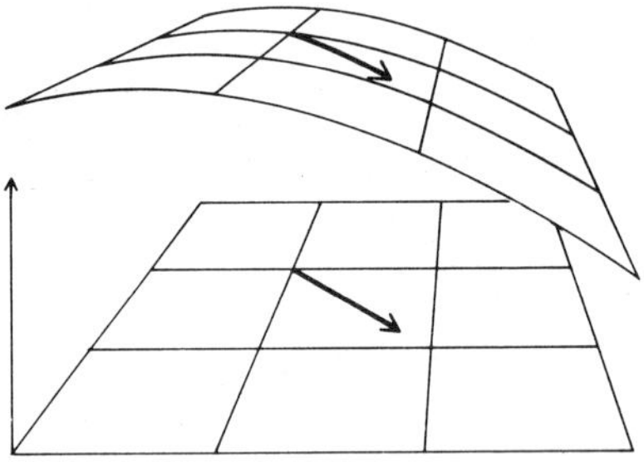

Fig. 11.9 Induced coordinate system

In the idealized version of the reduced gradient method for solving (66), the vector $\mathbf{x}$ is partitioned as $\mathbf{x} = (\mathbf{y}, \mathbf{z})$ where $\mathbf{y} \in E^m$, $\mathbf{z} \in E^{n-m}$. It is assumed that the $m \times m$ matrix $\nabla_{\mathbf{y}}\mathbf{h}(\mathbf{y}, \mathbf{z})$ is nonsingular throughout a given region of interest. (With respect to the more general problem, this region is a small neighborhood around the solution where it is not necessary to change the partition.) The vector $\mathbf{y}$ is regarded as an implicit function of $\mathbf{z}$ through the equation

$$\mathbf{h}(\mathbf{y}(\mathbf{z}), \mathbf{z}) = \mathbf{0}. \tag{67}$$

The ordinary method of steepest descent is then applied to the function $q(\mathbf{z}) = f(\mathbf{y}(\mathbf{z}), \mathbf{z})$. We note that the gradient $\mathbf{r}$ of this function is given by (63).

Since the method is really just the ordinary method of steepest descent with respect to $\mathbf{z}$, the rate of convergence is determined by the eigenvalues of the Hessian of the function q at the solution. We therefore turn to the question of evaluating this Hessian.

Denote by $\mathbf{Y}(\mathbf{z})$ the first derivatives of the implicit function $\mathbf{y}(\mathbf{z})$, that is, $\mathbf{Y}(\mathbf{z}) \triangle \nabla_{\mathbf{z}}\mathbf{y}(\mathbf{z})$. Explicitly,

$$\mathbf{Y}(\mathbf{z}) = -[\nabla_{\mathbf{y}}\mathbf{h}(\mathbf{y}, \mathbf{z})]^{-1}\nabla_{\mathbf{z}}\mathbf{h}(\mathbf{y}, \mathbf{z}). \tag{68}$$

For any $\boldsymbol{\lambda} \in E_m$ we have

$$q(\mathbf{z}) = f(\mathbf{y}(\mathbf{z}), \mathbf{z}) = f(\mathbf{y}(\mathbf{z}), \mathbf{z}) + \boldsymbol{\lambda}\mathbf{h}(\mathbf{y}(\mathbf{z}), \mathbf{z}). \tag{69}$$

Thus

$$\nabla q(\mathbf{z}) = [\nabla_{\mathbf{y}} f(\mathbf{y}, \mathbf{z}) + \boldsymbol{\lambda}\nabla_{\mathbf{y}}\mathbf{h}(\mathbf{y}, \mathbf{z})]\mathbf{Y}(\mathbf{z}) + \nabla_{\mathbf{z}} f(\mathbf{y}, \mathbf{z}) + \boldsymbol{\lambda}\nabla_{\mathbf{z}}\mathbf{h}(\mathbf{y}, \mathbf{z}). \tag{70}$$

Now if at a given point $\mathbf{x}^* = (\mathbf{y}^*, \mathbf{z}^*) = (\mathbf{y}(\mathbf{z}^*), \mathbf{z}^*)$ we let $\boldsymbol{\lambda}$ satisfy

$$\nabla_{\mathbf{y}} f(\mathbf{y}^*, \mathbf{z}^*) + \boldsymbol{\lambda}\nabla_{\mathbf{y}}\mathbf{h}(\mathbf{y}^*, \mathbf{z}^*) = \mathbf{0}; \tag{71}$$

then introducing the Lagrangian $l(\mathbf{y}, \mathbf{z}, \boldsymbol{\lambda}) = f(\mathbf{y}, \mathbf{z}) + \boldsymbol{\lambda}\mathbf{h}(\mathbf{y}, \mathbf{z})$, we obtain by differentiating (70)

$$\begin{aligned}\nabla^2 q(\mathbf{z}^*) = {} & \mathbf{Y}(\mathbf{z}^*)' \nabla_{\mathbf{yy}}^2 l(\mathbf{y}^*, \mathbf{z}^*)\mathbf{Y}(\mathbf{z}^*) + \nabla_{\mathbf{zy}}^2 l(\mathbf{y}^*, \mathbf{z}^*)\mathbf{Y}(\mathbf{z}^*) \\ & + \mathbf{Y}(\mathbf{z}^*)' \nabla_{\mathbf{yz}}^2 l(\mathbf{y}^*, \mathbf{z}^*) + \nabla_{\mathbf{zz}}^2 l(\mathbf{y}^*, \mathbf{z}^*).\end{aligned} \tag{72}$$

Or defining the $n \times (n-m)$ matrix

$$\mathbf{C} = \begin{bmatrix} \mathbf{Y}(\mathbf{z}^*) \\ \hdashline \mathbf{I} \end{bmatrix} \tag{73}$$

where $\mathbf{I}$ is the $(n-m) \times (n-m)$ identity, we have

$$\mathbf{Q} \triangleq \nabla^2 q(\mathbf{z}^*) = \mathbf{C}' \mathbf{L}(\mathbf{x}^*)\mathbf{C}. \tag{74}$$

The matrix $\mathbf{L}(\mathbf{x}^*)$ is the $n \times n$ Hessian of the Lagrangian at $\mathbf{x}^*$, and $\nabla^2 q(\mathbf{z}^*)$ is an $(n-m) \times (n-m)$ matrix that is a restriction of $\mathbf{L}(\mathbf{x}^*)$ to the tangent subspace M, but it is not the usual restriction. We summarize our conclusion with the following theorem.

Theorem. *Let* $\mathbf{x}^*$ *be a local solution of problem* (66). *Suppose that the idealized reduced gradient method produces a sequence* $\{\mathbf{x}_k\}$ *converging to* $\mathbf{x}^*$ *and that the partition* $\mathbf{x} = (\mathbf{y}, \mathbf{z})$ *is used throughout the tail of the sequence. Let* $\mathbf{L}$ *be the Hessian of the Lagrangian at* $\mathbf{x}^*$ *and define the matrix* $\mathbf{C}$ *by* (73) *and* (68). *Then the sequence of objective values* $\{f(\mathbf{x}_k)\}$ *converges to* $f(\mathbf{x}^*)$ *linearly with a ratio no greater than* $[(B-b)/(B+b)]^2$ *where* b *and* B *are, respectively, the smallest and largest eigenvalues of the matrix* $\mathbf{Q} = \mathbf{C}'\mathbf{L}\mathbf{C}$.

To compare the matrix $\mathbf{C}'\mathbf{L}\mathbf{C}$ with the usual restriction of $\mathbf{L}$ to M that determines the convergence rate of most methods, we note that the $n \times (n-m)$ matrix $\mathbf{C}$ maps $\Delta\mathbf{z} \in E^{n-m}$ into $(\Delta\mathbf{y}, \Delta\mathbf{z}) \in E^n$ lying in the tangent subspace M; that is, $\nabla_y \mathbf{h} \Delta \mathbf{y} + \nabla_z \mathbf{h} \Delta \mathbf{z} = \mathbf{0}$. Thus the columns of $\mathbf{C}$ form a basis for the subspace M. Next note that the columns of the matrix

$$\mathbf{E} = \mathbf{C}(\mathbf{C}'\mathbf{C})^{-1/2} \tag{75}$$

form an orthonormal basis for M, since each column of $\mathbf{E}$ is just a linear combination of columns of $\mathbf{C}$ and by direct calculation we see that $\mathbf{E}'\mathbf{E} = \mathbf{I}$. Thus by the procedure described in Section 10.4 we see that a representation for the usual restriction of $\mathbf{L}$ to M is

$$\mathbf{L}_M = (\mathbf{C}'\mathbf{C})^{-1/2}\mathbf{C}'\mathbf{L}\mathbf{C}(\mathbf{C}'\mathbf{C})^{-1/2}. \tag{76}$$

Comparing (76) with (74) we deduce that

$$\mathbf{Q} = (\mathbf{C}'\mathbf{C})^{1/2}\mathbf{L}_M(\mathbf{C}'\mathbf{C})^{1/2}. \tag{77}$$

This means that the Hessian matrix for the reduced gradient method is the restriction of $\mathbf{L}$ to M but pre- and post-multiplied by a positive definite symmetric matrix.

The eigenvalues of $\mathbf{Q}$ depend on the exact nature of $\mathbf{C}$ as well as $\mathbf{L}_M$. Thus, the rate of convergence rate of the reduced gradient method is not coordinate independent but depends strongly on just which variables are declared as independent at the final stage of the process. The convergence rate can be either faster or slower than that of the gradient projection method. In general, however, if $\mathbf{C}$ is well-behaved (that is, well-conditioned), the ratio of eigenvalues for the reduced gradient method can be expected to be the same order of magnitude as that of the gradient projection method. If, however, $\mathbf{C}$ should be ill-conditioned, as would arise in the case where the implicit equation $\mathbf{h}(\mathbf{y}, \mathbf{z}) = \mathbf{0}$ is itself ill-conditioned, then it can be shown that the eigenvalue ratio for the reduced gradient method will most likely be considerably worsened. This suggests that care should be taken to select a set of basic variables $\mathbf{y}$ that leads to a well-behaved $\mathbf{C}$ matrix.

11.8 VARIATIONS

It is possible to modify either the gradient projection method or the reduced gradient method so as to move in directions that are determined through additional considerations. For example, analogs of the conjugate gradient method, PARTAN, or any of the quasi-Newton methods can be applied to constrained problems by handling constraints through projection or reduction. The corresponding asymptotic rates of convergence for such methods are easily determined by applying the results for unconstrained problems on the $n - m$ dimensional surface of constraints, as illustrated in this chapter.

Although such generalizations can sometimes lead to substantial improvement in convergence rates, one must recognize that the detailed logic for a complicated generalization can become exhorbitant. Furthermore, some methods are not particularly well-suited to the projection or reduction philosophy. It has, for instance, sometimes been naively suggested that a problem with inequality constraints might be solved by application of a modified Davidon method, which projects the directions that would ordinarily be generated onto the tangent space of the active constraints. Since the set of active constraints will change as the method progresses, successive directions will not be conjugate, and the inherently poor eigenvalue structure of the Davidon method will dominate the convergence characteristics in the early stages of such a scheme. In general, then, it is usually best to consider only simple modifications of projection and reduction algorithms.

Convex Simplex Method

A popular modification of the reduced gradient method, termed the *convex simplex method*, most closely parallels the highly effective simplex method for solving linear programs. The major difference between this method and the reduced gradient method is that instead of moving all (or several) of the independent variables in the direction of the negative reduced gradient, only one independent variable is changed at a time. The selection of the one independent variable to change is made much as in the ordinary simplex method.

At a given feasible point, let $\mathbf{x} = (\mathbf{y}, \mathbf{z})$ be the partition of $\mathbf{x}$ into dependent and independent parts, and assume for simplicity that the bounds on $\mathbf{x}$ are $\mathbf{x} \geqslant \mathbf{0}$. Given the reduced gradient $\mathbf{r}$ at the current point, the component z_i to be changed is found from:

1. Let $r_{i_1} = \min_i \{r_i\}$
2. Let $r_{i_2} z_{i_2} = \max \{r_i z_i\}$

   If $r_{i_1} = r_{i_2} z_{i_2} = 0$, terminate. Otherwise:

   If $r_{i_1} \leqslant -|r_{i_2} z_{i_2}|$, increase z_{i_1}

   If $r_{i_1} \geqslant -|r_{i_2} z_{i_2}|$, decrease z_{i_2}.

The rule in step (2) amounts to selecting the variable that yields the best potential decrease in the cost function. The rule accounts for the non-negativity constraint on the independent variables by weighting the cost coefficients of those variables that are candidates to be decreased by their distance from zero. This feature ensures global convergence of the method.

The remaining details of the method are identical to those of the reduced gradient method. Once a particular component of $\mathbf{z}$ is selected for change, according to the above criterion, the corresponding $\mathbf{y}$ vector is computed as a function of the change in that component so as to continuously satisfy the constraints. The component of $\mathbf{z}$ is continuously changed until either a local minimum with respect to that component is attained or the boundary of one nonnegativity constraint is reached.

Just as in the discussion of the reduced gradient method, it is convenient, for purposes of convergence analysis, to view the problem as unconstrained with respect to the independent variables. The convex simplex method is then seen to be a coordinate descent procedure in the space of these $n - m$ variables. Indeed, since the component selected is based on the magnitude of the components of the reduced gradient, the method is merely an adaptation of the Gauss–Southwell scheme discussed in Section 7.8 to the constrained situation. Hence, although it is difficult to pin down precisely, we expect that it would take approximately $n - m$ steps of this coordinate descent method to make the progress of a single reduced gradient step. To be competitive with the reduced gradient method, therefore, the difficulties associated with a single step—line searching and constraint evaluation—must be approximately $n - m$ times simpler when only a single component is varied than when all $n - m$ are varied simultaneously. This is indeed the case for linear programs and for some quadratic programs but not for nonlinear problems that require the full line search machinery. Hence, in general, the convex simplex method is not a bargain.

11.9 SUMMARY

The concept of feasible direction methods is a straightforward and logical extension of the methods used for unconstrained problems but leads to some subtle difficulties. These methods are highly susceptible to *jamming* (lack of global convergence) because many simple direction finding mappings and the usual line search mapping are not closed. To rectify this, the concept of *uniformly feasible directions* is introduced and simple direction finding algorithms are modified so as to be closed and produce uniformly feasible directions.

Problems having nonlinear constraints introduce numerous additional difficulties arising from the problem of remaining in the feasible region. The most practical primal methods for such problems are the gradient projection methods and the reduced gradient method. Both of these basic methods can be regarded as the method steepest descent applied on the surface defined by the active constraints. The rate of convergence for the two methods can be expected to be approximately equal and is determined by the eigenvalues of the Hessian of the Lagrangian restricted to the subspace tangent to the active constraints. Of the two methods, the reduced gradient method seems to be best. It can be easily modified to ensure against jamming, while that is more difficult for the gradient projection method, and the reduced gradient method requires fewer computations per iterative step and therefore, for most problems, will probably converge in less time than the gradient projection method.

11.10 EXERCISES

1. Show that the problem of finding $\mathbf{d} = (d_1, d_2, \ldots, d_n)$ to

$$\begin{aligned} \text{minimize} \quad & \mathbf{cd} \\ \text{subject to} \quad & \mathbf{Ad} \leqslant \mathbf{0}, \\ & \sum_{i=1}^{n} |d_i| = 1 \end{aligned}$$

can be converted to a linear program.

2. Sometimes a different normalizing term is used in (4). Show that the problem of finding $\mathbf{d} = (d_1, d_2, \ldots, d_n)$ to

$$\begin{aligned} \text{minimize} \quad & \mathbf{cd} \\ \text{subject to} \quad & \mathbf{Ad} \leqslant \mathbf{0}, \\ & \max_i |d_i| = 1 \end{aligned}$$

can be converted to a linear program.

3. Perhaps the most natural normalizing term to use in (4) is one based on the Euclidean norm. This leads to the problem of finding $\mathbf{d} = (d_1, d_2, \ldots, d_n)$ to

$$\begin{aligned} \text{minimize} \quad & \mathbf{cd} \\ \text{subject to} \quad & \mathbf{Ad} \leqslant \mathbf{0}, \\ & \sum_{i=1}^{n} d_i^2 = 1. \end{aligned}$$

Find the Kuhn–Tucker necessary conditions for this problem and show how they can be solved by a modification of the simplex procedure.

4. Let $\Gamma \subset E^{2n}$ be a set of uniformly feasible direction vectors for Ω with feasibility constant δ. For $\varepsilon > 0$ define the map ${}^{\varepsilon}\mathbf{M}_{\delta}$ on Γ by

$$\begin{aligned} {}^{\varepsilon}\mathbf{M}_{\delta}(\mathbf{x}, \mathbf{d}) = \{\mathbf{y} : f(\mathbf{y}) \leqslant f(\mathbf{x} + \tau\mathbf{d}) + \varepsilon \\ \text{for all} \quad \tau, 0 \leqslant \tau \leqslant \delta, \mathbf{y} = \mathbf{x} + \alpha\mathbf{d} \\ \text{for some} \quad \alpha, 0 \leqslant \alpha < \infty, \mathbf{y} \in \Omega\}. \end{aligned}$$

The map ${}^{\varepsilon}\mathbf{M}_{\delta}$ corresponds to an "inaccurate" constrained line search. Show that this map is closed.

5. Show that the direction finding mapping of Example 3, Section 11.3, is closed.

6. For the problem

$$\begin{aligned} \text{minimize} \quad & f(\mathbf{x}) \\ \text{subject to} \quad & \mathbf{a}_1'\mathbf{x} \leqslant b_1 \\ & \mathbf{a}_2'\mathbf{x} \leqslant b_2 \\ & \cdot \\ & \cdot \\ & \cdot \\ & \mathbf{a}_m'\mathbf{x} \leqslant b_m \end{aligned}$$

consider selecting $\mathbf{d} = (d_1, d_2, \ldots, d_n)$ at a feasible point $\mathbf{x}$ by solving the problem

$$\begin{aligned} \text{minimize} \quad & \nabla f(\mathbf{x})\mathbf{d} \\ \text{subject to} \quad & \mathbf{a}_i'\mathbf{d} \leqslant (b_i - \mathbf{a}_i'\mathbf{x})M, \qquad i = 1, 2, \ldots, m \\ & \sum_{i=1}^{n} |d_i| = 1 \end{aligned}$$

where M is some given positive constant. For large M the ith inequality of this subsidiary problem will be active only if the corresponding inequality in the original problem is nearly active at $\mathbf{x}$ (indeed, note that $M \to \infty$ corresponds to Zoutendijk's method). Show that this direction finding mapping is closed and generates uniformly feasible directions with feasibility constant $1/M$.

7. Generalize the method of Exercise 6 so that it is applicable to non-linear inequalities.

8. An alternate, but equivalent, definition of the projected gradient $\mathbf{p}$ is that it is the vector solving

$$\begin{aligned} \text{minimize} \quad & |\mathbf{g} - \mathbf{p}|^2 \\ \text{subject to} \quad & \mathbf{A}_q\mathbf{p} = \mathbf{0}. \end{aligned}$$

Using the Kuhn–Tucker necessary conditions, solve this problem and thereby derive the formula for the projected gradient.

9. Show that finding the $\mathbf{d}$ that solves

$$\begin{aligned} \text{minimize} \quad & \mathbf{g}'\mathbf{d} \\ \text{subject to} \quad & \mathbf{A}_q\mathbf{d} = \mathbf{0}, \qquad |\mathbf{d}|^2 = 1 \end{aligned}$$

gives a vector $\mathbf{d}$ that has the same direction as the negative projected gradient.

10. Let $\mathbf{P}$ be a projection matrix. Show that $\mathbf{P}' = \mathbf{P}, \mathbf{P}^2 = \mathbf{P}$.

11. Suppose $\mathbf{A}_q = [\mathbf{a}', \mathbf{A}_{\bar{q}}]$ so that $\mathbf{A}_q$ is the matrix $\mathbf{A}_{\bar{q}}$ with the row $\mathbf{a}'$ adjoined. Show that $(\mathbf{A}_q\mathbf{A}_q')^{-1}$ can be found from $(\mathbf{A}_{\bar{q}}\mathbf{A}_{\bar{q}}')^{-1}$ from the formula

$$(\mathbf{A}_q\mathbf{A}_q')^{-1} = \begin{bmatrix} \varepsilon & -\varepsilon\mathbf{a}'\mathbf{A}_{\bar{q}}'(\mathbf{A}_{\bar{q}}\mathbf{A}_{\bar{q}}')^{-1} \\ -\varepsilon(\mathbf{A}_{\bar{q}}\mathbf{A}_{\bar{q}}')^{-1}\mathbf{A}_{\bar{q}}\mathbf{a} & (\mathbf{A}_{\bar{q}}\mathbf{A}_{\bar{n}}')^{-1}[\mathbf{I} + \mathbf{A}_{\bar{q}}\mathbf{a}\mathbf{a}'\mathbf{A}_{\bar{q}}'(\mathbf{A}_{\bar{q}}\mathbf{A}_{\bar{q}}')^{-1}] \end{bmatrix}$$

where

$$\varepsilon = \frac{1}{\mathbf{a}'\mathbf{a} - \mathbf{a}'\,\mathbf{A}_{\bar{q}}'\,(\mathbf{A}_{\bar{q}}\,\mathbf{A}_{\bar{q}}')^{-1}\,\mathbf{A}_{\bar{q}}\,\mathbf{a}}.$$

Develop a similar formula for $(\mathbf{A}_{\bar{q}}\,\mathbf{A}_{\bar{q}})^{-1}$ in terms of $(\mathbf{A}_q\,\mathbf{A}_q)^{-1}$.

12. Show that the gradient projection method will solve a linear program in a finite number of steps.

13. Suppose that the projected negative gradient $\mathbf{d}$ is calculated satisfying

$$-\mathbf{g} = \mathbf{d} + \mathbf{A}_q'\,\boldsymbol{\beta}$$

and that some component β_i of $\boldsymbol{\beta}$, corresponding to an inequality, is negative. Show that if the ith inequality is dropped, the projection $\mathbf{d}_i$ of the negative gradient onto the remaining constants is a feasible direction of descent.

14. Using the result of Exercise 13, it is possible to avoid the discontinuity at $\mathbf{d} = \mathbf{0}$ in the direction finding mapping of the simple gradient projection method. At a given point let $\gamma = -\min\{0, \beta_i\}$, with the minimum taken with respect to the indices i corresponding the active inequalities. The direction to be taken at this point is $\mathbf{d} = -\mathbf{Pg}$ if $|\mathbf{P}_g| \geqslant \gamma$, or $\bar{\mathbf{d}}$, defined by dropping the inequality i for which $\beta_i = -\gamma$, if $|\mathbf{P}_g| \leqslant \gamma$. (In case of equality either direction is selected.) Show that this direction finding map is closed over a region where the set of active inequalities does not change.

15. Find the geodesics of
    a) a two-dimensional plane
    b) a sphere.

16. Suppose that the problem

$$\text{minimize } f(\mathbf{x})$$
$$\text{subject to } \mathbf{h}(\mathbf{x}) = \mathbf{0}$$

is such that every point is a regular point. And suppose that the sequence of points $\{\mathbf{x}_k\}_{k=0}^{\infty}$ generated by geodesic descent is bounded. Prove that every limit point of the sequence satisfies the first-order necessary conditions for a constrained minimum.

17. Show that, for linear constraints, if at some point in the reduced gradient method $\Delta\mathbf{z}$ is zero, that point satisfies the Kuhn–Tucker first-order necessary conditions for a constrained minimum.

18. Consider the problem

$$\text{minimize } f(\mathbf{x})$$
$$\text{subject to } \mathbf{Ax} = \mathbf{b},$$
$$\mathbf{x} \geqslant \mathbf{0}$$

where $\mathbf{A}$ is $n \times m$. Assume $f \in C$, that the feasible set is bounded, and that the nondegeneracy assumption holds. Suppose a "modified" reduced gradient algorithm is defined following the procedure in Section 11.6 but with two modifications: (i) the basic variables are, at the beginning of an iteration, always taken as the m largest variables (ties are broken arbitrarily);

(ii) The formula for $\Delta\mathbf{z}$ is replaced by

$$\Delta z_i = \begin{cases} r_i & \text{if} \quad r_i \geqslant 0 \\ x_i r_i & \text{if} \quad r_i < 0. \end{cases}$$

Establish the global convergence of this algorithm.

19. Find the exact solution to the example presented in Section 11.4.
20. Find the direction of movement that would be taken by the gradient projection method if in the example of Section 11.4 the constraint $x_4 = 0$ were relaxed. Show that if the term $-3x_4$ in the objective function were replaced by $-x_4$, then both the gradient projection method and the reduced gradient method would move in identical directions.
21. Show that in terms of convergence characteristics, the reduced gradient method behaves like the gradient projection method applied to a scaled version of the problem.
22. Let r be the condition number of $\mathbf{L}_M$ and s the condition number of $\mathbf{C}'\mathbf{C}$. Show that the rate of convergence of the reduced gradient method is no worse than $[(s-r)/(s+r)]^2$.
23. Referring to Exercise 18 establish a global convergence result for the convex simplex method.

REFERENCES

11.2 Feasible direction methods of various types were originally suggested and developed by Zoutendijk [Z4].

11.3. The systematic study of the global convergence properties of feasible direction methods was begun by Topkis and Veinott [T2] and by Zangwill [Z2]. The algorithm given in Example 3 is due to Topkis and Veinott [T2]. The concept of the map $\mathbf{M}_\delta$ is new.

11.4. The gradient projection method was proposed and developed by Rosen [R1], [R2].

11.5. This material is taken from Luenberger [L11].

11.6–7. The reduced gradient method was originally proposed by Wolfe [W4] for problems with linear constraints and generalized to nonlinear constraints by Abadie and Carpentier [A1]. Wolfe [W3] presents an example of jamming in the reduced gradient method and McCormick [Mc1] presents a simple modification that prevents jamming. The convergence analysis given in this section is new.

11.8. The convex simplex method, for problems with linear constraints, together with a proof of its global convergence is due to Zangwill [Z2].

Chapter 12 PENALTY AND BARRIER METHODS

Penalty and barrier methods are procedures for approximating constrained optimization problems by unconstrained problems. The approximation is accomplished in the case of penalty methods by adding to the objective function a term that prescribes a high cost for violation of the constraints, and in the case of barrier methods by adding a term that favors points interior to the feasible region over those near the boundary. Associated with these methods is a parameter μ that determines the severity of the penalty or barrier and consequently the degree to which the unconstrained problem approximates the original constrained problem. As $\mu \to \infty$ the approximation becomes increasingly accurate.

There are two fundamental issues associated with these methods. The first has to do with how well the unconstrained problem approximates the constrained one. This is essential in examining whether, as the parameter μ is increased toward infinity, the solution of the unconstrained problem converges to a solution of the constrained problem. The other issue, most important from a practical viewpoint, is the question of how to solve a given unconstrained problem when its objective function contains a penalty or barrier term. It turns out that as μ is increased to yield a good approximating problem, the corresponding structure of the resulting unconstrained problem becomes increasingly unfavorable thereby slowing the convergence rate of many algorithms that might be applied. It is necessary, then, to devise acceleration procedures that circumvent this slow convergence phenomenon.

Penalty and barrier methods are of great interest to both the practitioner and the theorist. To the practitioner they offer a simple straightforward method for handling constrained problems that can be implemented without sophisticated computer programming and that possess much the same degree of generality as primal methods. The theorist, striving to make this approach practical by overcoming its inherently slow convergence, finds it appropriate to bring into play nearly all aspects of optimization theory; including Lagrange multipliers, necessary conditions, and many of the algorithms discussed earlier in this book. The canonical rate of convergence associated with the original constrained problem again asserts its fundamental role by essentially determining the natural *accelerated* rate of convergence for unconstrained penalty or barrier problems.

12.1 PENALTY METHODS

Consider the problem

$$\begin{aligned} &\text{minimize} \quad f(\mathbf{x}) \\ &\text{subject to} \quad \mathbf{x} \in S \end{aligned} \tag{1}$$

where f is a continuous function on E^n and S is a constraint set in E^n. In most applications S is defined implicitly by a number of functional constraints, but in this section the more general description in (1) can be handled. The idea of a penalty function method is to replace problem (1) by an unconstrained problem of the form

$$\text{minimize} \quad f(\mathbf{x}) + \mu P(\mathbf{x}) \tag{2}$$

where μ is a positive constant and P is a function on E^n satisfying: (i) P is continuous, (ii) $P(\mathbf{x}) \geqslant 0$ for all $\mathbf{x} \in E^n$, and (iii) $P(\mathbf{x}) = 0$ if and only if $\mathbf{x} \in S$.

Example 1. Suppose S is defined by a number of inequality constraints:

$$S = \{\mathbf{x} : g_i(\mathbf{x}) \leqslant 0, \quad i = 1, 2, \ldots, p\}.$$

A very useful penalty function in this case is

$$P(\mathbf{x}) = \sum_{i=1}^{p} (\max[0, g_i(\mathbf{x})])^2.$$

The function $\mu P(\mathbf{x})$ is illustrated in Fig. 12.1 for the one-dimensional case with $g_1(x) = x - b$, $g_2(x) = a - x$.

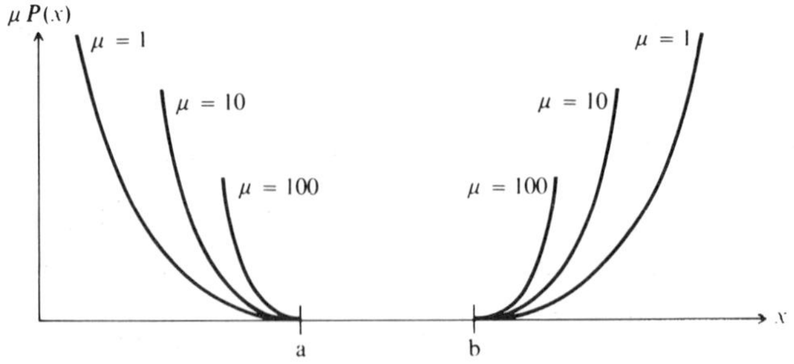

Fig. 12.1 Plot of $\mu P(x)$

For large μ it is clear that the minimum point of problem (2) will be in a region where P is small. Thus, for increasing μ it is expected that the corresponding solution points will approach the feasible region S and, subject to being close, will minimize f. Ideally then, as $\mu \to \infty$ the solution point of the penalty problem will converge to a solution of the constrained problem.

The Method

The procedure for solving problem (1) by the penalty function method is this: Let $\{\mu_k\}$, $k = 1, 2, \ldots$, be a sequence tending to infinity such that for each k, $\mu_k \geqslant 0$, $\mu_{k+1} > \mu_k$. Define the function

$$q(\mu, \mathbf{x}) = f(\mathbf{x}) + \mu P(\mathbf{x}). \tag{3}$$

For each k solve the problem

$$\text{minimize} \quad q(\mu_k, \mathbf{x}) \tag{4}$$

obtaining a solution point $\mathbf{x}_k$.

We assume here that, for each k, problem (4) has a solution. This will be true, for example, if $q(\mu, \mathbf{x})$ is unbounded as $|\mathbf{x}| \to \infty$. (Also see Exercise 2 to see that it is not necessary to obtain the minimum precisely.)

Convergence

The following lemma gives a set of inequalities that follow directly from the definition of $\mathbf{x}_k$ and the inequality $\mu_{k+1} > \mu_k$.

***Lemma* 1**

$$q(\mu_k, \mathbf{x}_k) \leqslant q(\mu_{k+1}, \mathbf{x}_{k+1}) \tag{5}$$

$$P(\mathbf{x}_k) \geqslant P(\mathbf{x}_{k+1}) \tag{6}$$

$$f(\mathbf{x}_k) \leqslant f(\mathbf{x}_{k+1}). \tag{7}$$

Proof.

$$\begin{aligned} q(\mu_{k+1}, \mathbf{x}_{k+1}) &= f(\mathbf{x}_{k+1}) + \mu_{k+1} P(\mathbf{x}_{k+1}) \geqslant f(\mathbf{x}_{k+1}) + \mu_k P(\mathbf{x}_{k+1}) \\ &\geqslant f(\mathbf{x}_k) + \mu_k P(\mathbf{x}_k) = q(\mu_k, \mathbf{x}_k). \end{aligned}$$

which proves (5).

We also have

$$f(\mathbf{x}_k) + \mu_k P(\mathbf{x}_k) \leqslant f(\mathbf{x}_{k+1}) + \mu_k P(\mathbf{x}_{k+1}) \tag{8}$$

$$f(\mathbf{x}_{k+1}) + \mu_{k+1} P(\mathbf{x}_{k+1}) \leqslant f(\mathbf{x}_k) + \mu_{k+1} P(\mathbf{x}_k). \tag{9}$$

Adding (8) and (9) yields

$$(\mu_{k+1} - \mu_k) P(\mathbf{x}_{k+1}) \leqslant (\mu_{k+1} - \mu_k) P(\mathbf{x}_k)$$

which proves (6).

Also

$$f(\mathbf{x}_{k+1}) + \mu_k P(\mathbf{x}_{k+1}) \geqslant f(\mathbf{x}_k) + \mu_k P(\mathbf{x}_k)$$

and hence using (6) we obtain (7). ▮

***Lemma* 2.** *Let* $\mathbf{x}^*$ *be a solution to problem* (1). *Then for each* k

$$f(\mathbf{x}^*) \geqslant q(\mu_k, \mathbf{x}_k) \geqslant f(\mathbf{x}_k).$$

Proof.

$$f(\mathbf{x}^*) = f(\mathbf{x}^*) + \mu_k P(\mathbf{x}^*) \geqslant f(\mathbf{x}_k) + \mu_k P(\mathbf{x}_k) \geqslant f(\mathbf{x}_k). \ ▮$$

Global convergence of the penalty method, or more precisely verification that any limit point of the sequence is a solution, follows easily from the two lemmas above.

Theorem. *Let* $\{\mathbf{x}_k\}$ *be a sequence generated by the penalty method. Then, any limit point of the sequence is a solution to* (1).

Proof. Suppose the subsequence $\{\mathbf{x}_k\}$, $k \in \mathscr{K}$ is a convergent subsequence of $\{\mathbf{x}_k\}$ having limit $\bar{\mathbf{x}}$. Then by the continuity of f, we have

$$\operatorname*{limit}_{k \in \mathscr{K}} f(\mathbf{x}_k) = f(\bar{\mathbf{x}}). \tag{10}$$

Let f^* be the optimal value associated with problem (1). Then according to Lemmas 1 and 2, the sequence of values $q(\mu_k, \mathbf{x}_k)$ is nondecreasing and bounded above by f^*. Thus

$$\operatorname*{limit}_{k \in \mathscr{K}} q(\mu_k, \mathbf{x}_k) = q^* \leqslant f^*. \tag{11}$$

Substracting (10) from (11) yields

$$\operatorname*{limit}_{k \in \mathscr{K}} \mu_k P(\mathbf{x}_k) = q^* - f(\bar{\mathbf{x}}). \tag{12}$$

Since $P(\mathbf{x}_k) \geqslant 0$ and $\mu_k \to \infty$, (12) implies

$$\operatorname*{limit}_{k \in \mathscr{K}} P(\mathbf{x}_k) = 0.$$

Using the continuity of P, this implies $P(\bar{\mathbf{x}}) = 0$. We therefore have shown that the limit point $\bar{\mathbf{x}}$ is feasible for (1).

To show that $\bar{\mathbf{x}}$ is optimal we note that from Lemma 2, $f(\mathbf{x}_k) \leqslant f^*$ and hence

$$f(\bar{\mathbf{x}}) = \operatorname*{limit}_{k \in \mathscr{K}} f(\mathbf{x}_k) \leqslant f^*. \ ▮$$

12.2 BARRIER METHODS

Barrier methods are applicable to problems of the form

$$\begin{array}{ll} \text{minimize} & f(\mathbf{x}) \\ \text{subject to} & \mathbf{x} \in S \end{array} \tag{13}$$

where the constraint set S has a nonempty interior that is arbitrarily close to any point of S. Intuitively, what this means is that the set has an interior and it is possible to get to any boundary point by approaching it from the interior. We shall refer to such a set as *robust*. Some examples of robust and nonrobust sets are shown in Fig. 10.2. This kind of set often arises in conjunction with inequality constraints, where S takes the form

$$S = \{\mathbf{x} : g_i(\mathbf{x}) \leqslant 0,\ i = 1, 2, \ldots, p\}.$$

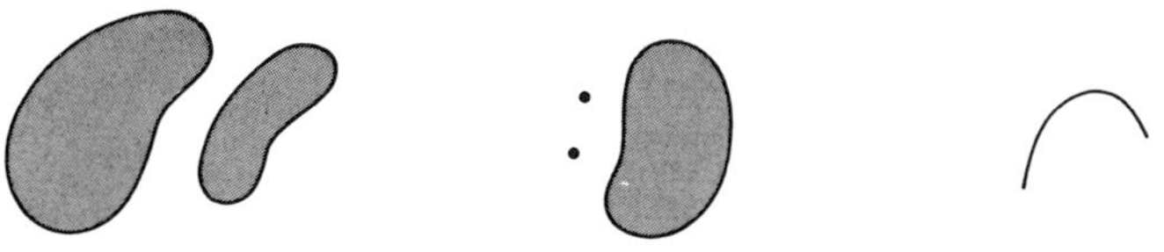

Fig. 12.2 Examples

Barrier methods work by establishing a barrier on the boundary of the feasible region that prevents a search procedure from leaving the region. A *barrier function* is a function B defined on the interior of S such that: (i) B is continuous, (ii) $B(\mathbf{x}) \geqslant 0$, (iii) $B(\mathbf{x}) \to \infty$ as $\mathbf{x}$ approaches the boundary of S.

Example 1. Let g_i, $i = 1, 2, \ldots, p$ be continuous functions on E^n. Suppose

$$S = \{\mathbf{x} : g_i(\mathbf{x}) \leqslant 0, \quad i = 1, 2, \ldots, p\}$$

is robust, and suppose the interior of S is the set of $\mathbf{x}$'s where $g_i(\mathbf{x}) < 0$, $i = 1, 2, \ldots, p$. Then the function

$$B(\mathbf{x}) = -\sum_{i=1}^{p} \frac{1}{g_i(\mathbf{x})}, \tag{14}$$

defined on the interior of S, is a barrier function. It is illustrated in one-dimension for $g_1 = x - a$, $g_2 = x - b$ in Fig. 12.3.

Corresponding to the problem (13), consider the approximate problem

$$\begin{array}{ll} \text{minimize} & f(\mathbf{x}) + \dfrac{1}{\mu} B(\mathbf{x}) \\ \text{subject to} & \mathbf{x} \in \text{interior of } S, \end{array} \tag{15}$$

where μ is a positive constant. This is a constrained problem, and indeed the constraint is somewhat more complicated than in the original problem (13). The advantage of this problem, however, is that it can be solved by using an unconstrained search technique. To find the solution one starts at an initial interior point and then searches from that point using steepest descent or some other iterative descent method applicable to unconstrained problems. Since the value of the objective function approaches infinity near the boundary of S, the search technique will automatically remain within the interior of S, and the constraint need not be accounted for explicitly. Thus, although problem (15) is from a formal viewpoint a constrained problem, from a computational viewpoint it is unconstrained.

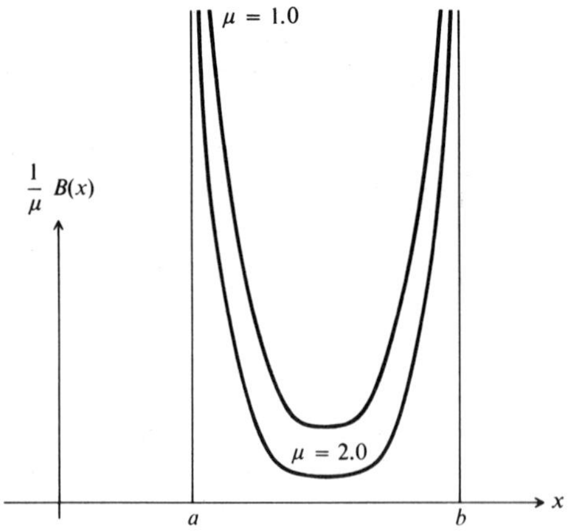

Fig. 12.3 Barrier function

The Method

The barrier method is quite analogous to the penalty method. Let $\{\mu_k\}$ be a sequence tending to infinity such that for each k, $k = 1, 2, 3 \ldots$, $\mu_k \geqslant 0$, $\mu_{k+1} > \mu_k$. Define the function

$$r(\mu, \mathbf{x}) = f(\mathbf{x}) + \frac{1}{\mu} B(\mathbf{x}).$$

For each k solve the problem

$$\begin{array}{ll} \text{minimize} & r(\mu_k, \mathbf{x}) \\ \text{subject to} & \mathbf{x} \in \text{interior of } S \end{array}$$

obtaining the point $\mathbf{x}_k$.

Convergence

Virtually the same convergence properties hold for the barrier method as for the penalty method. We leave to the reader the proof of the following result.

> ***Theorem.*** *Any limit point of a sequence* $\{\mathbf{x}_k\}$ *generated by the barrier method is a solution to problem* (13).

12.3 PROPERTIES OF PENALTY AND BARRIER FUNCTIONS

Penalty and barrier methods are applicable to nonlinear programming problems having a very general form of constraint set S. In most situations, however, this set is not given explicitly but is defined implicitly by a number of functional constraints. In these situations, the penalty or barrier function is invariably defined in terms of the constraint functions themselves, and although there are an unlimited number of ways in which this can be done, some important general implications follow from this kind of construction.

For economy of notation we consider problems of the form

$$\begin{aligned} &\text{minimize} \quad f(\mathbf{x}) \\ &\text{subject to} \quad g_i(\mathbf{x}) \leqslant 0, \qquad i = 1, 2, \ldots, p. \end{aligned} \tag{16}$$

For our present purposes, equality constraints are suppressed, at least notationally, by writing each of them as two inequalities. If the problem is to be attacked with a barrier method, then, of course, equality constraints are not present even in an unsuppressed version.

Penalty Functions

A penalty function for a problem expressed in the form (16) will most naturally be expressed in terms of the auxiliary constraint functions

$$g_i^{+}(\mathbf{x}) \triangleq \max\,[0, g_i(\mathbf{x})], \qquad i = 1, 2, \ldots, p. \tag{17}$$

This is because in the interior of the constraint region $P(\mathbf{x}) \triangleq 0$ and hence P should be a function only of violated constraints. Denoting by $\mathbf{g}^{+}(\mathbf{x})$ the p-dimensional vector made up of the $g_i^{+}(\mathbf{x})$'s, we consider the general class of penalty functions

$$P(\mathbf{x}) = \gamma(\mathbf{g}^{+}(\mathbf{x})) \tag{18}$$

where γ is a continuous function from E^p to the real numbers, defined in such a way that P satisfies the requirements demanded of a penalty function.

Example 1. Set

$$P(\mathbf{x}) = \sum_{i=1}^{p} g_i^{+}(\mathbf{x})^2 = |\mathbf{g}^{+}(\mathbf{x})|^2$$

which is without doubt the most popular penalty function. In this case γ is the identity quadratic form on E^p, that is, $\gamma(\mathbf{y}) = |\mathbf{y}|^2$.

Example 2. By letting

$$\gamma(\mathbf{y}) = \mathbf{y}'\boldsymbol{\Gamma}\mathbf{y}$$

where Γ is a symmetric positive definite $p \times p$ matrix, we obtain the penalty function

$$P(\mathbf{x}) = \mathbf{g}^+(\mathbf{x})'\boldsymbol{\Gamma}\mathbf{g}^+(\mathbf{x}).$$

Example 3. A general class of penalty functions is

$$P(\mathbf{x}) = \sum_{i=1}^{p} (g_i^+(\mathbf{x}))^\varepsilon$$

for some $\varepsilon > 0$.

Lagrange Multipliers

In the penalty method we solve, for various μ_k, the unconstrained problem

$$\text{minimize} \quad f(\mathbf{x}) + \mu_k P(\mathbf{x}). \tag{19}$$

Most algorithms require that the objective function has continuous first partial derivatives. Since we shall, as usual, assume that both f and $\mathbf{g} \in C^1$, it is natural to require, then, that the penalty function $P \in C^1$. We define

$$\nabla g_i^+(\mathbf{x}) = \begin{cases} \nabla g_i(\mathbf{x}) & \text{if} \quad g_i(\mathbf{x}) \geqslant 0 \\ \mathbf{0} & \text{if} \quad g_i(\mathbf{x}) < 0 \end{cases} \tag{20}$$

and, of course, $\nabla \mathbf{g}^+(\mathbf{x})$ is the $m \times n$ matrix whose rows are the ∇g_i^+'s. Unfortunately, $\nabla \mathbf{g}^+$ is usually discontinuous at points where $g_i^+(\mathbf{x}) = 0$ for some $i = 1, 2, \ldots, p$, and thus some restrictions must be placed on γ in order to guarantee $P \in C^1$. We assume that $\gamma \in C^1$ and that if $\mathbf{y} = (y_1, y_2, \ldots, y_n)$, $\nabla\gamma(\mathbf{y}) = (\nabla\gamma_1, \nabla\gamma_2, \ldots, \nabla\gamma_n)$, then

$$y_i = 0 \quad \text{implies} \quad \nabla\gamma_i = 0. \tag{21}$$

(In Example 3 above, for instance, this condition is satisfied only for $\varepsilon > 1$.) With this assumption, the derivative of $\gamma(\mathbf{g}^+(\mathbf{x}))$ with respect to $\mathbf{x}$ is continuous and can be written as $\nabla\gamma(\mathbf{g}^+(\mathbf{x}))\nabla\mathbf{g}(\mathbf{x})$. In this result $\nabla\mathbf{g}(\mathbf{x})$ legitimately replaces the discontinuous $\nabla\mathbf{g}^+(\mathbf{x})$, because it is premultiplied by $\nabla\gamma(\mathbf{g}^+(\mathbf{x}))$. Of course, these considerations are necessary only for inequality constraints. If equality constraints are treated directly, the situation is far simpler.

In view of this assumption, problem (19) will have its solution at a point $\mathbf{x}_k$ satisfying

$$\nabla f(\mathbf{x}_k) + \mu_k \nabla\gamma(\mathbf{g}^+(\mathbf{x}_k))\nabla\mathbf{g}(\mathbf{x}_k) = \mathbf{0}$$

which can be written as

$$\nabla f(\mathbf{x}_k) + \boldsymbol{\lambda}_k \nabla\mathbf{g}(\mathbf{x}_k) = \mathbf{0}, \tag{22}$$

where

$$\boldsymbol{\lambda}_k \stackrel{\Delta}{=} \mu_k \nabla\gamma(\mathbf{g}^+(\mathbf{x}_k)). \tag{23}$$

Thus, associated with every μ is a Lagrange multiplier vector that is determined after the unconstrained minimization is performed.

If a solution $\mathbf{x}^*$ to the original problem (16) is a regular point of the constraints, then there is a unique Lagrange multiplier vector $\boldsymbol{\lambda}^*$ associated with the solution. The result stated below says that $\boldsymbol{\lambda}_k \to \boldsymbol{\lambda}^*$.

Proposition. *Suppose that the penalty function method is applied to problem* (16) *using a penalty function of the form* (18) *with* $\gamma \in C^1$ *and satisfying* (21). *Corresponding to the sequence* $\{\mathbf{x}_k\}$ *generated by this method, define* $\boldsymbol{\lambda}_k = \mu_k \nabla\gamma(\mathbf{g}^+(\mathbf{x}_k))$. *If* $\mathbf{x}_k \to \mathbf{x}^*$, *a solution to* (16), *and this solution is a regular point, then* $\boldsymbol{\lambda}_k \to \boldsymbol{\lambda}^*$, *the Lagrange multiplier associated with problem* (16).

Proof. Left to the reader.

Example 4. For $P(\mathbf{x}) = |\mathbf{g}^+(\mathbf{x})|^2$ we have $\boldsymbol{\lambda}_k = 2\mu_k \mathbf{g}^+(\mathbf{x}_k)'$.

As a final observation we note that in general if $\mathbf{x}_k \to \mathbf{x}^*$, then since $\boldsymbol{\lambda}_k = \mu_k \nabla\gamma(\mathbf{g}^+(\mathbf{x}_k)) \to \boldsymbol{\lambda}^*$, the sequence $\mathbf{x}_k$ approaches $\mathbf{x}^*$ from outside the constraint region. Indeed, as $\mathbf{x}_k$ approaches $\mathbf{x}^*$ all constraints that are active at $\mathbf{x}^*$ and have positive Lagrange multipliers will be violated at $\mathbf{x}_k$ because the corresponding components of $\nabla\gamma(\mathbf{g}^+(\mathbf{x}_k))$ are positive. Thus, if we assume that the active constraints are nondegenerate (all Lagrange multipliers are strictly positive), every active constraint will be approached from the outside.

The Hessian Matrix

Since the penalty function method must, for various (large) values of μ, solve the unconstrained problem

$$\text{minimize} \quad f(\mathbf{x}) + \mu P(\mathbf{x}) \tag{24}$$

it is important, in order to evaluate the difficulty of such a problem, to determine the eigenvalue structure of the Hessian of this modified objective function. We show here that the structure becomes increasingly unfavorable as μ increases.

Although one usually insists for computational as well as theoretical purposes that the function $P \in C^1$, one usually does not insist that $P \in C^2$. In particular, the most popular penalty function $P(\mathbf{x}) = |\mathbf{g}^+(\mathbf{x})|^2$, illustrated in Fig. 12.1 for one component, has a discontinuity in its second derivative at any point where a component of $\mathbf{g}$ is zero. At first this might appear to be a serious drawback, since it means the Hessian is discontinuous at the boundary of the constraint region—right where, in general, the solution is expected to lie. However, as pointed out above, the penalty method generates points that approach a boundary solution from outside the constraint region. Thus, except for some possible chance occurrences, the sequence will, as $\mathbf{x}_k \to \mathbf{x}^*$, be at points where the Hessian is well-defined. Furthermore, in iteratively solving the unconstrained problem (24) with a fixed μ_k, a sequence will be generated that converges to $\mathbf{x}_k$ which is (for most values of k) a point where the Hessian is well-defined, and hence the standard type of analysis will be applicable to the tail of such a sequence.

Defining $q(\mu, \mathbf{x}) = f(\mathbf{x}) + \mu\gamma(\mathbf{g}^+(\mathbf{x}))$ we have for the Hessian, $\mathbf{Q}$, of q (with respect to $\mathbf{x}$)

$$\mathbf{Q}(\mu, \mathbf{x}) = \mathbf{F}(\mathbf{x}) + \mu\nabla\gamma(\mathbf{g}^+(\mathbf{x}))\mathbf{G}(\mathbf{x}) + \mu\nabla\mathbf{g}^+(\mathbf{x})'\mathbf{\Gamma}(\mathbf{g}^+(\mathbf{x}))\nabla\mathbf{g}^+(\mathbf{x})$$

where $\mathbf{F}$, $\mathbf{G}$, and $\mathbf{\Gamma}$ are, respectively, the Hessians of f, $\mathbf{g}$, and γ. For a mixed μ_k we use the definition of $\boldsymbol{\lambda}_k$ given by (23) and introduce the rather natural definition

$$\mathbf{L}_k(\mathbf{x}_k) = \mathbf{F}(\mathbf{x}_k) + \boldsymbol{\lambda}_k\mathbf{G}(\mathbf{x}_k) \tag{25}$$

which is the Hessian of the corresponding Lagrangian. Then we have

$$\mathbf{Q}(\mu_k, \mathbf{x}_k) = \mathbf{L}_k(\mathbf{x}_k) + \mu_k\nabla\mathbf{g}^+(\mathbf{x}_k)'\mathbf{\Gamma}(\mathbf{g}^+(\mathbf{x}_k))\nabla\mathbf{g}^+(\mathbf{x}_k) \tag{26}$$

which is the desired expression.

The first term on the right side of (26) converges to the Hessian of the Lagrangian of the original constrained problem as $\mathbf{x}_k \to \mathbf{x}^*$, and hence has a limit that is independent of μ_k. The second term is a matrix having rank equal to the rank of the active constraints and having a magnitude tending to infinity. (See Exercise 7.)

Example 5. For $P(\mathbf{x}) = |\mathbf{g}^+(\mathbf{x})|^2$ we have

$$\mathbf{\Gamma}(\mathbf{g}^+(\mathbf{x}_k)) = \begin{bmatrix} e_1 & 0 & \cdots & 0 \\ 0 & e_2 & & 0 \\ 0 & & \ddots & \vdots \\ \vdots & & \ddots & \vdots \\ \vdots & & & \vdots \\ 0 & \cdots & 0 & e_p \end{bmatrix}$$

where
$$e_i = \begin{cases} 1 & \text{if } g_i(\mathbf{x}_k) > 0 \\ 0 & \text{if } g_i(\mathbf{x}_k) < 0 \\ \text{undefined} & \text{if } g_i(\mathbf{x}_k) = 0. \end{cases}$$

Thus
$$\mu_k \nabla \mathbf{g}^+(\mathbf{x}_k)' \mathbf{\Gamma}(\mathbf{g}^+(\mathbf{x}_k)) \nabla \mathbf{g}^+(\mathbf{x}_k) = \mu_k \nabla \mathbf{g}^+(\mathbf{x}_k)' \nabla \mathbf{g}^+(\mathbf{x}_k)$$

which is μ_k times a matrix that approaches $\nabla \mathbf{g}^+(\mathbf{x}^*)' \nabla \mathbf{g}^+(\mathbf{x}^*)$. This matrix has rank equal to the rank of the active constraints at $\mathbf{x}^*$ (refer to (20)).

Assuming that there are r active constraints at the solution $\mathbf{x}^*$, then for well-behaved γ, the Hessian matrix $\mathbf{Q}(\mu_k, \mathbf{x}_k)$ has r eigenvalues that tend to infinity as $\mu_k \to \infty$, arising from the second term on the right side of (26). There will be $n - r$ other eigenvalues that, although varying with μ_k, tend to finite limits. These limits turn out to be, as is perhaps not too surprising at this point, the eigenvalues of $\mathbf{L}(\mathbf{x}^*)$ restricted to the tangent subspace M of the active constraints. The proof of this requires some further analysis.

***Lemma* 1.** *Let $\mathbf{A}(\mu)$ be a symmetric matrix written in partitioned form*

$$\mathbf{A}(\mu) = \begin{bmatrix} \mathbf{A}_1(\mu) & \mathbf{A}_2(\mu) \\ \mathbf{A}_2'(\mu) & \mathbf{A}_3(\mu) \end{bmatrix} \tag{27}$$

where $\mathbf{A}_1(\mu)$ tends to a positive definite matrix $\mathbf{A}_1$, $\mathbf{A}_2(\mu)$ tends to a finite matrix, and $\mathbf{A}_3(\mu)$ is a positive definite matrix tending to infinity with μ (that is, for any $s > 0$, $\mathbf{A}_3(\mu) - s\mathbf{I}$ is positive definite for sufficiently large μ). Then

$$\mathbf{A}^{-1}(\mu) \to \begin{bmatrix} \mathbf{A}_1^{-1} & \mathbf{0} \\ \mathbf{0} & \mathbf{0} \end{bmatrix} \tag{28}$$

as $\mu \to \infty$.

Proof. We have the identity

$$\begin{bmatrix} \mathbf{A}_1 & \mathbf{A}_2 \\ \mathbf{A}_2' & \mathbf{A}_3 \end{bmatrix}^{-1}$$
$$= \begin{bmatrix} (\mathbf{A}_1 - \mathbf{A}_2\mathbf{A}_3^{-1}\mathbf{A}_2')^{-1} & -(\mathbf{A}_1 - \mathbf{A}_2\mathbf{A}_3^{-1}\mathbf{A}_2')\mathbf{A}_2\mathbf{A}_3^{-1} \\ -\mathbf{A}_3^{-1}\mathbf{A}_2'(\mathbf{A}_1 - \mathbf{A}_2\mathbf{A}_3^{-1}\mathbf{A}_2')^{-1} & (\mathbf{A}_3 - \mathbf{A}_2'\mathbf{A}_1^{-1}\mathbf{A}_2)^{-1} \end{bmatrix}. \tag{29}$$

Using the fact that $\mathbf{A}_3^{-1}(\mu) \to \mathbf{0}$ gives the result. ∎

To apply this result to the Hessian matrix (26) we associate $\mathbf{A}$ with $\mathbf{Q}(\mu_k, \mathbf{x}_k)$ and let the partition of $\mathbf{A}$ correspond to the partition of the space E^n into the subspace M and the subspace N that is orthogonal to M; that is, N is the subspace spanned by the gradients of the active constraints. In this partition, $\mathbf{L}_M$, the restriction of $\mathbf{L}$ to M, corresponds to the matrix $\mathbf{A}_1$. We leave the details of the required continuity arguments to the reader. The important conclusion is that if $\mathbf{x}^*$ is a solution to (16), is a regular point, and has exactly r active constraints none of which are degenerate, then the Hessian matrices $\mathbf{Q}(\mu_k, \mathbf{x}_k)$ of a penalty function of form (18) have r eigenvalues tending to infinity as $\mu_k \to \infty$, and $n - r$ eigenvalues tending to the eigenvalues of $\mathbf{L}_M$.

This explicit characterization of the structure of penalty function Hessians is of great importance in the remainder of the chapter. The fundamental point is that virtually any choice of penalty function leads both to an ill-conditioned Hessian and to consideration of the ubiquitous Hessian of the Lagrangian restricted to M.

Barrier Functions

Essentially the same story holds for barrier function. If we consider for Problem (16) barrier functions of the form

$$B(\mathbf{x}) = \eta(\mathbf{g}(\mathbf{x})), \tag{30}$$

then Lagrange multipliers and ill-conditioned Hessians are again inevitable. Rather than parallel the earlier analysis of penalty functions, we illustrate the conclusions with an example.

Define

$$B(\mathbf{x}) = \sum_{i=1}^{p} -\frac{1}{g_i(\mathbf{x})}. \tag{31}$$

The barrier objective

$$r(\mu_k, \mathbf{x}) = f(\mathbf{x}) - \frac{1}{\mu_k} \sum_{i=1}^{p} \frac{1}{g_i(\mathbf{x})}$$

has its minimum at a point $\mathbf{x}_k$ satisfying

$$\nabla f(\mathbf{x}_k) + \frac{1}{\mu_k} \sum_{i=1}^{p} \frac{1}{g_i(\mathbf{x}_k)^2} \nabla g_i(\mathbf{x}_k) = \mathbf{0}. \tag{32}$$

Thus, we define $\boldsymbol{\lambda}_k$ to be the vector having ith component $\dfrac{1}{\mu_k} \cdot \dfrac{1}{g_i(\mathbf{x}_k)^2}$. Then (32) can be written as

$$\nabla f(\mathbf{x}_k) + \boldsymbol{\lambda}_k \nabla \mathbf{g}(\mathbf{x}_k) = \mathbf{0}. \tag{33}$$

Again, assuming $\mathbf{x}_k \to \mathbf{x}^*$, the solution of (16), we can show that $\boldsymbol{\lambda}_k \to \boldsymbol{\lambda}^*$, the Lagrange multiplier vector associated with the solution. This implies that if g_i is an active constraint,

$$\frac{1}{\mu_k g_i(\mathbf{x}_k)^2} \to \lambda_i^* < \infty. \tag{34}$$

Next, evaluating the Hessian $\mathbf{R}(\mu_k, \mathbf{x}_k)$ of $r(\mu_k, \mathbf{x}_k)$, we have

$$\mathbf{R}(\mu_k,\mathbf{x}_k) = \mathbf{F}(\mathbf{x}_k) + \frac{1}{\mu_k}\sum_{i=1}^{p}\frac{1}{g_i(\mathbf{x}_k)^2}\mathbf{G}_i(\mathbf{x}_k)$$

$$- \frac{1}{\mu_k}\sum_{i=1}^{p}\frac{2}{g_i(\mathbf{x}_k)^3}\nabla g_i(\mathbf{x}_k)'\nabla g_i(\mathbf{x}_k)$$

$$= \mathbf{L}(\mathbf{x}_k) - \frac{1}{\mu_k}\sum_{i=1}^{p}\frac{2}{g_i(\mathbf{x}_k)^3}\nabla g_i(\mathbf{x}_k)'\nabla g_i(\mathbf{x}_k).$$

As $\mu_k \to \infty$ we have

$$\frac{-1}{\mu_k g_i(\mathbf{x}_k)^3} \to \begin{cases} \infty & \text{if } g_i \text{ is active at } \mathbf{x}^* \\ 0 & \text{if } g_i \text{ is inactive at } \mathbf{x}^* \end{cases}$$

so that we may write, from (34),

$$\mathbf{R}(\mu_k, \mathbf{x}_k) \to \mathbf{L}(\mathbf{x}^*) + \sum_{i\in I} -\frac{\lambda_i^*}{g_i(\mathbf{x}^*)}\nabla g_i(\mathbf{x}^*)'\nabla g_i(\mathbf{x}^*) \tag{35}$$

where I is the set of indices corresponding to active constraints. Thus the Hessian of the barrier objective function has exactly the same structure as that of penalty objective functions.

*12.4 EXTRAPOLATION

Penalty and barrier methods are quite different than other algorithms discussed in this book in that they are not iterative in nature. The definition of $\mathbf{x}_{k+1}$ in no way depends on that of $\mathbf{x}_k$. From this point of view, if one decides to terminate the sequence at the Nth term corresponding to μ_N, obtaining $\mathbf{x}_N$, the calculation of the previous vectors $\mathbf{x}_1, \mathbf{x}_2, \ldots, \mathbf{x}_{N-1}$ is irrelevant, since $\mathbf{x}_N$ could have been calculated directly by solving a single unconstrained minimization problem. Indeed, this is generally the manner that penalty and barrier functions are employed; one selects a large value of μ, solves the corresponding unconstrained problem, and takes the resulting solution as the final approximate answer to the original problem.

It is sometimes recognized, however, that selecting a single large value of μ can lead to difficulty. First, exactly what is a large value relative to a given problem may not be known in advance and consequently an initial trial may produce a solution point that is not close enough to the feasible region—in which case μ must be increased. Second, large values of μ yield, as shown in the previous section, ill-conditioned Hessians which in turn imply slow convergence for many algorithms.

A partial remedy to these difficulties is obtained by noting that the search for $\mathbf{x}_{k+1}$ can be initiated from $\mathbf{x}_k$, a starting point that may be fairly close to

$\mathbf{x}_{k+1}$. Solution of the $k+1$th problem will then probably require less time than if the search were initiated from an arbitrary point $\mathbf{x}_0$. For this reason the penalty and barrier methods are often regarded as truly iterative algorithms. It has never been determined, however, that solving a sequence of unconstrained problems for increasing values of μ leads to a computational saving over just solving the problem corresponding to the largest value of μ directly. Indeed, indications are that it does not!

An extension of this idea, that does in some cases lead to significant improvement, is to use several past values of μ and their corresponding solution points to predict the solution point of the next problem. The search for the next solution point can then be initiated from this predicted point.

The predictions are based on polynomial extrapolation applied to the variable $s = 1/\mu$. Given the points $\mathbf{x}_1, \mathbf{x}_2, \ldots, \mathbf{x}_q$ corresponding to $s_1 > s_2 > \ldots > s_q$ we determine the polynomial

$$p_i(s) = a_0{}^i + a_1{}^i s + \cdots + a_{q-1}^i s^{q-1}, \qquad i = 1, 2, \ldots, n \tag{36}$$

that fits the values of the ith component of the $\mathbf{x}_k$'s to the corresponding s_k's, that is,

$$(\mathbf{x}_k)_i = p_i(s_k) \qquad \text{for} \qquad k = 1, 2, \ldots, q.$$

Collecting these polynomials we may write

$$\mathbf{x}_k = \sum_{j=0}^{q-1} \mathbf{a}_j s_k{}^j \tag{37}$$

where $\mathbf{a}_j$ is the vector of coefficients $a_j{}^1, a_j{}^2, \ldots, a_j{}^n$. Given any new value of $s > s_q$, (37) can serve to estimate the corresponding solution point by substituting s into the right-hand side. In particular, at any stage q both $s = s_{q+1}$ and $s = 0$ can be substituted to obtain estimates of both $\mathbf{x}_{q+1}$ and the solution point $\mathbf{x}^*$.

Although such extrapolation schemes seem to be of value in many circumstances, their success has not been overwhelming. This may be due to the fact that the first q vectors $\mathbf{x}_1, \mathbf{x}_2, \ldots, \mathbf{x}_q$ (and hence the vectors $\mathbf{a}_0, \mathbf{a}_1, \ldots, \mathbf{a}_{q-1}$) span only a q-dimensional space, whereas n dimensions are required to express a general point. This general idea, however, is certainly worthy of further investigation.

12.5 NEWTON'S METHOD AND PENALTY FUNCTIONS

In the remainder of this chapter we address the problem of efficiently solving the unconstrained problems associated with a penalty or barrier method. The main difficulty is the extremely unfavorable eigenvalue structure that, as explained in Section 12.3, always accompanies unconstrained problems derived in this way. Certainly straightforward application of the method of steepest descent is out of the question!

One method for avoiding slow convergence for these problems is to apply Newton's method (or one of its variations), since the order two convergence of Newton's method is unaffected by the poor eigenvalue structure. In applying the method, however, special care must be devoted to the manner by which the Hessian is inverted, since it is ill-conditioned. Nevertheless, if second-order information is easily available, Newton's method offers an extremely attractive and effective method for solving modest size penalty or barrier optimization problems. When such information is not readily available, or if data handling and storage requirements of Newton's method are excessive, attention naturally focuses on first order methods.

A simple modified Newton's method can often be quite effective for some penalty problems. For example, consider the problem having only equality constraints

$$\begin{aligned} &\text{minimize} \quad f(\mathbf{x}) \\ &\text{subject to} \quad \mathbf{h}(\mathbf{x}) = \mathbf{0} \end{aligned} \tag{38}$$

with $\mathbf{x} \in E^n$, $\mathbf{h}(\mathbf{x}) \in E^m$, $m < n$. Applying the standard quadratic penalty method we solve instead the unconstrained problem

$$\text{minimize} \quad f(\mathbf{x}) + \mu|\mathbf{h}(\mathbf{x})|^2 \tag{39}$$

for some large μ. Calling the penalty objective function $q(\mathbf{x})$ we consider the iterative process

$$\mathbf{x}_{k+1} = \mathbf{x}_k - \alpha_k(\mathbf{I} + \mu\nabla\mathbf{h}(\mathbf{x}_k)'\nabla\mathbf{h}(\mathbf{x}_k))^{-1}\nabla q(\mathbf{x}_k)' \tag{40}$$

where α_k is chosen to minimize $q(\mathbf{x}_{k+1})$. The matrix $\mathbf{I} + \mu\nabla\mathbf{h}(\mathbf{x}_k)'\nabla\mathbf{h}(\mathbf{x}_k)$ is positive definite and although quite ill-conditioned it can be inverted efficiently (see Exercise (11)).

According to the Modified Newton Method Theorem (Section 9.1) the rate of convergence of this method is determined by the eigenvalues of the matrix

$$[\mathbf{I} + \mu\nabla\mathbf{h}(\mathbf{x}_k)'\nabla\mathbf{h}(\mathbf{x}_k)]^{-1}\mathbf{Q}(\mathbf{x}_k) \tag{41}$$

where $\mathbf{Q}(\mathbf{x}_k)$ is the Hessian of q at $\mathbf{x}_k$. In view of (26), as $\mu \to \infty$ the matrix (41) will have m eigenvalues that approach unity, while the remaining $n - m$ eigenvalues approach the eigenvalues of $\mathbf{L}_M$ evaluated at the solution $\mathbf{x}^*$ of (38). Thus if the smallest and largest eigenvalues of $\mathbf{L}_M$, a and A, are located such that the interval $[a, A]$ contains unity, the convergence ratio of this modified Newton's method will be equal (in the limit of $\mu \to \infty$) to the canonical ratio $[(A - a)/(A + a)]^2$ for problem (38).

If the eigenvalues of $\mathbf{L}_M$ are not spread below and above unity, the convergence rate will be slowed. If a point in the interval containing the eigenvalues of $\mathbf{L}_M$ is known, a scalar factor can be introduced so that the canonical rate is achieved, but such information is often not easily available.

Inequalities

If there are inequality as well as equality constraints in the problem, the analogous procedure can be applied to the associated penalty objective function. The unusual feature of this case is that corresponding to an inequality constraint $g_i(\mathbf{x}) \leqslant 0$, the term $\nabla g_i^{+}(\mathbf{x})'\nabla g_i^{+}(\mathbf{x})$ used in the iteration matrix will suddenly appear if the constraint is violated. Thus, the iteration matrix is discontinuous with respect to $\mathbf{x}$, and as the method progresses its nature changes according to which constraints are violated. This discontinuity does not, however, imply that the method is subject to jamming, since the result of Exercise 4, Chapter 9 is applicable to this method.

12.6 CONJUGATE GRADIENTS AND PENALTY METHODS

The partial conjugate gradient method proposed and analyzed in Section 8.5 is ideally suited to penalty or barrier problems having only a few active constraints. If there are m active constraints, then taking cycles of $m + 1$ conjugate gradient steps will yield a rate of convergence that is independent of the penalty constant μ. For example, consider the problem having only equality constraints:

$$\begin{aligned} &\text{minimize} \quad f(\mathbf{x}) \\ &\text{subject to} \quad \mathbf{h}(\mathbf{x}) = \mathbf{0} \end{aligned} \tag{42}$$

where $\mathbf{x} \in E^n$, $\mathbf{h}(\mathbf{x}) \in E^m$, $m < n$. Applying the standard quadratic penalty method, we solve instead the unconstrained problem

$$\text{minimize} \quad f(\mathbf{x}) + \mu|\mathbf{h}(\mathbf{x})|^2 \tag{43}$$

for some large μ. The objective function of this problem has a Hessian matrix that has m eigenvalues that are of the order μ in magnitude, while the remaining $n - m$ eigenvalues are close to the eigenvalues of the matrix $\mathbf{L}_M$, corresponding to problem (42). Thus, letting $\mathbf{x}_{k+1}$ be determined from $\mathbf{x}_k$ by taking $m + 1$ steps of a (nonquadratic) conjugate gradient method, and assuming $\mathbf{x}_k \to \overline{\mathbf{x}}$, a solution to (43), the sequence $\{f(\mathbf{x}_k)\}$ converges linearly to $f(\overline{\mathbf{x}})$ with a convergence ratio equal to approximately

$$\left(\frac{A - a}{A + a}\right)^2 \tag{44}$$

where a and A are, respectively, the smallest and largest eigenvalues of $\mathbf{L}_M(\overline{\mathbf{x}})$.

This is an extremely effective technique when m is relatively small. The programming logic required is only slightly greater than that of steepest descent, and the time per iteration is only about $m + 1$ times as great as for steepest descent. The method can be used for problems having inequality constraints as well but it is advisable to change the cycle length, depending on the number of constraints active at the end of the previous cycle.

Example

$$\text{Minimize} \quad f(x_1, x_2, \ldots, x_{10}) = \sum_{k=1}^{10} k x_k^2$$

$$\begin{aligned}
\text{subject to} \quad 1.5x_1 + x_2 + x_3 + 0.5x_4 + 0.5x_5 &= 5.5 \\
2.0x_6 - 0.5x_7 - 0.5x_8 + x_9 - x_{10} &= 2.0 \\
x_1 + x_3 + x_5 + x_7 + x_9 &= 10 \\
x_2 + x_4 + x_6 + x_8 + x_{10} &= 15.
\end{aligned}$$

This problem was treated by the penalty function approach, and the resulting composite function was then solved for various values of μ by using various cycle lengths of a conjugate gradient algorithm. In the table below p is the number of conjugate gradient steps in a cycle. Thus $p = 1$ corresponds to ordinary steepest descent; $p = 5$ corresponds, by the theory of Section 8.5, to the smallest value of p for which the rate of convergence is independent of μ; and $p = 10$ is the standard conjugate gradient method. Note that for $p < 5$ the convergence rate does indeed depend on μ, while it is more or less

Table 1

	p (steps per cycle)	Number of cycles to convergence	No. of steps	Value of modified objective
$\mu = 10$	1	90	90	388.565
	3	8	24	388.563
	5	3	15	388.563
	7	3	21	388.563
$\mu = 100$	1	230*	230	488.607
	3	21	63	487.446
	5	4	20	487.438
	7	2	14	487.433
$\mu = 1000$	1	260*	260	525.238
	3	45*	135	503.550
	5	3	15	500.910
	7	3	21	500.882

* Program not run to convergence due to excessive time.

constant for $p \geqslant 5$. The value of μ's selected are not artificially large, since for $\mu = 100$ the constraints are satisfied only to within 0.5 percent of their right-hand sides.

12.7 NORMALIZATION OF PENALTY FUNCTIONS

There is a good deal of freedom in the selection of penalty or barrier functions that can be exploited to accelerate convergence. We propose here a simple normalization procedure that together with a two-step cycle of conjugate gradients yields the canonical rate of convergence. Again for simplicity we illustrate the technique for the penalty method applied to the problem

$$\begin{aligned} \text{minimize} \quad & f(\mathbf{x}) \\ \text{subject to} \quad & \mathbf{h}(\mathbf{x}) = \mathbf{0} \end{aligned} \tag{45}$$

as in Sections 12.5 and 12.6, but the idea is easily extended to other penalty or barrier situations.

Corresponding to (45) we consider the family of quadratic penalty functions

$$P(\mathbf{x}) = \mathbf{h}(\mathbf{x})'\mathbf{\Gamma}\mathbf{h}(\mathbf{x}) \tag{46}$$

where $\mathbf{\Gamma}$ is a symmetric positive definite $m \times m$ matrix. We ask what the best choice of $\mathbf{\Gamma}$ might be.

Letting

$$q(\mu, \mathbf{x}) = f(\mathbf{x}) + \mu P(\mathbf{x}) \tag{47}$$

the Hessian of q turns out to be, using (26),

$$\mathbf{Q}(\mu, \mathbf{x}_k) = \mathbf{L}(\mathbf{x}_k) + \mu \nabla \mathbf{h}(\mathbf{x}_k)'\mathbf{\Gamma}\nabla \mathbf{h}(\mathbf{x}_k). \tag{48}$$

The m large eigenvalues are due to the second term on the right. The observation we make is that although the m large eigenvalues are all proportional to μ, they are not necessarily all equal. Indeed, for very large μ these eigenvalues are determined almost exclusively by the second term, and are therefore μ times the nonzero eigenvalues of the matrix $\nabla \mathbf{h}(\mathbf{x}_k)'\mathbf{\Gamma}\nabla \mathbf{h}(\mathbf{x}_k)$. We would like to select $\mathbf{\Gamma}$ so that these eigenvalues are not spread out but are nearly equal to one another. An ideal choice for the kth iteration would be

$$\mathbf{\Gamma} = [\nabla \mathbf{h}(\mathbf{x}_k)\nabla \mathbf{h}(\mathbf{x}_k)']^{-1} \tag{49}$$

since then all nonzero eigenvalues would be exactly equal. However, we do not allow $\mathbf{\Gamma}$ to change at each step, and therefore compromise by setting

$$\mathbf{\Gamma} = [\nabla \mathbf{h}(\mathbf{x}_0)\nabla \mathbf{h}(\mathbf{x}_0)']^{-1} \tag{50}$$

where $\mathbf{x}_0$ is the initial point of the iteration.

Using this penalty function, the corresponding eigenvalue structure will at any point look approximately like that shown in Fig. 12.4. The eigenvalues are bunched into two separate groups. As μ is increased the smaller eigenvalues move into the interval $[a, A]$ where a and A are, as usual, the smallest and largest eigenvalues of $\mathbf{L}_M$ at the solution to (45). The larger eigenvalues move forward to the right and spread further apart.

Fig. 12.4 Eigenvalue distributions

Using the result of Exercise 11, Chapter 8, we see that if $\mathbf{x}_{k+1}$ is determined from $\mathbf{x}_k$ by two conjugate gradient steps, the rate of convergence will be linear at a ratio determined by the widest of the two eigenvalue groups. If our normalization is sufficiently accurate, the large-valued group will have the lesser width. In that case convergence of this scheme is approximately that of the canonical rate for the original problem. Thus, by proper normalization it is possible to obtain the canonical rate of convergence for only about twice the time per iteration as required by steepest descent.

There are, of course, numerous variations of this method that can be used in practice. $\mathbf{\Gamma}$ can, for example, be allowed to vary at each step, or it can be occasionally updated.

Table 2

	p (steps per cycle)	Number of cycles to convergence	No. of steps	Value of modified objective
$\mu = 10$	1	28	28	251.2657
	2	9	18	251.2657
	3	5	15	251.2657
$\mu = 100$	1	153	153	379.5955
	2	13	26	379.5955
	3	11	33	379.5955
$\mu = 1000$	1	261*	261	402.0903
	2	14	28	400.1687
	3	13	39	400.1687

* Program not run to completion due to excessive time.

Example. The example problem presented in the previous section was also solved by the normalization method presented above. The results for various values of μ and for cycle lengths of one, two, and three are presented in Table 2. (All runs were initiated from the zero vector.)

12.8 PENALTY FUNCTIONS AND GRADIENT PROJECTION

The penalty function method can be combined with the idea of the gradient projection method to yield an attractive general purpose procedure for solving constrained optimization problems. The proposed combination method can be viewed either as a way of accelerating the rate of convergence of the penalty function method by eliminating the effect of the large eigenvalues, or as a technique for efficiency handling the delicate and usually cumbersome requirement in the gradient projection method that each point be feasible. The combined method converges at the canonical rate (the same as does the gradient projection method), is globally convergent (unlike the gradient projection method), and avoids much of the computational difficulty associated with staying feasible.

Underlying Concept

The basic theoretical result that motivates the development of this algorithm is the Combined Steepest Descent and Newton's Method Theorem of Section 9.6. The idea is to apply this combined method to a penalty problem. For simplicity we first consider the equality constrained problem

$$\begin{aligned} &\text{minimize} \quad f(\mathbf{x}) \\ &\text{subject to} \quad \mathbf{h}(\mathbf{x}) = \mathbf{0} \end{aligned} \tag{51}$$

where $\mathbf{x} \in E^n$, $\mathbf{h}(\mathbf{x}) \in E^m$. The associated unconstrained penalty problem that we consider is

$$\text{minimize} \quad q(\mathbf{x}) \tag{52}$$

where

$$q(\mathbf{x}) = f(\mathbf{x}) + \mu|\mathbf{h}(\mathbf{x})|^2.$$

At any point $\mathbf{x}_k$ let $M(\mathbf{x}_k)$ be the subspace tangent to the surface $S_k = \{\mathbf{x} : \mathbf{h}(\mathbf{x}) = \mathbf{h}(\mathbf{x}_k)\}$. This is a slight extension of the tangent subspaces that we have considered before, since $M(\mathbf{x}_k)$ is defined even for points that are not feasible. If the sequence $\{\mathbf{x}_k\}$ converges to a solution $\mathbf{x}_\mu$ of problem (52), then we expect that $M(\mathbf{x}_k)$ will in some sense converge to $M(\mathbf{x}_\mu)$. The orthogonal complement of $M(\mathbf{x}_k)$ is the space generated by the gradients of the constraint functions evaluated at $\mathbf{x}_k$. Let us denote this space by $N(\mathbf{x}_k)$. The idea of the algorithm is to take N as the subspace over which Newton's method is applied, and M as the space over which the gradient method is applied. A cycle of the algorithm would be as follows:

1. Given $\mathbf{x}_k$, apply one step of Newton's method over the subspace $N(\mathbf{x}_k)$ to obtain a point $\mathbf{w}_k$ of the form

$$\mathbf{w}_k = \mathbf{x}_k + \nabla \mathbf{h}(\mathbf{x}_k)'\mathbf{u}_k$$
$$\mathbf{u}_k \in E^m.$$

2. From $\mathbf{w}_k$, take an ordinary steepest descent step to obtain $\mathbf{x}_{k+1}$.

Of course, we must show how step (1) can be easily executed, and this is done below, but first, without drawing out the details, let us examine the general structure of this algorithm.

The process is illustrated in Fig. 12.5. The first step is analogous to the step in the gradient projection method that returns to the feasible surface; except that here the criterion is reduction of the objective function rather than satisfaction of constraints. To interpret the second step, suppose for the moment that the original problem (51) has a quadratic objective and linear constraints; so that, consequently, the penalty problem (52) has a quadratic objective and $N(\mathbf{x})$, $M(\mathbf{x})$ and $\nabla \mathbf{h}(\mathbf{x})$ are independent of $\mathbf{x}$. In that case the first (Newton) step would exactly minimize q with respect to N, so that the gradient of q at $\mathbf{w}_k$ would be orthogonal to N; that is, the gradient would lie in the subspace M. Furthermore, since $\nabla q(\mathbf{w}_k) = \nabla f(\mathbf{w}_k) + 2\mu \mathbf{h}(\mathbf{w}_k)\nabla \mathbf{h}(\mathbf{w}_k)$, we see that $\nabla q(\mathbf{w}_k)$ would in that case be equal to the projection of the gradient of f onto M. Hence, the second step is, in the quadratic case exactly, and in the general case approximately, a move in the direction of the projected negative gradient of the original objective function.

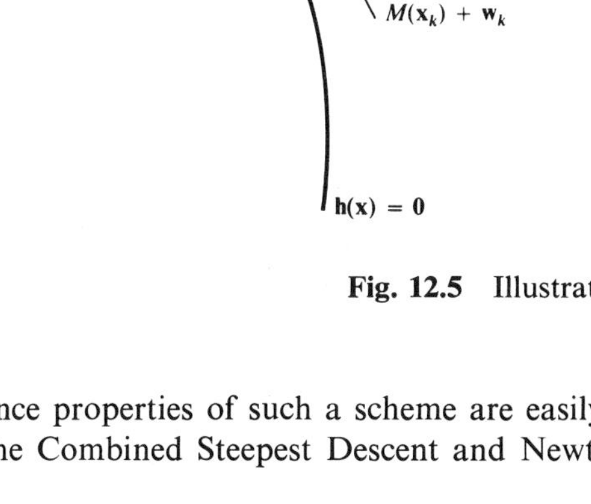

Fig. 12.5 Illustration of the method

The convergence properties of such a scheme are easily predicted from the theorem on the Combined Steepest Descent and Newton's Method, in

Section 9.6, and our analysis of the structure of the Hessian of the penalty objective function given by (26). As $\mathbf{x}_k \to \mathbf{x}_\mu$ the rate will be determined by the ratio of largest to smallest eigenvalues of the Hessian restricted to $M(\mathbf{x}_\mu)$. This leads, however, by what was shown in Section 11.3, to approximately the canonical rate for problem (51). Thus, this combined method will yield again the canonical rate as $\mu \to \infty$.

Implementing the First Step

To implement the first step of the algorithm suggested above it is necessary to show how a Newton step can be taken in the subspace $N(\mathbf{x}_k)$. We show that, again for large values of μ, this can be accomplished easily.

At the point $\mathbf{x}_k$ the function c, defined by

$$c(\mathbf{u}) = q(\mathbf{x}_k + \nabla\mathbf{h}(\mathbf{x}_k)'\mathbf{u}) \tag{53}$$

for $\mathbf{u} \in E^m$, measures the variations in q with respect to displacements in $N(\mathbf{x}_k)$. We shall, for simplicity, assume that at each point, $\mathbf{x}_k$, $\nabla\mathbf{h}(\mathbf{x}_k)$ has rank m. We can immediately calculate the gradient with respect to $\mathbf{u}$,

$$\nabla c(\mathbf{u}) = \nabla q(\mathbf{x}_k + \nabla\mathbf{h}(\mathbf{x}_k)'\mathbf{u})\nabla\mathbf{h}(\mathbf{x}_k)' \tag{54}$$

and the $m \times n$ Hessian with respect to $\mathbf{u}$ at $\mathbf{u} = \mathbf{0}$,

$$\mathbf{C} = \nabla\mathbf{h}(\mathbf{x}_k)\mathbf{Q}(\mathbf{x}_k)\nabla\mathbf{h}(\mathbf{x}_k)' \tag{55}$$

where $\mathbf{Q}$ is the $n \times n$ Hessian of q with respect to $\mathbf{x}$. From (26) we have that at $\mathbf{x}_k$

$$\mathbf{Q}(\mathbf{x}_k) = \mathbf{L}_k(\mathbf{x}_k) + 2\mu\nabla\mathbf{h}(\mathbf{x}_k)'\nabla\mathbf{h}(\mathbf{x}_k). \tag{56}$$

And given $\mathbf{C}$, the direction for the Newton step in N would be

$$\begin{aligned} \mathbf{d}_k &= -\nabla\mathbf{h}(\mathbf{x}_k)'\mathbf{C}^{-1}\nabla c(\mathbf{0})' \\ &= -\nabla\mathbf{h}(\mathbf{x}_k)'\mathbf{C}^{-1}\nabla\mathbf{h}(\mathbf{x}_k)\nabla q(\mathbf{x}_k)'. \end{aligned} \tag{57}$$

It is clear from (55) and (56) that exact evaluation of the Newton step requires knowledge of $\mathbf{L}(\mathbf{x}_k)$ which usually is costly to obtain. For large values of μ, however, $\mathbf{C}$ can be approximated by

$$\mathbf{C} \simeq 2\mu[\nabla\mathbf{h}(\mathbf{x}_k)\nabla\mathbf{h}(\mathbf{x}_k)']^2 \tag{58}$$

and hence a good approximation to the Newton direction is

$$\mathbf{d}_k = -\frac{1}{2\mu}\nabla\mathbf{h}(\mathbf{x}_k)'[\nabla\mathbf{h}(\mathbf{x}_k)\nabla\mathbf{h}(\mathbf{x}_k)']^{-2}\nabla\mathbf{h}(\mathbf{x}_k)\nabla q(\mathbf{x}_k)'. \tag{59}$$

Thus, a suitable implementation of one cycle of the algorithm is:

1. Calculate

$$\mathbf{d}_k = -\frac{1}{2\mu}\nabla\mathbf{h}(\mathbf{x}_k)'[\nabla\mathbf{h}(\mathbf{x}_k)\nabla\mathbf{h}(\mathbf{x}_k)']^{-2}\nabla\mathbf{h}(\mathbf{x}_k)\nabla q(\mathbf{x}_k)'$$

2. Find β_k to minimize $q(\mathbf{x}_k + \beta \mathbf{d}_k)$ (using $\beta_k = 1$ as an initial search point), and set $\mathbf{w}_k = \mathbf{x}_k + \beta_k \mathbf{d}_k$.
3. Calculate $\mathbf{p}_k = -\nabla q(\mathbf{w}_k)'$.
4. Find α_k to minimize $q(\mathbf{w}_k + \alpha \mathbf{p}_k)$, and set $\mathbf{x}_{k+1} = \mathbf{w}_k + \alpha_k \mathbf{p}_k$.

It is interesting to compare the Newton step of this version of the algorithm with the step for returning to the feasible region used in the ordinary gradient projection method. We have

$$\nabla q(\mathbf{x}_k)' = \nabla f(\mathbf{x}_k)' + 2\mu \nabla \mathbf{h}(\mathbf{x}_k)'\mathbf{h}(\mathbf{x}_k). \tag{60}$$

If we neglect $\nabla f(\mathbf{x}_k)'$ on the right (as would be valid if we are a long distance from the constraint boundary, then the vector $\mathbf{d}_k$ reduces to

$$\mathbf{d}_k = -\nabla \mathbf{h}(\mathbf{x}_k)'[\nabla \mathbf{h}(\mathbf{x}_k)\nabla \mathbf{h}(\mathbf{x}_k)']^{-1}\mathbf{h}(\mathbf{x}_k)$$

which is precisely the first estimate used to return to the boundary in the gradient projection method. The scheme developed in this section can therefore be regarded as one which corrects this estimate by accounting for the variation in f.

An important advantage of the present method is that it is not necessary to carry out the search in detail. If $\beta = 1$ yields an improved value for the penalty objective, no further search is required. If not, one need search only until some improvement is obtained. At worst, if this search is poorly performed, the method degenerates to steepest descent. When one finally gets close to the solution, however, $\beta = 1$ is bound to yield an improvement and terminal convergence will progress at nearly the canonical rate.

Inequality Constraints

The procedure is conceptually the same for problems with inequality constraints. The only difference is that at the beginning of each cycle the subspace $M(\mathbf{x}_k)$ is calculated on the basis of those constraints that are either active or violated at $\mathbf{x}_k$, the others being ignored. The resulting technique is a descent algorithm in that the penalty objective function decreases at each cycle; it is globally convergent because of the pure gradient step taken at the end of each cycle; its rate of convergence approaches the canonical rate for the original constrained problem as $\mu \to \infty$; and there are no feasibility tolerances or subroutine iterations required.

12.9 SUMMARY

Penalty methods approximate a constrained problem by an unconstrained problem that assigns high cost to points that are far from the feasible region. As the approximation is made more exact (by letting the parameter μ tend to infinity) the solution of the unconstrained penalty problem approaches the solution to the original constrained problem from outside the active constraints. Barrier methods, on the other hand, approximate a constrained problem by an (essentially) unconstrained problem that assigns high cost to being near the boundary of the feasible region, but unlike penalty methods, these methods are applicable only to problems having a robust feasible region. As the approximation is made more exact, the solution of the unconstrained barrier problem approaches the solution to the original constrained problem from inside the feasible region.

The objective functions of all penalty and barrier methods have ill-conditioned Hessians. In all methods, as $\mu \to \infty$ the Hessian (at the solution) is equal to the sum of $\mathbf{L}$, the Hessian of the Lagrangian associated with the original constrained problem, and a matrix of rank r that tends to infinity (where r is the number of active constraints). This is a fundamental property of these methods.

Effective exploitation of these methods requires that schemes be devised that eliminate the effect of the associated large eigenvalues. For this purpose the three general principles developed in earlier chapters, The Partial Conjugate Gradient Method, The Modified Newton Method, and The Combination of Steepest Descent and Newton's Method, when creatively applied, all yield methods that converge at approximately the canonical rate associated with the original constrained problem.

It is necessary to add a point of qualification with respect to some of the algorithms introduced in this chapter, lest it be inferred that they are offered as panaceas for the general programming problem. As has been repeatedly emphasized, the ideal study of convergence is a careful blend of analysis, good sense, and experimentation. The rate of convergence does not always tell the whole story, although it is often a major component of it. Although some of the algorithms presented in this chapter asymptotically achieve the canonical rate of convergence (at least approximately), for large μ the points may have to be quite close to the solution before this rate characterizes the process. In other words, for large μ the process may converge slowly in its initial phase, and, to obtain a truly representative analysis, one must look beyond the first-order convergence properties of these methods. This subject cannot, however, be further explored in an introductory text such as this.

Nevertheless, in spite of the qualification expressed above, our analysis techniques have taken us a considerable distance with respect to penalty functions. We have identified a major obstacle to fast convergence and developed several techniques for overcoming this obstacle. The resulting algorithms converge at the canonical rate.

12.10 EXERCISES

1. Show that if $q(\mu, \mathbf{x})$ is continuous (with respect to $\mathbf{x}$) and $q(\mu, \mathbf{x}) \to \infty$ as $|\mathbf{x}| \to \infty$, then $q(\mu, \mathbf{x})$ has a minimum.

2. Suppose problem (1), with f continuous, is approximated by the penalty problem (2), and let $\{\mu_k\}$ be an increasing sequence of positive constants tending to infinity. Define $q(\mu, \mathbf{x}) = f(\mathbf{x}) + P(\mathbf{x})$, and fix $\varepsilon > 0$. For each k let $\mathbf{x}_k$ be determined satisfying

$$q(\mu_k, \mathbf{x}_k) \leqslant \left[\min_{\mathbf{x}} q(\mu_k, \mathbf{x})\right] + \varepsilon.$$

Show that if $\mathbf{x}^*$ is a solution to (1), any limit point, $\bar{\mathbf{x}}$, of the sequence $\{\mathbf{x}_k\}$ is feasible and satisfies $f(\bar{\mathbf{x}}) \leqslant f(\mathbf{x}^*) + \varepsilon$.

3. Construct an example problem and a penalty function such that, as $\mu \to \infty$, the solution to the penalty problem diverges to infinity.

4. *Combined Penalty and Barrier Method.* Consider a problem of the form

$$\begin{aligned} &\text{minimize} \quad f(\mathbf{x}) \\ &\text{subject to} \quad \mathbf{x} \in S \cap T \end{aligned}$$

and suppose P is a penalty function for S and B is a barrier function for T. Define

$$c(\mu, \mathbf{x}) = f(\mathbf{x}) + \mu\, P(\mathbf{x}) + \frac{1}{\mu} B(\mathbf{x}).$$

Let $\{\mu_k\}$ be a sequence $\mu_k \to \infty$, and for $k = 1, 2, \ldots$ let $\mathbf{x}_k$ be a solution to

$$\text{minimize } c(\mu_k, \mathbf{x})$$

subject to $\mathbf{x} \in$ interior of T. Assume all functions are continuous, T is compact (and robust), the original problem has a solution $\mathbf{x}^*$, and that $S \cap$ [interior of T] is not empty. Show that

a) $\lim_{\mu \to \infty} c(\mu_k, \mathbf{x}_k) = f(\mathbf{x}^*)$.

b) $\lim_{k \to \infty} \mu_k P(\mathbf{x}_k) = 0$.

c) $\lim_{k \to \infty} \frac{1}{\mu_k} B(\mathbf{x}_k) = 0$.

5. Prove the Theorem at the end of Section 12.2.

6. Parallel the analysis (31)–(35) for the barrier function

$$B(\mathbf{x}) = -\sum_{i=1}^{p} \log\left(-g_i(\mathbf{x})\right)$$

(Assume for the sake of consistency that $B(\mathbf{x}) \geqslant 0$ in the feasible region.)

7. Consider a penalty function for the equality constraints

$$\mathbf{h}(\mathbf{x}) = \mathbf{0}, \qquad \mathbf{h}(\mathbf{x}) \in E^m,$$

having the form

$$P(\mathbf{x}) = \gamma(\mathbf{h}(\mathbf{x})) = \sum_{i=1}^{m} w(h_i(\mathbf{x}))$$

where w is a function whose derivative w' is analytic and has a zero of order $s \geqslant 1$ at zero.

a) Show that corresponding to (26) we have

$$\mathbf{Q}(\mu_k, \mathbf{x}_k) = L_k(\mathbf{x}_k) + \mu_k \sum_{i=1}^{m} \{w''(h_i(\mathbf{x}_k))\} \nabla h_i(\mathbf{x}_k)' \nabla h_i(\mathbf{x}_k).$$

b) Show that as $\mu_k \to \infty$, m eigenvalues of $\mathbf{Q}(\mu_k, \mathbf{x}_k)$ have magnitude on the order of $(\mu_k)^{1/s}$.

8. Corresponding to the problem

$$\begin{aligned} &\text{minimize} \quad f(\mathbf{x}) \\ &\text{subject to} \quad g(\mathbf{x}) \leqslant 0, \end{aligned}$$

consider the sequence of unconstrained problems

$$\text{minimize} \quad f(\mathbf{x}) + [g^+(\mathbf{x}) + 1]^k - 1$$

and suppose $\mathbf{x}_k$ is the solution to the kth problem.

a) Find an appropriate definition of a Lagrange multiplier λ_k to associate with $\mathbf{x}_k$.

b) Find the limiting form of the Hessian of the associated objective function, and determine how fast the largest eigenvalues tend to infinity.

9. Repeat Exercise 8 for the sequence of unconstrained problems

$$\text{minimize } f(\mathbf{x}) + [(g(\mathbf{x}) + 1)^+]^k.$$

10. *Morrison's Method.* Suppose the problem

$$\begin{aligned} &\text{minimize} \quad f(\mathbf{x}) \\ &\text{subject to} \quad \mathbf{h}(\mathbf{x}) = \mathbf{0} \end{aligned} \tag{65}$$

has solution $\mathbf{x}^*$. Let M be an optimistic estimate of $f(\mathbf{x}^*)$, that is, $M \leqslant f(\mathbf{x}^*)$. Define $v(M, \mathbf{x}) = [f(\mathbf{x}) - M]^2 + |\mathbf{h}(\mathbf{x})|^2$ and define the unconstrained problem

$$\text{minimize } v(M, \mathbf{x}). \tag{66}$$

Given $M_k \leqslant f(\mathbf{x}^*)$, a solution $\mathbf{x}_{M_k}$ to the corresponding problem (66) is found, then M_k is updated through

$$M_{k+1} = M_k + [v(M_k, \mathbf{x}_{M_k})]^{1/2} \tag{67}$$

and the process repeated.

a) Show that if $M = f(\mathbf{x}^*)$, a solution to (66) is a solution to (65).

b) Show that if $\mathbf{x}_M$ is a solution to (66), then $f(\mathbf{x}_M) \leqslant f(\mathbf{x}^*)$.

c) Show that if $M_k \leqslant f(\mathbf{x}^*)$ then M_{k+1} determined by (67) satisfies $M_{k+1} \leqslant f(\mathbf{x}^*)$.

d) Show that $M_k \to f(\mathbf{x}^*)$.

e) Find the Hessian of $v(M, \mathbf{x})$ (with respect to $\mathbf{x}^*$). Show that, to within a scale factor, it is identical to that associated with the standard penalty function method.

11. Let $\mathbf{A}$ be an $m \times n$ matrix of rank m. Prove the matrix identity

$$[\mathbf{I} + \mathbf{A}'\,\mathbf{A}]^{-1} = \mathbf{I} - \mathbf{A}'\,[\mathbf{I} + \mathbf{A}\mathbf{A}']^{-1}\,\mathbf{A}$$

and discuss how it can be used in conjunction with the method of Section 12.5.

12. Show that in the limit of large μ, a single cycle of the normalization method of Section 12.7 is exactly the same as a single cycle of the combined penalty function and gradient projection method of Section 12.8.

13. Suppose that at some step k of the combined penalty function and gradient projection method, the $m \times n$ matrix $\nabla \mathbf{h}(\mathbf{x}_k)$ is not of rank m. Show how the method can be continued by temporarily executing the Newton step over a subspace of dimension less than m.

14. For a problem with equality constraints, show that in the combined penalty function and gradient projection method the second step (the steepest descent step) can be replaced by a step in the direction of the negative projected gradient (projected onto M_k) without destroying the global convergence property and without changing the rate of convergence.

15. Show, using the change of coordinates (64) and for simplicity ignoring the inequality constraint $\mathbf{x} \geqslant \mathbf{0}$, that according to the combined method of steepest descent and Newton's method, step (3) of the algorithm in Section 12.9 should be

$$\Delta \mathbf{z} = \nabla_{\mathbf{z}} q(\mathbf{w}_k)' - \nabla_{\mathbf{z}} \mathbf{h}(\mathbf{x}_k)'\,[\nabla_{\mathbf{y}} \mathbf{h}(\mathbf{x}_k)']^{-1}\,\nabla_{\mathbf{y}} q(\mathbf{w}_k)'$$
$$\Delta \mathbf{y} = \nabla_{\mathbf{y}} q(\mathbf{w}_k)' - \nabla_{\mathbf{y}} \mathbf{h}(\mathbf{x}_k)^{-1}\,\nabla_{\mathbf{z}} \mathbf{h}(\mathbf{x}_k)\,\Delta \mathbf{z}.$$

Show that if $\nabla_{\mathbf{y}} q(\mathbf{w}_k) = \mathbf{0}$, the above becomes identical to that given in the text. Note that if the Newton step found the exact minimum over $N(\mathbf{x}_k)$, then $\nabla_{\mathbf{y}} q(\mathbf{w}_k) = \mathbf{0}$, and hence justify the use of the reduced gradient. Compare with Exercise 14.

16. Develop a method that is analogous to that of Section 12.8, but which is a combination of penalty functions and the reduced gradient method. Establish that the rate of convergence of the method is identical to that of the reduced gradient method.

REFERENCES

12.1. The penalty approach to constrained optimization is generally attributed to Courant [C5]. For more details than presented here see Butler and Martin [B5] or Zangwill [Z1]. The penalty method has an interesting geometrical interpretation which is presented in Luenberger [L6].

12.2. The barrier method is due to Carroll [C2], but was developed and popularized by Fiacco and McCormick [F3] who proved the general effectiveness of the method.

12.3. It has long been known that penalty problems are solved slowly by steepest descent, and the difficulty has been traced to the ill-conditioning of the Hessian. The explicit characterization given here is a generalization of that in Luenberger [L10].

12.4. The idea of this section is taken from Fiacco and McCormick [F3] where some further related analysis can be found.

12.5. Most previous successful implementations of penalty or barrier methods have employed Newton's method to solve the unconstrained problems and thereby have largely avoided the effects of the ill-conditioned Hessian. See Fiacco and McCormick [F3] for some suggestions. The technique at the end of the section is new.

12.6. This method was first presented in Luenberger [L10].

12.7. This method is new.

12.8. See Luenberger [L8], for further analysis of this method.

12.10. For analysis along the lines of Exercise 7 see Lootsma [L5]. For the functions suggested in Exercises 8 and 9 see Levitin and Polyak [L3]. For the method of Exercise 10 see Morrison [M1].

Chapter 13 CUTTING PLANE AND DUAL METHODS

The two general classes of methods explained and analyzed in this chapter, cutting plane and dual methods, both require that the problem being attacked possess some sort of convexity property; and hence, these methods, at least in their purest forms, are applicable only to a subclass of the general nonlinear programming problem. It can be convincingly argued, however, that the required convexity assumptions are satisfied in a sufficient percentage of practical situations to amply justify careful consideration of these special algorithms. Indeed, there are important classes of problems for which these methods are the best available.

Cutting plane algorithms, exceedingly elementary in principle, develop a series of ever-improving approximating linear programs, whose solutions converge to the solution of the original problem. The methods differ only in the manner by which an improved approximating problem is constructed once a solution to the old approximation is known. The theory associated with these algorithms is, unfortunately, scant and their convergence properties are not particularly attractive. They are, however, often very easy to implement.

Dual methods are based on the viewpoint that it is the Lagrange multipliers which are the fundamental unknowns associated with a constrained problem; once these multipliers are known determination of the solution point is simple (at least in some situations). Dual methods, therefore, do not attack the original constrained problem directly but instead attack an alternate problem, the dual problem, whose unknowns are the Lagrange multipliers of the first problem. Because Lagrange multipliers measure sensitivities and hence often have meaningful intuitive interpretations as prices associated with constraint resources, searching for these multipliers, is often, in the context of a given practical problem, as appealing as searching for the values of the original problem variables.

The study of dual methods, and more particularly the introduction of the dual problem, precipitates some extensions of earlier concepts. Thus, perhaps the most interesting feature of this chapter is the calculation of the Hessian of the dual problem and the discovery of a *dual canonical convergence ratio* associated with a constrained problem that governs the convergence of steepest ascent applied to the dual.

13.1 CUTTING PLANE METHODS

Cutting plane methods are applied to problems having the general form

$$\begin{aligned} &\text{minimize} && \mathbf{cx} \\ &\text{subject to} && \mathbf{x} \in S \end{aligned} \tag{1}$$

where $S \subset E^n$ is a closed convex set. Problems that involve minimization of a convex function over a convex set, such as the problem

$$\begin{aligned} &\text{minimize} && f(\mathbf{y}) \\ &\text{subject to} && \mathbf{y} \in R \end{aligned} \tag{2}$$

where $R \subset E^{n-1}$ is a convex set and f is a convex function, can be easily converted to the form (1) by writing (2) equivalently as

$$\begin{aligned} &\text{minimize} && r \\ &\text{subject to} && f(\mathbf{y}) - r \leqslant 0 \\ & && \mathbf{y} \in R \end{aligned} \tag{3}$$

which, with $\mathbf{x} = (r, \mathbf{y})$, is a special case of (1).

General Form of Algorithm

The general form of a cutting-plane algorithm for problem (1) is as follows:

Given a polytope $P_k \supset S$

Step 1. Minimize $\mathbf{cx}$ over P_k obtaining a point $\mathbf{x}_k$ in P_k. If $\mathbf{x}_k \in S$, stop; $\mathbf{x}_k$ is optimal. Otherwise,

Step 2. Find a hyperplane H_k separating the point $\mathbf{x}_k$ from S, that is, find $\mathbf{a}_k \in E_n$, $b_k \in E^1$ such that $S \subset \{\mathbf{x}\colon \mathbf{a}_k\mathbf{x} \leqslant b_k\}$, $\mathbf{x}_k \in \{\mathbf{x}\colon \mathbf{a}_k\mathbf{x} > b_k\}$. Update P_k to obtain P_{k+1} including as a constraint $\mathbf{a}_k\mathbf{x} \leqslant b_k$.

The process is illustrated in Fig. 13.1.

Specific algorithms differ mainly in the manner in which the hyperplane that separates the current point $\mathbf{x}_k$ from the constraint set S is selected. This selection is, of course, the most important aspect of the algorithm, since it is the deepness of the cut associated with the separating hyperplane, the distance of the hyperplane from the current point, that governs how much improvement there is in the approximation to the constraint set, and hence how fast the method converges.

Specific algorithms also differ somewhat with respect to the manner by which the polytope is updated once the new hyperplane is determined. The most straightforward procedure is to simply adjoin the linear inequality associated with that hyperplane to the ones determined previously. This yields the best possible updated approximation to the constraint set but tends to produce, after a large number of iterations, an unwieldy number of inequalities expressing the approximation. Thus in some algorithms, older inequalities that are not binding at the current point are discarded from further consideration.

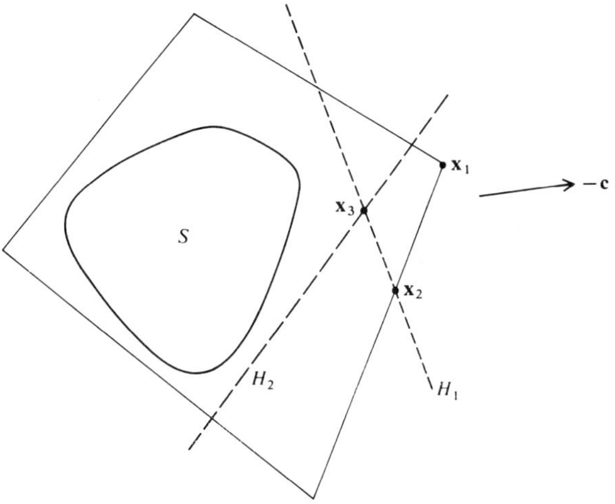

Fig. 13.1 Cutting plane method

Duality

The general cutting plane algorithm can be regarded as an extended application of duality in linear programming, and although this viewpoint does not particularly aid in the analysis of the method, it reveals the basic interconnection between cutting plane and dual methods. The foundation of this viewpoint is the fact that S can be written as the intersection of all the half-spaces that contain it; thus

$$S = \{\mathbf{x}:\ \mathbf{a}_i\mathbf{x} \leqslant b_i,\ i \in I\}$$

where I is an (infinite) index set corresponding to all half-spaces containing S. With S viewed in this way problem (1) can be thought of as an (infinite) linear programming problem.

Corresponding to this linear program there is (at least formally) the dual problem

$$\begin{aligned} \text{maximize} \quad & \sum_{i \in I} \lambda_i b_i \\ \text{subject to} \quad & \sum_{i \in I} \lambda_i \mathbf{a}_i = \mathbf{c} \\ & \lambda_i \geqslant 0, \; i \in I. \end{aligned} \tag{4}$$

Selecting a finite subset of I, say $\bar{I}$, and forming

$$P = \{\mathbf{x}\colon \; \mathbf{a}_i \mathbf{x} \leqslant b_i, \; i \in \bar{I}\}$$

gives a polytope that contains S. Minimizing $\mathbf{cx}$ over this polytope yields a point and a corresponding subset of active constraints I_A. The dual problem with the additional restriction $\lambda_i = 0$ for $i \notin I_A$ will then have a feasible solution, but this solution will in general not be optimal. Thus, a solution to a polytope problem corresponds to a feasible but non-optimal solution to the dual. For this reason the cutting plane method can be regarded as working toward optimality of the (infinite dimensional) dual.

13.2 KELLEY'S CONVEX CUTTING PLANE ALGORITHM

The convex cutting plane method was developed to solve convex programming problems of the form

$$\begin{aligned} \text{minimize} \quad & f(\mathbf{x}) \\ \text{subject to} \quad & g_i(\mathbf{x}) \leqslant 0, \qquad i = 1, 2, \ldots, p, \end{aligned} \tag{5}$$

where $\mathbf{x} \in E^n$ and f and the g_i's are differentiable convex functions. As indicated in the last section, it is sufficient to consider the case where the objective function is linear; thus, we consider the problem

$$\begin{aligned} \text{minimize} \quad & \mathbf{cx} \\ \text{subject to} \quad & \mathbf{g}(\mathbf{x}) \leqslant \mathbf{0} \end{aligned} \tag{6}$$

where $\mathbf{x} \in E^n$ and $\mathbf{g}(\mathbf{x}) \in E^p$ is convex and differentiable.

For $\mathbf{g}$ convex and differentiable we have the fundamental inequality

$$\mathbf{g}(\mathbf{x}) \geqslant \mathbf{g}(\mathbf{w}) + \nabla \mathbf{g}(\mathbf{w})(\mathbf{x} - \mathbf{w}) \tag{7}$$

for any $\mathbf{x}$, $\mathbf{w}$. We use this equation to determine the separating hyperplane. Specifically, the algorithm is as follows:

Let $S = \{\mathbf{x}\colon \; \mathbf{g}(\mathbf{x}) \leqslant \mathbf{0}\}$ and let P be an initial polytope containing S and such that $\mathbf{cx}$ is bounded on P. Then

Step 1. Minimize $\mathbf{cx}$ over P obtaining the point $\mathbf{x} = \mathbf{w}$. If $\mathbf{g}(\mathbf{w}) \leqslant \mathbf{0}$, stop; $\mathbf{w}$ is an optimal solution. Otherwise

Step 2. Let i be an index maximizing $g_i(\mathbf{w})$. Clearly $g_i(\mathbf{w}) > 0$. Define the new approximating polytope to be the old one intersected with the half-space

$$\{\mathbf{x}:\ g_i(\mathbf{w}) + \nabla g_i(\mathbf{w})(\mathbf{x} - \mathbf{w}) \leqslant 0\}. \tag{8}$$

Return to step 1.

The set defined by (8) is actually a half-space if $\nabla g_i(\mathbf{w}) \neq \mathbf{0}$. However, $\nabla g_i(\mathbf{w}) = \mathbf{0}$ would imply that $\mathbf{w}$ minimizes g_i which is impossible if S is nonempty. Furthermore, the half-space given by (8) contains S, since if $\mathbf{g}(\mathbf{x}) = 0$ then by (7) $g_i(\mathbf{w}) + \nabla g_i(\mathbf{w})(\mathbf{x} - \mathbf{w}) \leqslant g_i(\mathbf{x}) \leqslant 0$. The half-space does not contain the point $\mathbf{w}$ since $g_i(\mathbf{w}) > 0$. This method for selecting the separating hyperplane is illustrated in Fig. 13.2 for the one-dimensional case. Note that in one dimension, the procedure reduces to Newton's method.

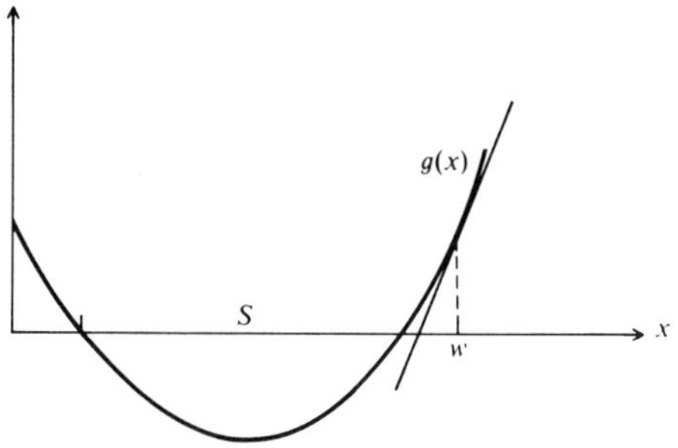

Fig. 13.2 Convex cutting plane

Calculation of the separating hyperplane is exceedingly simple in this algorithm, and hence the method really amounts to the solution of a series of linear programming problems. It should be noted that this algorithm, valid for any convex programming problem, does not involve any line searches. In that respect it is also similar to Newton's method applied to a convex function.

Convergence

Under fairly mild assumptions on the convex function, the convex cutting plane method is globally convergent. It is possible to apply the general convergence theorem to prove this, but somewhat easier, in this case, to prove it directly.

> ***Theorem.*** *Let the convex functions g_i, $i = 1, 2, \ldots, p$ be continuously differentiable, and suppose the convex cutting plane algorithm generates the sequence of points $\{\mathbf{w}_k\}$. Any limit point of this sequence is a solution to problem* (6).

Proof. Suppose $\{\mathbf{w}_k\}$, $k \in \mathscr{K}$ is a subsequence of $\{\mathbf{w}_k\}$ converging to $\mathbf{w}$. By taking a further subsequence of this, if necessary, we may assume that the index i corresponding to step 2 of the algorithm is fixed throughout the subsequence. Now if $k \in \mathscr{K}$, $k' \in \mathscr{K}$ and $k' > k$, then we must have

$$g_i(\mathbf{w}_k) + \nabla g_i(\mathbf{w}_k)(\mathbf{w}_{k'} - \mathbf{w}_k) \leqslant 0$$

which implies that

$$g_i(\mathbf{w}_k) \leqslant |\nabla g_i(\mathbf{w}_k)|\,|\mathbf{w}_{k'} - \mathbf{w}_k|. \tag{9}$$

Since $|\nabla g_i(\mathbf{w}_k)|$ is bounded with respect to $k \in \mathscr{K}$, the right-hand side of (9) goes to zero as k and k' go to infinity. The left-hand side goes to $g_i(\mathbf{w})$. Thus $g_i(\mathbf{w}) \leqslant 0$ and we see that $\mathbf{w}$ is feasible for problem (6).

If f^* is the optimal value of problem (6), we have $\mathbf{c}\mathbf{w}_k \leqslant f^*$ for each k since $\mathbf{w}_k$ is obtained by minimizing over a set containing S. Thus, by continuity, $\mathbf{c}\mathbf{w} \leqslant f^*$ and hence $\mathbf{w}$ is an optimal solution. ▮

As with most algorithms based on linear programming concepts, the rate of convergence of cutting plane algorithms has not yet been satisfactorily analyzed. Preliminary research shows that these algorithms converge arithmetically, that is, if $\mathbf{x}^*$ is optimal, then $|\mathbf{x}_k - \mathbf{x}^*|^2 \leqslant c/k$ for some constant c. This is an exceedingly poor type of convergence. This estimate, however, may not be the best possible and indeed there are indications that the convergence is actually geometric but with a ratio that goes to unity as the dimension of the problem increases.

13.3 MODIFICATIONS

In this section we describe the supporting hyperplane algorithm (an alternative method for determining a cutting plane) and examine the possibility of dropping from consideration some old hyperplanes so that the linear programs do not grow too large.

The Supporting Hyperplane Algorithm

The convexity requirements are less severe for this algorithm. It is applicable to problems of the form

$$\begin{aligned} &\text{minimize} \quad \mathbf{c}\mathbf{x} \\ &\text{subject to} \quad \mathbf{g}(\mathbf{x}) \leqslant \mathbf{0} \end{aligned}$$

where $\mathbf{x} \in E^n$, $\mathbf{g}(\mathbf{x}) \in E^p$, the g_i's are continuously differentiable, and the constraint region S defined by the inequalities is convex. Note that convexity of the functions themselves is not required. We also assume the existence of a point interior to the constraint region, that is, we assume the existence of a point $\mathbf{y}$ such that $\mathbf{g}(\mathbf{y}) < \mathbf{0}$, and we assume that on the constraint boundary $g_i(\mathbf{x}) = 0$ implies $\nabla g_i(x) \neq \mathbf{0}$. The algorithm is as follows:

Start with an initial polytope P containing S and such that $\mathbf{cx}$ is bounded below on S. Then

Stel 1. Determine $\mathbf{w} = \mathbf{x}$ to minimize $\mathbf{cx}$ over P. If $\mathbf{w} \in S$, stop. Otherwise,

Step 2. Find the point $\mathbf{u}$ on the line joining $\mathbf{y}$ and $\mathbf{w}$ that lies on the boundary of S. Let i be an index for which $g_i(\mathbf{u}) = 0$ and define the half-space $H = \{\mathbf{x}\colon \nabla g_i(\mathbf{u})(\mathbf{x} - \mathbf{u}) \leqslant 0\}$. Update P by intersecting with H. Return to step 1.

The algorithm is illustrated in Fig. 13.3.

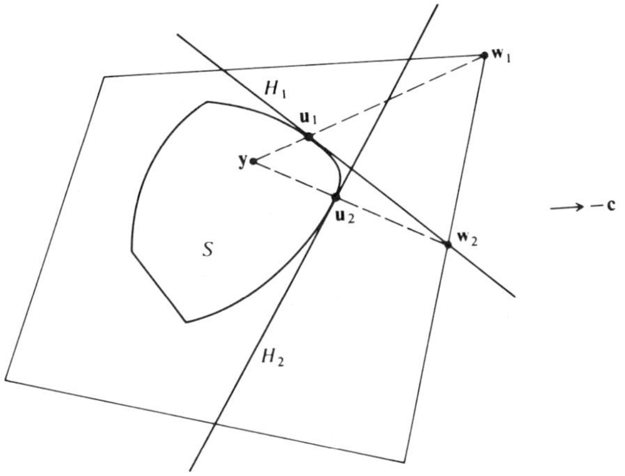

Fig. 13.3 Supporting hyperplane algorithm

The price paid for the generality of this method over the convex cutting plane method is that an interpolation along the line joining $\mathbf{y}$ and $\mathbf{w}$ must be executed to find the point $\mathbf{u}$. This is analogous to the line search for a minimum point required by most programming algorithms.

Dropping Nonbinding Constraints

In all cutting plane algorithms nonbinding constraints can be dropped from the approximating set of linear inequalities so as to keep the complexity of the approximation manageable. Indeed, since n linearly independent hyperplanes determine a single point in E^n, the algorithm can be arranged, by discarding the nonbinding constraints at the end of each step, so that the polytope consists of exactly n linear inequalities at every stage.

Global convergence is not destroyed by this process, since the sequence of objective values will still be monotonically increasing. It is not known, however, what effect this has on the speed of convergence.

13.4 LOCAL DUALITY

In recent years duality has come to play a central role in the development and unification of the theory of optimization. From its unique viewpoint has sprung new and valuable interpretations of the ubiquitous Lagrange multipliers and the evolution of new algorithms effective for wide classes of practical problems. Early forms of the theory were stated in terms of derivatives of the functions involved, and were local rather than global in character. These theories, although of significant theoretical impact, were not well suited to computation. Later, by imposing convexity assumptions, a complete global theory, stated only in terms of the functions themselves, rather than their derivatives, and well-suited to computation, was developed. To be broadly applicable, however, a theory of duality should require a minimum of convexity assumptions. This leads, quite naturally, to the concept of requiring only local convexity, and thereby to a local duality theory. We present such a theory in this section, since it is in keeping with the spirit of the earlier chapters and is perhaps the simplest way to develop computationally useful duality results.

As often done before for convenience, we again consider nonlinear programming problems of the form

$$\begin{aligned} &\text{minimize} \quad f(\mathbf{x}) \\ &\text{subject to} \quad \mathbf{h}(\mathbf{x}) = \mathbf{0} \end{aligned} \tag{10}$$

where $\mathbf{x} \in E^n$, $\mathbf{h}(\mathbf{x}) \in E^m$ and $f, \mathbf{h} \in C^2$. Everything we do can be easily extended to problems having inequality as well as equality constraints, for the price of a somewhat more involved notation.

We focus attention on a local solution $\mathbf{x}^*$ of (10). Assuming that $\mathbf{x}^*$ is a regular point of the constraints, then, as we know, there will be a corresponding Lagrange multiplier (row) vector $\boldsymbol{\lambda}^*$ such that

$$\nabla f(\mathbf{x}^*) + \boldsymbol{\lambda}^* \nabla \mathbf{h}(\mathbf{x}^*) = \mathbf{0} \tag{11}$$

and the Hessian of the Lagrangian

$$\mathbf{L}(\mathbf{x}^*) = \mathbf{F}(\mathbf{x}^*) + \boldsymbol{\lambda}^* \mathbf{H}(\mathbf{x}^*) \tag{12}$$

must be positive semidefinite on the tangent subspace

$$M = \{\mathbf{x} : \nabla \mathbf{h}(\mathbf{x}^*)\mathbf{x} = \mathbf{0}\}.$$

At this point we introduce the special local convexity assumption necessary for the development of the local duality theory. Specifically, we assume that the Hessian $\mathbf{L}(\mathbf{x}^*)$ is positive definite. Of course, it should be emphasized that by this we mean $\mathbf{L}(\mathbf{x}^*)$ is positive definite on the whole space E^n, not just on the subspace M. The assumption guarantees that the Lagrangian $l(\mathbf{x}) = f(\mathbf{x}) + \boldsymbol{\lambda}^* \mathbf{h}(\mathbf{x})$ is locally convex at $\mathbf{x}^*$.

With this assumption, the point $\mathbf{x}^*$ is not only a local solution to the constrained problem (10); it is also a local solution to the unconstrained problem

$$\text{minimize} \quad f(\mathbf{x}) + \boldsymbol{\lambda}^*\mathbf{h}(\mathbf{x}) \tag{13}$$

since it satisfies the first- and second-order sufficiency conditions for a local minimum point. Furthermore, for any $\boldsymbol{\lambda}$ sufficiently close to $\boldsymbol{\lambda}^*$ the function $f(\mathbf{x}) + \boldsymbol{\lambda}\mathbf{h}(\mathbf{x})$ will have a local minimum point at a point $\mathbf{x}$ near $\mathbf{x}^*$. This follows by noting that, by the Implicit Function Theorem, the equation

$$\nabla f(\mathbf{x}) + \boldsymbol{\lambda}\nabla\mathbf{h}(\mathbf{x}) = \mathbf{0} \tag{14}$$

has a solution $\mathbf{x}$ near $\mathbf{x}^*$ when $\boldsymbol{\lambda}$ is near $\boldsymbol{\lambda}^*$, because $\mathbf{L}^*$ is nonsingular; and by the fact that, at this solution $\mathbf{x}$, the Hessian $\mathbf{F}(\mathbf{x}) + \boldsymbol{\lambda}\mathbf{H}(\mathbf{x})$ is positive definite. Thus, locally there is a unique correspondence between $\boldsymbol{\lambda}$ and $\mathbf{x}$ through solution of the unconstrained problem

$$\text{minimize} \quad f(\mathbf{x}) + \boldsymbol{\lambda}\mathbf{h}(\mathbf{x}). \tag{15}$$

Furthermore, this correspondence is continuously differentiable.

Near $\boldsymbol{\lambda}^*$ we define the *dual function* ϕ by the equation

$$\phi(\boldsymbol{\lambda}) = \text{minimum} \quad [f(\mathbf{x}) + \boldsymbol{\lambda}\mathbf{h}(\mathbf{x})] \tag{16}$$

where here it is understood that the minimum is taken locally with respect to $\mathbf{x}$ near $\mathbf{x}^*$. We are then able to show (and will do so below) that locally the original constrained problem (10) is equivalent to unconstrained local maximization of the dual function ϕ with respect to $\boldsymbol{\lambda}$. Hence, we establish an equivalence between a constrained problem in $\mathbf{x}$ and an unconstrained problem in $\boldsymbol{\lambda}$.

To establish the duality relation we must prove two important lemmas. In the statements below we denote by $\mathbf{x}(\boldsymbol{\lambda})$ the unique solution to (15) in the neighborhood of $\mathbf{x}^*$.

***Lemma* 1.** *The dual function ϕ has gradient*

$$\nabla\phi(\boldsymbol{\lambda}) = \mathbf{h}(\mathbf{x}(\boldsymbol{\lambda}))'. \tag{17}$$

Proof. We have explicitly, from (16),

$$\phi(\boldsymbol{\lambda}) = f(\mathbf{x}(\boldsymbol{\lambda})) + \boldsymbol{\lambda}\mathbf{h}(\mathbf{x}(\boldsymbol{\lambda})).$$

Thus

$$\nabla\phi(\boldsymbol{\lambda}) = [\nabla f(\mathbf{x}(\boldsymbol{\lambda})) + \boldsymbol{\lambda}\nabla\mathbf{h}(\mathbf{x}(\boldsymbol{\lambda}))]\nabla\mathbf{x}(\boldsymbol{\lambda}) + \mathbf{h}(\mathbf{x}(\boldsymbol{\lambda}))'.$$

Since the first term on the right vanishes by definition of $\mathbf{x}(\boldsymbol{\lambda})$, we obtain (7). ▌

Lemma 1 is of extreme practical importance, since it shows that the gradient of the dual function is simple to calculate. Once the dual function itself is evaluated, by minimization with respect to $\mathbf{x}$, the corresponding $\mathbf{h}(\mathbf{x})'$, which is the gradient, can be evaluated without further calculation.

Lemma 2. *The Hessian of the dual function is*

$$\mathbf{\Phi}(\boldsymbol{\lambda}) = -\nabla\mathbf{h}(\mathbf{x}(\boldsymbol{\lambda}))\mathbf{L}^{-1}(\mathbf{x}(\boldsymbol{\lambda}))\nabla\mathbf{h}(\mathbf{x}(\boldsymbol{\lambda}))'. \tag{18}$$

Proof. The Hessian is the derivative of the gradient. Thus by Lemma 1,

$$\mathbf{\Phi}(\boldsymbol{\lambda}) = \nabla\mathbf{h}(\mathbf{x}(\boldsymbol{\lambda}))\nabla\mathbf{x}(\boldsymbol{\lambda}). \tag{19}$$

By definition we have

$$\nabla f(\mathbf{x}(\boldsymbol{\lambda})) + \boldsymbol{\lambda}\nabla\mathbf{h}(\mathbf{x}(\boldsymbol{\lambda})) = \mathbf{0},$$

and differentiating this with respect to $\boldsymbol{\lambda}$ we obtain

$$\mathbf{L}(\mathbf{x}(\boldsymbol{\lambda}))\nabla\mathbf{x}(\boldsymbol{\lambda}) + \nabla\mathbf{h}(\mathbf{x}(\boldsymbol{\lambda}))' = \mathbf{0}.$$

Solving for $\nabla\mathbf{x}(\boldsymbol{\lambda})$ and substituting in (19) we obtain (18). ∎

Since $\mathbf{L}^{-1}(\mathbf{x}(\boldsymbol{\lambda}))$ is positive definite, and since $\nabla\mathbf{h}(\mathbf{x}(\boldsymbol{\lambda}))$ is of full rank near $\mathbf{x}^*$, we have as an immediate consequence of Lemma 2 that the $m \times m$ Hessian of ϕ is negative definite. As might be expected, this Hessian plays a dominant role in the analysis of dual methods.

Local Duality Theorem. *Suppose that the problem*

$$\begin{aligned} &\textit{minimize} \quad f(\mathbf{x}) \\ &\textit{subject to} \quad \mathbf{h}(\mathbf{x}) = \mathbf{0} \end{aligned} \tag{20}$$

has a local solution at $\mathbf{x}^*$ with corresponding value r^* and Lagrange multiplier $\boldsymbol{\lambda}^*$. Suppose also that $\mathbf{x}^*$ is a regular point of the constraints and that the corresponding Hessian of the Lagrangian $\mathbf{L}^* = \mathbf{L}(\mathbf{x}^*)$ is positive definite. Then the dual problem

$$\textit{maximize} \quad \phi(\boldsymbol{\lambda}) \tag{21}$$

has a local solution at $\boldsymbol{\lambda}^*$ with corresponding value r^* and $\mathbf{x}^*$ as the point corresponding to $\boldsymbol{\lambda}^*$ in the definition of ϕ.

Proof. It is clear that $\mathbf{x}^*$ corresponds to $\boldsymbol{\lambda}^*$ in the definition of ϕ. Now at $\boldsymbol{\lambda}^*$ we have by Lemma 1

$$\nabla\phi(\boldsymbol{\lambda}^*) = \mathbf{h}(\mathbf{x}^*)' = \mathbf{0}$$

and by Lemma 2 the Hessian of ϕ is negative definite. Thus $\boldsymbol{\lambda}^*$ satisfies the first- and second-order sufficiency conditions for an unconstrained maximum point of ϕ. The corresponding value $\phi(\lambda^*)$ is found from the definition of ϕ to be r^*. ∎

Example 1. Consider the problem in two variables

$$\begin{aligned} &\text{minimize} \quad -xy \\ &\text{subject to} \quad (x-3)^2 + y^2 = 5. \end{aligned}$$

The first-order necessary conditions are

$$-y + (2x - 6)\lambda = 0$$
$$-x + 2y\lambda = 0$$

together with the constraint. These equations have a solution at

$$x = 4, \qquad y = 2, \qquad \lambda = 1.$$

The Hessian of the corresponding Lagrangian is

$$\mathbf{L} = \begin{bmatrix} 2 & -1 \\ -1 & 2 \end{bmatrix}.$$

Since this is positive definite, we conclude that the solution obtained is a local minimum. (It can be shown, however, that it is not the global solution.)

Since $\mathbf{L}$ is positive definite, we can apply the local duality theory near this solution. We define

$$\phi(\lambda) = \min\{-xy + \lambda[(x-3)^2 + y^2 - 5]\}$$

which leads to

$$\phi(\lambda) = \frac{4\lambda + 4\lambda^3 - 80\lambda^5}{(4\lambda^2 - 1)^2}$$

valid for $\lambda > \frac{1}{2}$. It can be verified that ϕ has a local maximum at $\lambda = 1$.

Inequality Constraints

For problems having inequality constraints as well as equality constraints the above development requires only minor modification. Consider the problem

$$\begin{aligned} &\text{minimize} \quad f(\mathbf{x}) \\ &\text{subject to} \quad \mathbf{h}(\mathbf{x}) = \mathbf{0} \\ &\qquad\qquad\quad \mathbf{g}(\mathbf{x}) \leqslant \mathbf{0} \end{aligned} \tag{22}$$

where $\mathbf{g}(\mathbf{x}) \in E^p$, $\mathbf{g} \in C^2$ and everything else is as before. Suppose $\mathbf{x}^*$ is a local solution of (22) and is a regular point of the constraints. Then, as we know, there are Lagrange multipliers $\boldsymbol{\lambda}^*$ and $\boldsymbol{\mu}^* \geqslant \mathbf{0}$ such that

$$\nabla f(\mathbf{x}^*) + \boldsymbol{\lambda}^* \nabla \mathbf{h}(\mathbf{x}^*) + \boldsymbol{\mu}^* \nabla \mathbf{g}(\mathbf{x}^*) = \mathbf{0} \tag{23}$$

$$\boldsymbol{\mu}^* \mathbf{g}(\mathbf{x}^*) = 0. \tag{24}$$

We impose the local convexity assumptions that the Hessian of the Lagrangian

$$\mathbf{L}(\mathbf{x}^*) = \mathbf{F}(\mathbf{x}^*) + \boldsymbol{\lambda}^*\mathbf{H}(\mathbf{x}^*) + \boldsymbol{\mu}^*\mathbf{G}(\mathbf{x}^*) \tag{25}$$

is positive definite (on the whole space).

For $\boldsymbol{\lambda}$ and $\boldsymbol{\mu} \geqslant \mathbf{0}$ near $\boldsymbol{\lambda}^*$ and $\boldsymbol{\mu}^*$ we define the dual function

$$\phi(\boldsymbol{\lambda}, \boldsymbol{\mu}) = \min \left[f(\mathbf{x}) + \boldsymbol{\lambda}\mathbf{h}(\mathbf{x}) + \boldsymbol{\mu}\mathbf{g}(\mathbf{x})\right] \tag{26}$$

where the minimum is taken locally near $\mathbf{x}^*$. Then, it is easy to show, paralleling the development above for equality constraints, that ϕ achieves a local maximum with respect to $\boldsymbol{\lambda}$, $\boldsymbol{\mu} \geqslant \mathbf{0}$ at $\boldsymbol{\lambda}^*$, $\boldsymbol{\mu}^*$.

Convex Duality

If in addition to our other assumptions we suppose that the functions $\mathbf{f}$ and $\mathbf{g}$ are convex and $\mathbf{h}$ is affine (linear plus a constant), then (22) is a convex programming problem. In that case $\mathbf{x}^*$ is a global, as well as local, solution (see Section 6.4). Furthermore, the Lagrangian $f(\mathbf{x}) + \boldsymbol{\lambda}\mathbf{h}(\mathbf{x}) + \boldsymbol{\mu}\mathbf{g}(\mathbf{x})$ is convex for any $\boldsymbol{\lambda}$, $\boldsymbol{\mu} \geqslant \mathbf{0}$, and hence the minimization (26) can be taken to be global minimization. Finally, it can be shown (Exercise 5) that ϕ is concave, and hence any local maximum of it will be a global maximum. All of this together implies the duality results are all valid in this case with global operations replacing the local ones.

Partial Duality

It is not necessary to include the Lagrange multipliers of all the constraints of a problem in the definition of the dual function. In general, if the local convexity assumption holds, local duality can be defined with respect to any subset of functional constraints. Thus, for example, in the problem

$$\begin{aligned} \text{minimize} \quad & f(\mathbf{x}) \\ \text{subject to} \quad & \mathbf{h}(\mathbf{x}) = \mathbf{0} \\ & \mathbf{g}(\mathbf{x}) \leqslant \mathbf{0} \end{aligned} \tag{27}$$

we might define the dual function with respect to only the equality constraints. In this case we would define

$$\phi(\boldsymbol{\lambda}) = \min_{\mathbf{g}(\mathbf{x}) \leqslant \mathbf{0}} \{f(\mathbf{x}) + \boldsymbol{\lambda}\mathbf{h}(\mathbf{x})\} \tag{28}$$

where the minimum is taken locally near the solution $\mathbf{x}^*$ but constrained by the remaining constraints $\mathbf{g}(\mathbf{x}) \leqslant \mathbf{0}$. Again, the dual function defined in this way will achieve a local maximum at the optimal Lagrange multiplier $\boldsymbol{\lambda}^*$.

13.5 DUAL CANONICAL CONVERGENCE RATE

Constrained problems satisfying the local convexity assumption can be solved by solving the associated unconstrained dual problem, and any of the standard algorithms discussed in Chapters 6 through 9 can be used for this purpose. Of course, the method that suggests itself immediately is the method of steepest ascent. It can be implemented by noting that, according to Lemma 1. Section 13.4, the gradient of ϕ is available almost without cost once ϕ itself is evaluated. Without some special properties, however, the method as a whole can be extremely costly to execute, since every evaluation of ϕ requires the solution of an unconstrained problem in the unknown $\mathbf{x}$. Nevertheless, as shown in the next section, many important problems do have a structure which is suited to this approach.

The method of steepest ascent, and other gradient-based algorithms, when applied to the dual problem will have a convergence rate governed by the eigenvalue structure of the Hessian of the dual function ϕ. At the Lagrange multiplier $\boldsymbol{\lambda}^*$ corresponding to a solution $\mathbf{x}^*$ this Hessian is (according to Lemma 2, Section 13.4)

$$\mathbf{\Phi} = -\nabla\mathbf{h}(\mathbf{x}^*)(\mathbf{L}^*)^{-1}\nabla\mathbf{h}(\mathbf{x}^*)'.$$

This expression shows that $\mathbf{\Phi}$ is in some sense a restriction of the matrix $(\mathbf{L}^*)^{-1}$ to the subspace spanned by the gradients of the constraint functions, which is the orthogonal complement of the tangent subspace M. This restriction is not the orthogonal restriction of $(\mathbf{L}^*)^{-1}$ onto the complement of M since the particular representation of the constraints affects the structure of the Hessian. We see, however, that while the convergence of primal methods is governed by the restriction of $\mathbf{L}^*$ to M, the convergence of dual methods is governed by a restriction of $(\mathbf{L}^*)^{-1}$ to the orthogonal complement of M.

The *dual-canonical convergence rate* associated with the original constrained problem, which is the rate of convergence of steepest ascent applied to the dual, is $(B - b)^2/(B + b)^2$ where b and B are, respectively, the smallest and largest eigenvalues of

$$-\mathbf{\Phi} = \nabla\mathbf{h}(\mathbf{x}^*)(\mathbf{L}^*)^{-1}\nabla\mathbf{h}(\mathbf{x}^*)'.$$

For locally convex programming problems, this rate is as important as the primal canonical rate.

Scaling

We conclude this section by pointing out a kind of complementarity that exists between the primal and dual rates. Suppose one calculates the primal and dual canonical rates associated with the locally convex constrained problem

$$\begin{aligned}&\text{minimize} \quad f(\mathbf{x})\\ &\text{subject to} \quad \mathbf{h}(\mathbf{x}) = \mathbf{0}.\end{aligned}$$

If a change of primal variables $\mathbf{x}$ is introduced, the primal rate will in general change but the dual rate will not. On the other hand, if the constraints are transformed (by replacing them by $\mathbf{Th}(\mathbf{x}) = \mathbf{0}$ where $\mathbf{T}$ is a nonsingular $m \times m$ matrix), the dual rate will change but the primal rate will not.

13.6 SEPARABLE PROBLEMS

A structure that arises frequently in mathematical programming applications is that of the separable problem:

$$\text{minimize} \quad \sum_{i=1}^{q} f_i(\mathbf{x}_i) \tag{29}$$

$$\text{subject to} \quad \sum_{i=1}^{q} \mathbf{h}_i(\mathbf{x}_i) = \mathbf{0} \tag{30}$$

$$\sum_{i=1}^{p} \mathbf{g}_i(\mathbf{x}_i) \leqslant \mathbf{0}. \tag{31}$$

In this formulation the components of the n-vector $\mathbf{x}$ are partitioned into q disjoint groups, $\mathbf{x} = (\mathbf{x}_1, \mathbf{x}_2, \ldots, \mathbf{x}_q)$ where the groups may or may not have the same number of components. Both the objective function and the constraints separate into sums of functions of the individual groups. For each i, the functions f_i, $\mathbf{h}_i$, and $\mathbf{g}_i$ are twice continuously differentiable functions of dimension 1, m, and p, respectively.

Example 1. Suppose that we have a fixed budget of, say, A dollars that may be allocated among n activities. If x_i dollars is allocated to the ith activity, then there will be a benefit (measured in some units) of $f_i(x_i)$. To obtain the maximum benefit within our budget, we solve the separable problem

$$\begin{aligned}&\text{maximize} \quad \sum_{i=1}^{n} f_i(x_i)\\ &\text{subject to} \quad \sum_{i=1}^{n} x_i \leqslant A \qquad (32)\\ &\qquad\qquad x_i \geqslant 0.\end{aligned}$$

In this example $\mathbf{x}$ is partitioned into its individual components.

Example 2. Problems involving a series of decisions made at distinct times are often separable. For illustration, consider the problem of scheduling water release through a dam to produce as much electric power as possible over a given time interval while satisfying constraints on acceptable water levels. A discrete-time model of this problem is to

$$\text{maximize} \quad \sum_{k=1}^{N} f(y(k), u(k)) \tag{33}$$

$$\text{subject to} \quad y(k) = y(k-1) - u(k) + s(k), \qquad k = 1, \ldots, N \tag{34}$$

$$c \leqslant y(k) \leqslant d, \qquad k = 1, \ldots, N \tag{35}$$

$$0 \leqslant u(k), \qquad k = 1, \ldots, N. \tag{36}$$

Here $y(k)$ represents the water volume behind the dam at the end of period k, $u(k)$ represents the volume flow through the dam during period k, and $s(k)$ is the volume flowing into the lake behind the dam during period k from upper streams. The function f gives the power generation, and c and d are bounds on lake volume. The initial volume $y(0)$ is given.

In this example we consider $\mathbf{x}$ as the $2N$ dimensional vector of unknowns $y(k)$, $u(k)$, $k = 1, 2, \ldots, N$. This vector is partitioned into the pairs $\mathbf{x}_k = \big(y(k), u(k)\big)$. The objective function is then clearly in separable form. The constraints (34) can be viewed as being in the form (30) with $\mathbf{h}_k(\mathbf{x}_k)$ having dimension N and such that $\mathbf{h}_k(\mathbf{x}_k)$ is identically zero except in the k and $k+1$ components.

Decomposition

Separable problems are ideally suited to dual methods, because the required unconstrained minimization decomposes into small subproblems. To see this we recall that the generally most difficult aspect of a dual method is evaluation of the dual function. For a separable problem, if we associate $\boldsymbol{\lambda}$ with the equality constraints (30) and $\boldsymbol{\mu} \geqslant \mathbf{0}$ with the inequality constraints (31), the required dual function is

$$\phi(\lambda, \mu) = \min \sum_{i=1}^{q} f_i(\mathbf{x}_i) + \boldsymbol{\lambda}\mathbf{h}_i(\mathbf{x}_i) + \boldsymbol{\mu}\mathbf{g}_i(\mathbf{x}_i). \tag{37}$$

This minimization problem decomposes into the q separate problems

$$\min_{x_i} f_i(\mathbf{x}_i) + \boldsymbol{\lambda}\mathbf{h}(\mathbf{x}_i) + \boldsymbol{\mu}\mathbf{g}_i(\mathbf{x}_i). \tag{38}$$

The solution of these subproblems can usually be accomplished relatively efficiently, since they are of smaller dimension than the original problem.

Example 3. In Example 1 using duality with respect to the budget constraint, the ith subproblem becomes, for $\mu > 0$

$$\max_{x_i \geqslant 0} f_i(x_i) - \mu x_i$$

which is only a one-dimensional problem. It can be interpreted as setting a benefit value μ for dollars and then maximizing total benefit from activity i, accounting for the dollar expenditure.

Example 4. In Example 2 using duality with respect to the constraints (34) we denote the dual variables by $\lambda(k)$, $k = 1, 2, \ldots, N$. The kth subproblem becomes

$$\max_{\substack{c \leqslant y(k) \leqslant d \\ 0 \leqslant u(k)}} \{f(y(k), u(k)) + [\lambda(k+1) - \lambda(k)]y(k) - \lambda(k)[u(k) - s(k)]\}$$

which is a two-dimensional optimization problem. Selection of $\boldsymbol{\lambda} \in E^N$ decomposes the problem into separate problems for each time period. The variable $\lambda(k)$ can be regarded as a value, measured in units of power, for water at the beginning of period k. The kth subproblem can then be interpreted as that faced by an entrepreneur who leased the dam for one period. He can buy water for the dam at the beginning of the period at price $\lambda(k)$ and sell what he has left at the end of the period at price $\lambda(k+1)$. His problem is to determine $y(k)$ and $u(k)$ so that his net profit, accruing from sale of generated power and purchase and sale of water, is maximized.

13.7 DUALITY AND PENALTY FUNCTIONS

There are a number of interesting and important relationships between duality and penalty functions. This section shows how the two concepts can be combined in a very complementary manner to yield a promising optimization procedure.

The combination is based on an extremely simple observation. We present the idea for the problem having only equality constraints, although as usual everything easily generalizes to problems having inequalities as well. Consider

$$\begin{aligned} &\text{minimize} \quad f(\mathbf{x}) \\ &\text{subject to} \quad \mathbf{h}(\mathbf{x}) = \mathbf{0} \end{aligned} \tag{39}$$

where $\mathbf{x} \in E^n$, $\mathbf{h}(\mathbf{x}) \in E^m$. Let us suppose that $\mathbf{x}^*$ is a local solution point for this problem and that $\mathbf{x}^*$ is a regular point. Associated with this solution point is the Lagrange multiplier $\boldsymbol{\lambda}^*$ vector satisfying

$$\nabla f(\mathbf{x}^*) + \boldsymbol{\lambda}^* \nabla \mathbf{h}(\mathbf{x}^*) = \mathbf{0}. \tag{40}$$

Also there is the Hessian of the Lagrangian

$$\mathbf{L}(\mathbf{x}^*) = \mathbf{F}(\mathbf{x}^*) + \boldsymbol{\lambda}^*\mathbf{H}(\mathbf{x}^*) \tag{41}$$

which is positive semidefinite on the tangent subspace $M = \{\mathbf{x}: \nabla\mathbf{h}(\mathbf{x}^*)\mathbf{x} = \mathbf{0}\}$. If the second-order sufficiency conditions hold at $\mathbf{x}^*$, then $\mathbf{L}(\mathbf{x}^*)$ is positive definite on M. The entire matrix $\mathbf{L}(\mathbf{x}^*)$ cannot in general, however, be assumed to be positive definite, and thus duality theory cannot be applied.

The observation we make is that problem (39) is equivalent to the problem

$$\begin{aligned} &\text{minimize} \quad f(\mathbf{x}) + \mu|\mathbf{h}(\mathbf{x})|^2 \\ &\text{subject to} \quad \mathbf{h}(\mathbf{x}) = \mathbf{0} \end{aligned} \tag{42}$$

for any $\mu > 0$. The first-order necessary conditions satisfied at $\mathbf{x}^*$ for problem (42) are identical to those for (39). However, the Hessian of the Lagrangian is, for problem (42),

$$\mathbf{L}(\mathbf{x}^*) + 2\mu[\nabla\mathbf{h}(\mathbf{x}^*)' \, \nabla\mathbf{h}(\mathbf{x}^*)]. \tag{43}$$

Thus, since $\mathbf{L}(\mathbf{x}^*)$ is positive definite on M and $[\nabla\mathbf{h}(\mathbf{x}^*)' \, \nabla\mathbf{h}(\mathbf{x}^*)]$ is positive definite on the complement of M, their sum will, for large enough μ, be positive definite everywhere. Thus, local duality theory is applicable to problem (42) for sufficiently large μ.

To apply the dual method to problem (42) we define the dual function

$$\phi(\boldsymbol{\lambda}) = \min\,\{f(\mathbf{x}) + \boldsymbol{\lambda}\mathbf{h}(\mathbf{x}) + \mu|\mathbf{h}(\mathbf{x})|^2\}. \tag{44}$$

If $\mathbf{x}$ is the vector minimizing the right-hand side, then as we have seen in Section 13.5, $\mathbf{h}(\mathbf{x})'$ is the gradient of ϕ. The Hessian of ϕ is

$$\nabla\mathbf{h}(\mathbf{x})[\mathbf{F}(\mathbf{x}) + \boldsymbol{\lambda}\mathbf{H}(\mathbf{x}) + 2\mu[\nabla\mathbf{h}(\mathbf{x})' \, \nabla\mathbf{h}(\mathbf{x})]]^{-1}\nabla\mathbf{h}(\mathbf{x})' \tag{45}$$

which for large μ approaches $[1/(2\mu)]\mathbf{I}$, where $\mathbf{I}$ is the $m \times m$ identity matrix.

Now, if steepest ascent is applied to the dual, we see, from our knowledge that the Hessian (45) approaches $[1/(2\mu)]\mathbf{I}$, that the rate of convergence will be extremely fast since the ratios of the eigenvalues of the Hessian are nearly unity. Thus, not only is local duality applicable for large μ, but it is also very attractive.

We may carry our knowledge of the approximate structure of the Hessian one step further to develop a modified Newton's method for solving the dual problem. Thus, using $[1/(2\mu)]\mathbf{I}$ as the approximate Hessian and $\mathbf{h}(\mathbf{x})'$ for the gradient, we are led to the iterative scheme

$$\boldsymbol{\lambda}_{k+1} = \boldsymbol{\lambda}_k + 2\mu\mathbf{h}(\mathbf{x}_k)' \tag{46}$$

where $\mathbf{x}_k$ is the minimizer of

$$f(\mathbf{x}) + \boldsymbol{\lambda}_k\mathbf{h}(\mathbf{x}) + \mu|\mathbf{h}(\mathbf{x})|^2.$$

For large μ this iterative process converges to $\boldsymbol{\lambda}^*$ arbitrarily fast.

The technique suggested above is an excellent supplement to general penalty methods. Starting with a given (moderately large μ) a penalty problem can be solved by one of the techniques of Chapter 12 that has convergence rate independent of μ. If the solution thus obtained is not close enough to the constraint region, a new problem with the same μ but with a Lagrange multiplier obtained from (46) can be solved. In practice, since the dual iteration has fast convergence, a few total cycles will usually be sufficient.

We might summarize the observations of this section by the following statement relating primal and dual convergence rates. If a penalty term is incorporated into a problem, the condition number of the primal problem becomes increasingly poor as $\mu \to \infty$ but the condition number of the dual becomes increasingly good. To apply the dual method, however, an unconstrained penalty problem of poor condition number must be solved at each step.

13.8 EXERCISES

1. Establish global convergence for the supporting hyperplane algorithm.
2. Establish global convergence for an imperfect version of the supporting hyperplane algorithm that in interpolating to find the boundary point $\mathbf{u}$ actually finds a point somewhere on the segment joining $\mathbf{u}$ and $\frac{1}{2}\mathbf{u} + \frac{1}{2}\mathbf{w}$ and establishes a hyperplane there.
3. Prove that the convex cutting plane method is still globally convergent if it is modified by discarding from the definition of the polytope at each stage hyperplanes corresponding to inactive linear inequalities.
4. Find the global maximum of the dual function of Example 1, Section 13.4.
5. Show that the function ϕ defined for $\boldsymbol{\lambda}$, $\boldsymbol{\mu}$, ($\boldsymbol{\mu} \geqslant \mathbf{0}$), by $\phi(\boldsymbol{\lambda}, \boldsymbol{\mu}) = \min_{\mathbf{x}} [f(\mathbf{x}) + \boldsymbol{\lambda}\mathbf{h}(\mathbf{x}) + \boldsymbol{\mu}\mathbf{g}(\mathbf{x})]$ is concave over any convex region where it is finite.
6. Prove that the dual canonical rate of convergence is not affected by a change of variables in $\mathbf{x}$.
7. Corresponding to the dual function (26):
   a) Find its gradient.
   b) Find its Hessian.
   c) Verify that it has a local maximum at $\boldsymbol{\lambda}^*$, $\boldsymbol{\mu}^*$.
8. Find the Hessian of the dual function for a separable problem.
9. Verify that the Hessian (45) is $[1/(2\mu)]\mathbf{I} + o(1/\mu)$.

REFERENCES

13.1–3. Cutting plane methods were first introduced by Kelley [K2] who developed the convex cutting plane method. The supporting hyperplane algorithm was suggested by Veinott [V1]. To see how global convergence of cutting plane algorithms can be established from the general convergence theorem see Zangwill [Z2]. For some results on the convergence rates of cutting plane algorithms consult Topkis [T1], Eaves and Zangwill [E1], and Wolfe [W5].

13.4–5. An important early differential form of duality was developed by Wolfe [W2]. The convex theory can be traced to the Legendre transformation used in the calculus of variations but it owes its main heritage to Fenchel [F2]. This line was further developed by Karlin [K1] and Hurwicz [H7]. Also see Luenberger [L6]. The version treated here is new.

13.6. The solution of separable problems by dual methods in this manner was pioneered by Everett [E2].

13.7. The idea presented here was originally suggested by Hestenes [H5] although its relation to duality and Newton's method has not been previously explored. An equivalent method motivated from a different viewpoint has been proposed by Powell [P2].

Appendix A MATHEMATICAL REVIEW

The purpose of this appendix is to set down for reference and review some basic definitions, notation, and relations that are used frequently in the text.

A.1 SETS

If x is a member of the set S, we write $x \in S$. We write $y \notin S$ if y is not a member of S.

A set S may be specified by listing its elements between braces; such as, for example, $S = \{1, 2, 3, 4\}$. Alternatively, a set can be specified in the form $S = \{x : P(x)\}$ as the set of elements satisfying property P; such as $S = \{x : 1 \leqslant x \leqslant 4, x \text{ integer}\}$.

The *union* of two sets S and T is denoted $S \cup T$ and is the set consisting of the elements that belong to either S or T. The *intersection* of two sets S and T is denoted $S \cap T$ and is the set consisting of the elements that belong to both S and T. If S is a *subset* of T, that is, if every member of S is also a member of T, we write $S \subset T$ or $T \supset S$.

There are two ways that operations such as minimization over a set are represented. Specifically we write either

$$\min_{x \in S} f(x) \qquad \text{or} \qquad \min \{f(x) : x \in S\}$$

to denote the minimum value of f over the set S.

Sets of Real Numbers

If a and b are real numbers, $[a, b]$ denotes the set of real numbers x satisfying $a \leqslant x \leqslant b$. A rounded, instead of square, bracket denotes strict inequality in the definition. Thus $(a, b]$ denotes all x satisfying $a < x \leqslant b$.

If S is a set of real numbers bounded above, then there is a smallest real number y such that $x \leqslant y$ for all $x \in S$. The number y is called the *least upper bound* or *supremum* of S and is denoted

$$\sup_{x \in S} (x) \qquad \text{or} \qquad \sup \{x : x \in S\}.$$

Similarly, the *greatest lower bound* or *infimum* of a set S is denoted

$$\inf_{x \in S}(x) \qquad \text{or} \qquad \inf\{x : x \in S\}.$$

A.2 MATRIX NOTATION

A *matrix* is a rectangular array of numbers, called *elements.* The matrix itself is denoted by a boldface letter. When specific numbers are not used, the elements are denoted by italicized lower-case letters, having a double subscript. Thus we write

$$\mathbf{A} = \begin{bmatrix} a_{11} & a_{12} & \cdots & a_{1n} \\ a_{21} & a_{22} & \cdots & a_{2n} \\ \cdot & & & \\ \cdot & & & \\ \cdot & & & \\ a_{m1} & a_{m2} & \cdots & a_{mn} \end{bmatrix}$$

for a matrix $\mathbf{A}$ having m rows and n columns. Such a matrix is referred to as an $m \times n$ matrix. If we wish to specify a matrix by defining a general element, we use the notation $\mathbf{A} = [a_{ij}]$.

An $m \times n$ matrix all of whose elements are zero is called a *zero matrix* and denoted $\mathbf{0}$. A *square* matrix (a matrix with $m = n$) whose elements $a_{ij} = 0$ for $i \neq j$, and $a_{ii} = 1$ for $i = 1, 2, \ldots, n$ is said to be an *identity matrix* and denoted $\mathbf{I}$.

The *sum* of two $m \times n$ matrices $\mathbf{A}$ and $\mathbf{B}$ is written $\mathbf{A} + \mathbf{B}$ and is the matrix whose elements are the sum of the corresponding elements in $\mathbf{A}$ and $\mathbf{B}$. The *product* of a matrix $\mathbf{A}$ and a scalar λ, written $\lambda\mathbf{A}$ or $\mathbf{A}\lambda$, is obtained by multiplying each element of $\mathbf{A}$ by λ. The *product* $\mathbf{AB}$ of an $m \times n$ matrix $\mathbf{A}$ and an $n \times p$ matrix $\mathbf{B}$ is the $m \times p$ matrix $\mathbf{C}$ with elements $c_{ij} = \sum_{k=1}^{n} a_{ik}b_{kj}$.

The *transpose* of an $m \times n$ matrix $\mathbf{A}$ is the $n \times m$ matrix $\mathbf{A}'$ with elements $a'_{ij} = a_{ji}$. A (square) matrix $\mathbf{A}$ is *symmetric* if $\mathbf{A}' = \mathbf{A}$. A square matrix $\mathbf{A}$ is *nonsingular* if there is a matrix $\mathbf{A}^{-1}$, called the *inverse* of $\mathbf{A}$, such that $\mathbf{A}^{-1}\mathbf{A} = \mathbf{I} = \mathbf{A}\mathbf{A}^{-1}$. The *determinant* of a square matrix $\mathbf{A}$ is denoted by det ($\mathbf{A}$). The determinant is nonzero if and only if the matrix is nonsingular. Two square $n \times n$ matrices $\mathbf{A}$ and $\mathbf{B}$ are *similar* if there is a nonsingular matrix $\mathbf{S}$ such that $\mathbf{B} = \mathbf{S}^{-1}\mathbf{AS}$.

Matrices having a single row are referred to as *row vectors*; matrices having a single column are referred to as *column vectors. Vectors* of either type are usually denoted by lower-case boldface letters. To economize page space, row vectors are written $\mathbf{a} = [a_1, a_2, \ldots, a_n]$ and column vectors are written $\mathbf{a} = (a_1, a_2, \ldots, a_n)$. Since column vectors are used frequently, this notation avoids the necessity to display numerous columns. To further distinguish rows from columns, we write $\mathbf{a} \in E^n$ if $\mathbf{a}$ is a column vector with n components, and we write $\mathbf{b} \in E_n$ if $\mathbf{b}$ is a row vector with n components.

It is often convenient to partition a matrix into submatrices. This is indicated by drawing partitioning lines through the matrix, as for example,

$$\mathbf{A} = \left[\begin{array}{cc|cc} a_{11} & a_{12} & a_{13} & a_{14} \\ \hline a_{21} & a_{22} & a_{23} & a_{24} \\ a_{31} & a_{32} & a_{33} & a_{34} \end{array}\right] = \begin{bmatrix} \mathbf{A}_{11} & \mathbf{A}_{12} \\ \mathbf{A}_{21} & \mathbf{A}_{22} \end{bmatrix}$$

The resulting submatrices are usually denoted $\mathbf{A}_{ij}$, as illustrated.

A matrix can be partitioned into either column or row vectors, in which case a special notation is convenient. Denoting the columns of an $m \times n$ matrix $\mathbf{A}$ by $\mathbf{a}_j$, $j = 1, 2, \ldots, n$, we write $\mathbf{A} = [\mathbf{a}_1, \mathbf{a}_2, \ldots, \mathbf{a}_n]$. Similarly, denoting the rows of $\mathbf{A}$ by $\mathbf{a}^i$, $i = 1, 2, \ldots, m$, we write $\mathbf{A} = (\mathbf{a}^1, \mathbf{a}^2, \ldots, \mathbf{a}^m)$. Following the same pattern, we often write $\mathbf{A} = [\mathbf{B}, \mathbf{C}]$ for the partitioned matrix $\mathbf{A} = [\mathbf{B} \,|\, \mathbf{C}]$.

A.3 SPACES

We consider the n-component vectors $\mathbf{x} = (x_1, x_2, \ldots, x_n)$ as elements of a vector space. The space itself, n-dimensional Euclidean space, is denoted E^n. Vectors in the space can be added or multiplied by a scalar, by performing the corresponding operations on the components. We write $\mathbf{x} \geqslant \mathbf{0}$ if each component of $\mathbf{x}$ is nonnegative.

The *line segment* connecting two vectors $\mathbf{x}$ and $\mathbf{y}$ is denoted $[\mathbf{x}, \mathbf{y}]$ and consists of all vectors of the form $\alpha\mathbf{x} + (1 - \alpha)\mathbf{y}$ with $0 \leqslant \alpha \leqslant 1$.

The *scalar product* of two vectors $\mathbf{x} = (x_1, x_2, \ldots, x_n)$ and $\mathbf{y} = (y_1, y_2, \ldots, y_n)$ is defined as $\mathbf{x}'\mathbf{y} = \mathbf{y}'\mathbf{x} = \sum_{i=1}^{n} x_i y_i$. The vectors $\mathbf{x}$ and $\mathbf{y}$ are said to be *orthogonal* if $\mathbf{x}'\mathbf{y} = 0$. The *magnitude* of a vector $\mathbf{x}$ is $|\mathbf{x}| = (\mathbf{x}'\mathbf{x})^{1/2}$. For any two vectors $\mathbf{x}$ and $\mathbf{y}$ in E^n, the *Cauchy Schwarz Inequality* holds: $|\mathbf{x}'\mathbf{y}| \leqslant |\mathbf{x}| \cdot |\mathbf{y}|$.

A set of vectors $\mathbf{a}_1, \mathbf{a}_2, \ldots, \mathbf{a}_k$ is said to be *linearly dependent* if there are scalars $\lambda_1, \lambda_2, \ldots, \lambda_k$, not all zero, such that $\sum_{i=1}^{k} \lambda_i \mathbf{a}_i = \mathbf{0}$. If no such set of scalars exists, the vectors are said to be *linearly independent*. A *linear combination* of the vectors $\mathbf{a}_1, \mathbf{a}_2, \ldots, \mathbf{a}_k$ is a vector of the form $\sum_{i=1}^{k} \lambda_i \mathbf{a}_i$. The set of vectors that are linear combinations of $\mathbf{a}_1, \mathbf{a}_2, \ldots, \mathbf{a}_k$ is the set *spanned* by the vectors. A linearly independent set of vectors that span E^n is said to be a *basis* for E^n. Every basis for E^n contains exactly n vectors.

The *rank* of a matrix $\mathbf{A}$ is equal to the maximum number of linearly independent columns in $\mathbf{A}$. This number is also equal to the maximum number of linearly independent rows in $\mathbf{A}$. The $m \times n$ matrix $\mathbf{A}$ is said to be of *full rank* if the rank of $\mathbf{A}$ is equal to the minimum of m and n.

A *subspace* M of E^n is a subset that is closed under the operations of vector addition and scalar multiplication, that is, if $\mathbf{a}$ and $\mathbf{b}$ are vectors in M, then $\lambda \mathbf{a} + \mu \mathbf{b}$ is also in M for every pair of scalars λ, μ. The dimension of a subspace M is equal to the maximum number of linearly independent vectors in M. If M is a subspace of E^n, the *orthogonal complement* of M, denoted $M^{\perp}$, consists of all vectors that are orthogonal to every vector in M. The orthogonal complement of M is easily seen to be a subspace, and together M and $M^{\perp}$ span E^n in the sense that every vector $\mathbf{x} \in E^n$ can be written uniquely in the form $\mathbf{x} = \mathbf{a} + \mathbf{b}$ with $\mathbf{a} \in M$, $\mathbf{b} \in M^{\perp}$. In this case $\mathbf{a}$ and $\mathbf{b}$ are said to be the *orthogonal projections* of $\mathbf{x}$ onto the subspaces M and $M^{\perp}$, respectively.

A correspondence $\mathbf{A}$ that associates with each point in a space X a point in a space Y is said to be a *mapping from X to Y*. For convenience this situation is symbolized by $\mathbf{A}: X \to Y$. The mapping $\mathbf{A}$ may be either linear or nonlinear.

A.4 EIGENVALUES AND QUADRATIC FORMS

Corresponding to an $n \times n$ square matrix $\mathbf{A}$, a scalar λ and a nonzero vector $\mathbf{x}$ satisfying the equation $\mathbf{Ax} = \lambda \mathbf{x}$ are said to be, respectively, an eigenvalue and eigenvector of $\mathbf{A}$. In order that λ be an eigenvalue it is clear that it is necessary and sufficient for $\mathbf{A} - \lambda \mathbf{I}$ to be singular, and hence $\det(\mathbf{A} - \lambda \mathbf{I}) = 0$. This last result, when expanded, yields an nth order polynomial equation which can be solved for n (possibly nondistinct) complex roots λ which are the eigenvalues of $\mathbf{A}$.

Now, for the remainder of this section, assume that $\mathbf{A}$ is symmetric. Then the following properties hold:

i) The eigenvalues of $\mathbf{A}$ are real.

ii) Eigenvectors associated with distinct eigenvalues are orthogonal.

iii) There is an orthogonal basis for E^n, each element of which is an eigenvector of $\mathbf{A}$.

If the basis $\mathbf{u}_1, \mathbf{u}_2, \ldots, \mathbf{u}_n$ in (iii) is normalized so that each element has magnitude unity, then defining the matrix $\mathbf{Q} = [\mathbf{u}_1, \mathbf{u}_2, \ldots, \mathbf{u}_n]$ we note that $\mathbf{Q}'\mathbf{Q} = \mathbf{I}$ and hence $\mathbf{Q}' = \mathbf{Q}^{-1}$. A matrix with this property is said to be an *orthogonal* matrix. Also, we observe, in this case, that

$$\mathbf{Q}^{-1}\mathbf{AQ} = \mathbf{Q}'\mathbf{AQ} = \mathbf{Q}'[\mathbf{Au}_1, \mathbf{Au}_2, \ldots, \mathbf{Au}_n] = \mathbf{Q}'[\lambda_1 \mathbf{u}_1, \lambda_2 \mathbf{u}_2, \ldots, \lambda_n \mathbf{u}_n].$$

Thus

$$\mathbf{Q}^{-1}\mathbf{AQ} = \begin{bmatrix} \lambda_1 & & & & & \\ & \lambda_2 & & & & \\ & & \cdot & & & \\ & & & \cdot & & \\ & & & & \cdot & \\ & & & & & \lambda_n \end{bmatrix}$$

and therefore $\mathbf{A}$ is similar to a diagonal matrix.

A symmetric matrix $\mathbf{A}$ is said to be *positive definite* if the *quadratic form* $\mathbf{x}'\mathbf{A}\mathbf{x}$ is positive for all nonzero vectors $\mathbf{x}$. Similarly, we define *positive semidefinite*, *negative definite*, and *negative semidefinite* if $\mathbf{x}'\mathbf{A}\mathbf{x} \geqslant 0$, < 0, or $\leqslant 0$ for all $\mathbf{x}$. The matrix $\mathbf{A}$ is *indefinite* if $\mathbf{x}'\mathbf{A}\mathbf{x}$ is positive for some $\mathbf{x}$ and negative for others.

It is easy to obtain a connection between definiteness and the eigenvalues of $\mathbf{A}$. For any $\mathbf{x}$ let $\mathbf{y} = \mathbf{Q}^{-1}\mathbf{x}$ where $\mathbf{Q}$ is defined as above. Then $\mathbf{x}'\mathbf{A}\mathbf{x} = \mathbf{y}'\mathbf{Q}'\mathbf{A}\mathbf{Q}\mathbf{y} = \sum_{i=1}^{n} \lambda_i y_i^2$. Since the y_i's are arbitrary (since $\mathbf{x}$ is), it is clear that $\mathbf{A}$ is positive definite (or positive semidefinite) if and only if all eigenvalues of $\mathbf{A}$ are positive (or nonnegative).

Through diagonalization we can also easily show that a positive semidefinite matrix $\mathbf{A}$ has a positive semidefinite (symmetric) square root $\mathbf{A}^{\frac{1}{2}}$ satisfying $\mathbf{A}^{\frac{1}{2}} \cdot \mathbf{A}^{\frac{1}{2}} = \mathbf{A}$. For this we use $\mathbf{Q}$ as above and define

$$\mathbf{A}^{\frac{1}{2}} = \mathbf{Q} \begin{bmatrix} \lambda_1^{\frac{1}{2}} & & & & \\ & \lambda_1^{\frac{1}{2}} & & & \\ & & \cdot & & \\ & & & \cdot & \\ & & & & \cdot \\ & & & & & \lambda_n^{\frac{1}{2}} \end{bmatrix} \mathbf{Q}'$$

which is easily verified to have the desired properties.

A.5 TOPOLOGICAL CONCEPTS

A sequence of vectors $\mathbf{x}_0, \mathbf{x}_1, \ldots, \mathbf{x}_k, \ldots$, denoted $\{\mathbf{x}_k\}_{k=0}^{\infty}$, or if the index set is understood, by simply $\{\mathbf{x}_k\}$, is said to *converge* to the limit $\mathbf{x}$ if $|\mathbf{x}_k - \mathbf{x}| \to 0$ as $k \to \infty$ (that is, if given $\varepsilon > 0$, there is a N such that $k \geqslant N$ implies $|\mathbf{x}_k - \mathbf{x}| < \varepsilon$). If $\{\mathbf{x}_k\}$ converges to $\mathbf{x}$, we write $\mathbf{x}_k \to \mathbf{x}$ or limit $\mathbf{x}_k = \mathbf{x}$.

A point $\mathbf{x}$ is a *limit point* of the sequence $\{\mathbf{x}_k\}$ if there is a subsequence of $\{\mathbf{x}_k\}$ convergent to $\mathbf{x}$. Thus $\mathbf{x}$ is a limit point of $\{\mathbf{x}_k\}$ if there is a subset $\mathscr{K}$ of the positive integers such that $\{\mathbf{x}_k\}_{k \in \mathscr{K}}$ is convergent to $\mathbf{x}$.

A *sphere around* $\mathbf{x}$ is a set of the form $\{\mathbf{y} : |\mathbf{y} - \mathbf{x}| < \varepsilon\}$ for some $\varepsilon > 0$. Such a sphere is also referred to as the *neighborhood* of $\mathbf{x}$ of radius ε.

A subset S of E^n is *open* if around every point in S there is a sphere that is contained in S. Equivalently, S is open if given $\mathbf{x} \in S$ there is an $\varepsilon > 0$ such that $|\mathbf{y} - \mathbf{x}| < \varepsilon$ implies $\mathbf{y} \in S$. Thus the sphere $\{\mathbf{x} : |\mathbf{x}| < 1\}$ is open. In general, open sets can be characterized as sets having no sharp boundaries. The *interior* of any set S in E^n is the set of points $\mathbf{x} \in S$ which are the center of some sphere contained in S. The interior of a set is always open; indeed it is the largest open set contained in S. The interior of the set $\{\mathbf{x} : |\mathbf{x}| \leqslant 1\}$ is the sphere $\{\mathbf{x} : |\mathbf{x}| < 1\}$.

A set P is *closed* if every point that is arbitrarily close to the set P is a member of P. Equivalently, P is closed if $\mathbf{x}_k \to \mathbf{x}$ with $\mathbf{x}_k \in P$ implies $\mathbf{x} \in P$.

Thus the set $\{\mathbf{x} : |\mathbf{x}| \leqslant 1\}$ is closed. The *closure* of any set P in E^n is the smallest closed set containing P. The *boundary* of a set is that part of the closure that is not in the interior.

A set is *compact* if it is both closed and bounded (that is, if it is closed and is contained within some sphere of finite radius). An important result, due to Weierstrass, is that if S is a compact set and $\{\mathbf{x}_k\}$ is a sequence each member of which belongs to S, then $\{\mathbf{x}_k\}$ has a limit point in S (that is, there is subsequence converging to a point in S).

Corresponding to a bounded sequence $\{r_k\}_{k=0}^{\infty}$ of real numbers, if we let $s_k = \sup \{r_i : i \geqslant k\}$ then $\{s_k\}$ converges to some real number s_0. This number is called the *limit superior* of $\{r_k\}$ and is denoted $\overline{\lim\limits_{k \to \infty}} (r_k)$.

A.6 FUNCTIONS

A real-valued function f defined on a subset of E^n is said to be *continuous* at $\mathbf{x}$ if $\mathbf{x}_k \to \mathbf{x}$ implies $f(\mathbf{x}_k) \to f(\mathbf{x})$. Equivalently, f is continuous at $\mathbf{x}$ if given $\varepsilon > 0$ there is a $\delta > 0$ such that $|\mathbf{y} - \mathbf{x}| < \delta$ implies $|f(\mathbf{y}) - f(\mathbf{x})| < \varepsilon$. An important result connected with continuous functions is a *theorem of Weierstrass*: A continuous function f defined on a compact set S has a minimum point in S; that is, there is an $\mathbf{x}^* \in S$ such that for all $\mathbf{x} \in S$, $f(\mathbf{x}) \geqslant f(\mathbf{x}^*)$.

A set of real-valued functions $f_1, f_2, \ldots, f_m$ on E^n can be regarded as a single vector function $\mathbf{f} = (f_1, f_2, \ldots, f_m)$. This function assigns a vector $\mathbf{f}(\mathbf{x}) = (f_1(\mathbf{x}), f_2(\mathbf{x}), \ldots, f_m(\mathbf{x}))$ in E^m to every vector $\mathbf{x} \in E^n$. Such a vector-valued function is said to be *continuous* if each of its component functions is continuous.

If each component of $\mathbf{f} = (f_1, f_2, \ldots, f_m)$ is continuous on some open set of E^n, then we write $\mathbf{f} \in C$. If in addition, each component function has first partial derivatives which are continuous on this set, we write $\mathbf{f} \in C^1$. In general, if the component functions have continuous partial derivatives of order p, we write $\mathbf{f} \in C^p$.

If $f \in C^1$ is a real-valued function on E^n, $f(\mathbf{x}) = f(x_1, x_2, \ldots, x_n)$, we define the *gradient* of f to be the vector

$$\nabla f(\mathbf{x}) = \left[\frac{\partial f(\mathbf{x})}{\partial x_1}, \frac{\partial f(\mathbf{x})}{\partial x_2}, \ldots, \frac{\partial f(\mathbf{x})}{\partial x_n}\right].$$

We sometimes use the alternative notation $f_{\mathbf{x}}(\mathbf{x})$ for $\nabla f(\mathbf{x})$. In matrix calculations the gradient is considered to be a row vector.

If $f \in C^2$ then we define the *Hessian* of f at $\mathbf{x}$ to be the $n \times n$ matrix denoted $\nabla^2 f(\mathbf{x})$ or $\mathbf{F}(\mathbf{x})$ as

$$\mathbf{F}(\mathbf{x}) = \left[\frac{\partial^2 f(\mathbf{x})}{\partial x_i \partial x_j}\right].$$

Since

$$\frac{\partial^2 f}{\partial x_i \partial x_j} = \frac{\partial^2 f}{\partial x_j \partial x_j}$$

it is easily seen that the Hessian is symmetric.

For a vector-valued function $\mathbf{f} = (f_1, f_2, \ldots, f_m)$ the situation is similar. If $\mathbf{f} \in C^1$, the first derivative is defined as the $m \times n$ matrix

$$\nabla \mathbf{f}(\mathbf{x}) = \left[\frac{\partial f_i(\mathbf{x})}{\partial x_j}\right].$$

If $\mathbf{f} \in C^2$ it is possible to define the m Hessians $\mathbf{F}_1(\mathbf{x}), \mathbf{F}_2(\mathbf{x}), \ldots, \mathbf{F}_m(\mathbf{x})$ corresponding to the m component functions. The second derivative itself, for a vector function, is a third order tensor but we do not require its use explicitly. Given any $\boldsymbol{\lambda} = [\lambda_1, \lambda_2, \ldots, \lambda_m] \in E_m$, we note, however, that the real-valued function $\boldsymbol{\lambda}\mathbf{f}$ has gradient equal to $\boldsymbol{\lambda}\nabla\mathbf{f}(\mathbf{x})$ and Hessian, denoted $\boldsymbol{\lambda}\mathbf{F}(\mathbf{x})$, equal to

$$\boldsymbol{\lambda}\mathbf{F}(\mathbf{x}) = \sum_{i=1}^{m} \lambda_i \mathbf{F}_i(\mathbf{x}).$$

Taylor's Theorem

A group of results that are used frequently in analysis are referred to under the general heading of Taylor's Theorem or Mean Value Theorems. If $f \in C^1$ in a region containing the line segment $[\mathbf{x}_1, \mathbf{x}_2]$, then there is a θ, $0 \leqslant \theta \leqslant 1$ such that

$$f(\mathbf{x}_2) = f(\mathbf{x}_1) + \nabla f(\theta\mathbf{x}_1 + (1-\theta)\mathbf{x}_2)(\mathbf{x}_2 - \mathbf{x}_1).$$

Furthermore, if $f \in C^2$ then there is a θ, $0 \leqslant \theta \leqslant 1$ such that

$$f(\mathbf{x}_2) = f(\mathbf{x}_1) + \nabla f(\mathbf{x}_1)(\mathbf{x}_2 - \mathbf{x}_1) + \tfrac{1}{2}(\mathbf{x}_2 - \mathbf{x}_1)' \mathbf{F}(\theta\mathbf{x}_1 + (1-\theta)\mathbf{x}_2)(\mathbf{x}_2 - \mathbf{x}_1)$$

where $\mathbf{F}$ denotes the Hessian of f.

Implicit Function Theorem

Suppose we have a set of m equations in n variables

$$h_i(\mathbf{x}) = 0, \qquad i = 1, 2, \ldots, m.$$

The implicit function theorem addresses the question as to whether if $n - m$ of the variables are fixed, the equations can be solved for the remaining m equations. Thus, selecting m variables, say $x_1, x_2, \ldots, x_m$, we wish to determine if these may be expressed in terms of the remaining variables in the form

$$x_i = \phi_i(x_{m+1}, x_{m+2}, \ldots, x_n), \qquad i = 1, 2, \ldots, m.$$

The functions ϕ_i, if they exist, are called *implicit* functions.

Theorem. *Let* $\mathbf{x}^0 = (x_1^0, x_2^0, \ldots, x_n^0)$ *be a point in* E^n *satisfying the properties:*

i) *The functions* $h_i \in C^p$, $i = 1, 2, \ldots, m$ *in some neighborhood of* $\mathbf{x}^0$, *for some* $p \geqslant 1$.

ii) $h_i(\mathbf{x}^0) = 0, \quad i = 1, 2, \ldots, m.$

iii) *The* $m \times m$ Jacobian *matrix*

$$\mathbf{J} = \begin{bmatrix} \dfrac{\partial h_1(\mathbf{x}^0)}{\partial x_1} & \cdots & \dfrac{\partial h_1(\mathbf{x}^0)}{\partial x_m} \\ \vdots & & \vdots \\ \dfrac{\partial h_m(\mathbf{x}^0)}{\partial x_1} & \cdots & \dfrac{\partial h_m(\mathbf{x}^0)}{\partial x_m} \end{bmatrix}$$

is nonsingular.

Then there is a neighborhood of $\hat{\mathbf{x}}^0 = (x_{m+1}^0, x_{m+2}^0, \ldots, x_n^0) \in E^{n-m}$ *such that for* $\hat{\mathbf{x}} = (x_{m+1}, x_{m+2}, \ldots, x_n)$ *in this neighborhood there are functions* $\phi_i(\hat{\mathbf{x}})$, $i = 1, 2, \ldots, m$ *such that*

i) $\phi_i \in C^p$.

ii) $x_i^0 = \phi_i(\hat{\mathbf{x}}^0), \qquad i = 1, 2, \ldots, m.$

iii) $h_i(\phi_i(\hat{\mathbf{x}}), \phi_2(\hat{\mathbf{x}}), \ldots, \phi_m(\hat{\mathbf{x}}), \hat{\mathbf{x}}) = 0, \qquad i = 1, 2, \ldots, m.$

Example 1. Consider the equation $x_1^2 + x_2 = 0$. A solution is $x_1 = 0$, $x_2 = 0$. However, in a neighborhood of this solution there is no function ϕ such that $x_1 = \phi(x_2)$. At this solution condition (iii) of the implicit function theorem is violated. At any other solution, however, such a ϕ exists.

Example 2. Let $\mathbf{A}$ be an $m \times n$ matrix ($m < n$) and consider the system of linear equations $\mathbf{Ax} = \mathbf{b}$. If $\mathbf{A}$ is partitioned as $\mathbf{A} = [\mathbf{B}, \mathbf{C}]$ where $\mathbf{B}$ is $m \times m$ then condition (iii) is satisfied if and only if $\mathbf{B}$ is nonsingular. This condition corresponds, of course, exactly with what the theory of linear equations tells us. In view of this example, the implicit function can be regarded as a nonlinear generalization of the linear theory.

o, *O* Notation

If g is a real-valued function of a real variable, the notation $g(x) = O(x)$ means that $g(x)$ goes to zero at least as fast as x does. More precisely, it means that there is a $K \geqslant 0$ such that

$$\left|\frac{g(x)}{x}\right| \leqslant K \quad \text{as} \quad x \to 0.$$

The notation $g(x) = o(x)$ means that $g(x)$ goes to zero faster than x does; or equivalently, that K, above, is zero.

Appendix B CONVEX SETS

B.1 BASIC DEFINITIONS

Concepts related to convex sets so dominate the theory of optimization that it is essential for a student of optimization to have knowledge of their most fundamental properties. In this appendix is compiled a brief summary of the most important of these properties.

Definition. A set C in E^n is said to be *convex* if for every $\mathbf{x}_1, \mathbf{x}_2 \in C$ and every real number α, $0 < \alpha < 1$, the point $\alpha\mathbf{x}_1 + (1 - \alpha)\mathbf{x}_2 \in C$.

This definition can be interpreted geometrically as stating that a set is convex if, given two points in the set, every point on the line segment joining these two points is also a member of the set. This is illustrated in Fig. B.1.

Fig. B.1 Convexity

The following proposition shows that certain familiar set operations preserve convexity.

***Proposition* 1.** *Convex sets in E^n satisfy the following relations:*

i) *If C is a convex set and β is a real number, the set*

$$\beta C = \{\mathbf{x} : \mathbf{x} = \beta \mathbf{c},\ \mathbf{c} \in C\}$$

is convex.

ii) *If C and D are convex sets, then the set*

$$C + D = \{\mathbf{x} : \mathbf{x} = \mathbf{c} + \mathbf{d},\ \mathbf{c} \in C,\ \mathbf{d} \in D\}$$

is convex.

iii) *The intersection of any collection of convex sets is convex.*

The proofs of these three properties follow directly from the definition of a convex set and are left to the reader. The properties themselves are illustrated in Fig. B.2.

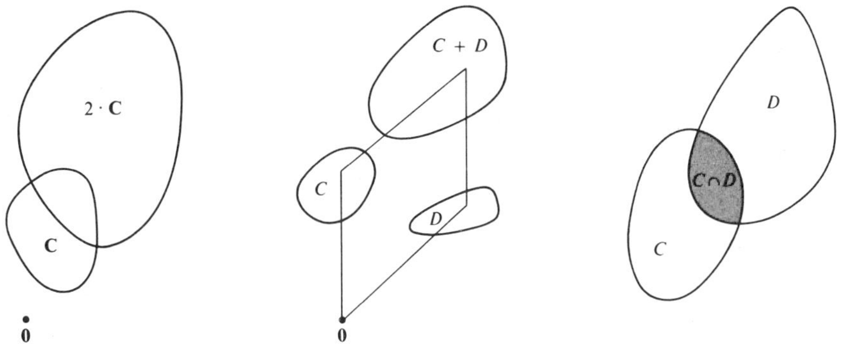

Fig. B.2 Properties of convex sets

Another important concept is that of forming the smallest convex set containing a given set.

Definition. Let S be a subset of E^n. The *convex hull* of S, denoted co(S), is the set which is the intersection of all convex sets containing S. The *closed convex hull* of S is defined as the closure of co(S).

Finally, we conclude this section by defining a *cone* and a *convex cone.* A convex cone is a special kind of convex set that arises quite frequently.

Definition. A set C is a *cone* if $\mathbf{x} \in C$ implies $\alpha\mathbf{x} \in C$ for all $\alpha > 0$. A cone that is also convex is a *convex cone.*

Some cones are shown in Fig. B.3. Their basic property is that if a point $\mathbf{x}$ belongs to a cone, then the entire half line from the origin through the point (but not the origin itself) also must belong to the cone.

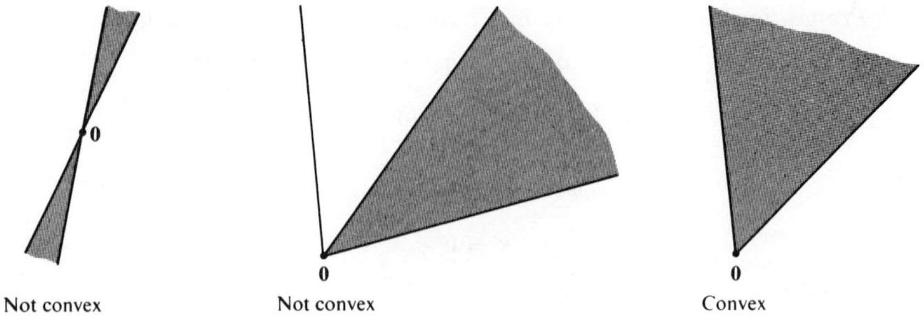

Fig. B.3 Cones

B.2 HYPERPLANES AND POLYTOPES

The most important type of convex set (aside from single points) is the hyperplane. Hyperplanes dominate the entire theory of optimization, appearing under the guise of Lagrange multipliers, duality theory, or gradient calculations.

The most natural definition of a hyperplane is the logical generalization of the geometric properties of a plane in three dimensions. We start by giving this geometric definition. For computations and for a concrete description of hyperplanes, however, there is an equivalent algebraic definition that is more useful. A major portion of this section is devoted to establishing this equivalence.

Definition. A set V in E^n is said to be a *linear variety*, if, given $\mathbf{x}_1, \mathbf{x}_2 \in V$, we have $\lambda \mathbf{x}_1 + (1 - \lambda)\mathbf{x}_2 \in V$ for all real numbers λ.

Note that the only difference between the definition of a linear variety and a convex set is that in a linear variety the entire line passing through any two points, rather than simply the line segment between them, must lie in the set. Thus in three dimensions the nonempty linear varieties are points, lines, two-dimensional planes, and the whole space. In general, it is clear that we may speak of the dimension of a linear variety. Thus, for example, a point is a linear variety of dimension zero and a line is a linear variety of dimension one. In the general case, the dimension of a linear variety in E^n can be found by translating it (moving it) so that it contains the origin and then determining the dimension of the resulting set, which is then a subspace of E^n.

Definition. A *hyperplane* in E^n is an $n - 1$ dimensional linear variety.

We see that hyperplanes generalizes the concept of a two-dimensional plane in three-dimensional space. They can be regarded as the largest linear varieties in a space, other than the entire space itself.

We now relate this abstract geometric definition to an algebraic one.

***Proposition* 2.** *Let* **a** *be a nonzero n-dimensional row vector, and let c be a real number. The set*

$$H = \{\mathbf{x} \in E^n : \mathbf{a}\mathbf{x} = c\}$$

is a hyperplane in E^n.

Proof. It follows directly from the linearity of the equation $\mathbf{a}\mathbf{x} = c$ that H is a linear variety. Let $\mathbf{x}_1$ be any vector in H. Translating by $-\mathbf{x}_1$ we obtain the set $M = H - \mathbf{x}_1$ which is a linear subspace of E^n. This subspace consists of all vectors $\mathbf{x}$ satisfying $\mathbf{a}\mathbf{x} = 0$; in other words, all vectors orthogonal to $\mathbf{a}$. This is clearly an $n - 1$ dimensional subspace. ▮

***Proposition* 3.** *Let H be a hyperplane in* E^n. *Then there is a nonzero n-dimensional row vector* **a** *and a constant c such that*

$$H = \{\mathbf{x} \in E^n : \mathbf{a}\mathbf{x} = c\}.$$

Proof. Let $\mathbf{x}_1 \in H$ and translate by $-\mathbf{x}_1$ obtaining the set $M = H - \mathbf{x}_1$. Since H is a hyperplane, M is an $n - 1$ dimensional subspace. Let $\mathbf{a}$ be any nonzero row vector that is orthogonal to this subspace, that is, $\mathbf{a}'$ belongs to the one-dimensional subspace $M^{\perp}$. Clearly $M = \{\mathbf{x} : \mathbf{a}\mathbf{x} = 0\}$. Letting $c = \mathbf{a}\mathbf{x}_1$ we see that if $\mathbf{x}_2 \in H$ we have $\mathbf{x}_2 - \mathbf{x}_1 \in M$ and thus $\mathbf{a}\mathbf{x}_2 - \mathbf{a}\mathbf{x}_1 = 0$ which implies $\mathbf{a}\mathbf{x}_2 = c$. Thus $H \subset \{\mathbf{x} : \mathbf{a}\mathbf{x} = c\}$. Since H is, by definition, of dimension $n - 1$ and $\{\mathbf{x} : \mathbf{a}\mathbf{x} = c\}$ is of dimension $n - 1$ by Proposition 2, these two sets must be equal. ▮

Combining Propositions 2 and 3, we see that a hyperplane is the set of solutions to a single linear equation. This is illustrated in Fig. B.4. We now use hyperplanes to build up other important classes of convex sets.

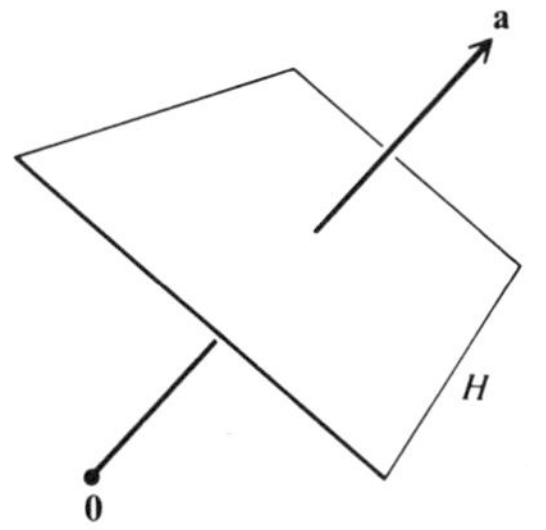

Fig. B.4

Definition. Let $\mathbf{a}$ be a nonzero row vector in E^n and let c be a real number. Corresponding to the hyperplane $H = \{\mathbf{x} : \mathbf{ax} = c\}$ are the *positive and negative closed half spaces*

$$H_+ = \{\mathbf{x} : \mathbf{ax} \geqslant c\}$$
$$H_- = \{\mathbf{x} : \mathbf{ax} \leqslant c\}$$

and the *positive* and *negative open half spaces*

$$\overset{\circ}{H}_+ = \{\mathbf{x} : \mathbf{ax} > c\}$$
$$\overset{\circ}{H}_- = \{\mathbf{x} : \mathbf{ax} < c\}.$$

It is easy to see that half spaces are convex sets and that the union of H_+ and H_- is the whole space.

Definition. A set which can be expressed as the intersection of a finite number of closed half spaces is said to be a *convex polytope.*

We see that convex polytopes are the sets obtained as the family of solutions to a set of linear inequalities of the form

$$\begin{aligned} \mathbf{a}^1\mathbf{x} &\leqslant b_1 \\ \mathbf{a}^2\mathbf{x} &\leqslant b_2 \\ &\vdots \\ \mathbf{a}^m\mathbf{x} &\leqslant b_m \end{aligned}$$

since each individual inequality defines a half space and the solution family is the intersection of these half spaces. (If some $\mathbf{a}^i = \mathbf{0}$, the resulting set can still, as the reader may verify, be expressed as the intersection of a finite number of half spaces.)

Several polytopes are illustrated in Fig. B.5. We note that a polytope may be empty, bounded, or unbounded. The case of a nonempty bounded polytope is of special interest and we distinguish this case by the following.

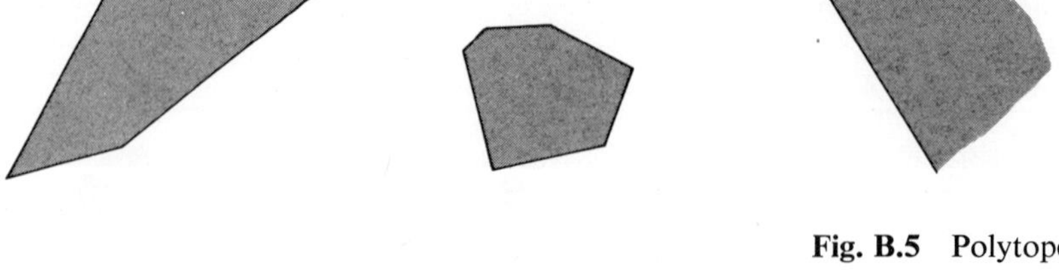

Fig. B.5 Polytopes

Definition. A nonempty bounded polytope is called a *polyhedron*.

B.3 SEPARATING AND SUPPORTING HYPERPLANES

The two theorems in this section are perhaps the most important results related to convexity. Geometrically, the first states that given a point outside a convex set, a hyperplane can be passed through the point that does not touch the convex set. The second, which is a limiting case of the first, states that given a boundary point of a convex set, there is a hyperplane that contains the boundary point and contains the convex set on one side of it.

Theorem 1. *Let C be a convex set and let* $\mathbf{y}$ *be a point exterior to the closure of C. Then there is a (row) vector* $\mathbf{a}$ *such that* $\mathbf{ay} < \inf_{\mathbf{x} \in C} \mathbf{ax}$.

Proof. Let

$$\delta = \inf_{\mathbf{x} \in C} |\mathbf{x} - \mathbf{y}| > 0.$$

There is an $\mathbf{x}_0$ on the boundary of C such that $|\mathbf{x}_0 - \mathbf{y}| = \delta$. This follows because the continuous function $f(\mathbf{x}) = |\mathbf{x} - \mathbf{y}|$ achieves its minimum over any closed and bounded set and it is clearly only necessary to consider $\mathbf{x}$ in the intersection of the closure of C and the sphere of radius 2δ centered at $\mathbf{y}$.

We shall show that setting $\mathbf{a}' = \mathbf{x}_0 - \mathbf{y}$ satisfies the conditions of the theorem. Let $\mathbf{x} \in C$. For any α, $0 \leqslant \alpha \leqslant 1$, the point $\mathbf{x}_0 + \alpha(\mathbf{x} - \mathbf{x}_0) \in \bar{C}$ and thus

$$|\mathbf{x}_0 + \alpha(\mathbf{x} - \mathbf{x}_0) - \mathbf{y}|^2 \geqslant |\mathbf{x}_0 - \mathbf{y}|^2.$$

Expanding,

$$2\alpha(\mathbf{x}_0 - \mathbf{y})'(\mathbf{x} - \mathbf{x}_0) + \alpha^2|\mathbf{x} - \mathbf{x}_0|^2 \geqslant 0.$$

Thus, considering this as $\alpha \to 0+$, we obtain

$$(\mathbf{x}_0 - \mathbf{y})'(\mathbf{x} - \mathbf{x}_0) \geqslant 0$$

or,

$$\begin{aligned}(\mathbf{x}_0 - \mathbf{y})'\mathbf{x} \geqslant (\mathbf{x}_0 - \mathbf{y})'\mathbf{x}_0 &= (\mathbf{x}_0 - \mathbf{y})'\mathbf{y} + (\mathbf{x}_0 - \mathbf{y})'(\mathbf{x}_0 - \mathbf{y}) \\ &= (\mathbf{x}_0 - \mathbf{y})'\mathbf{y} + \delta^2.\end{aligned}$$

Setting $\mathbf{a}' = \mathbf{x}_0 - \mathbf{y}$ proves the theorem. ∎

The geometrical interpretation of Theorem 1 is that, given a convex set C and a point $\mathbf{y}$ exterior to the closure of C, there is a hyperplane containing $\mathbf{y}$ that contains C in one of its open half spaces. We can easily extend this theorem to include the case where $\mathbf{y}$ is a boundary point of C.

Theorem 2. *Let C be a convex set and let* $\mathbf{y}$ *be a boundary point of C. Then there is a hyperplane containing* $\mathbf{y}$ *and containing C in one of its closed half spaces.*

Proof. Let $\{\mathbf{y}_k\}$ be a sequence of vectors, exterior to the closure of C, converging to $\mathbf{y}$. Let $\{\mathbf{a}^k\}$ be the sequence of corresponding vectors constructed according to Theorem 1, normalized so that $|\mathbf{a}^k| = 1$, such that

$$\mathbf{a}^k\mathbf{y} < \inf_{\mathbf{x} \in C} \mathbf{a}^k\mathbf{x}.$$

Since $\{\mathbf{a}^k\}$ is a bounded sequence, it has a convergent subsequence $\{\mathbf{a}^k\}$, $k \in \mathscr{K}$ with limit $\mathbf{a}$. For this vector we have for any $\mathbf{x} \in C$

$$\mathbf{a}\mathbf{y} = \lim_{k \in \mathscr{K}} \mathbf{a}^k\mathbf{y} \leqslant \lim_{k \in \mathscr{K}} \mathbf{a}^k\mathbf{x} = \mathbf{a}\mathbf{x}. \blacksquare$$

Definition. A hyperplane containing a convex set C in one of its closed half spaces and containing a boundary point of C is said to be a *supporting hyperplane* of C.

In terms of this definition, Theorem 2 says that, given a convex set C and a boundary point $\mathbf{y}$ of C, there is a hyperplane supporting C at $\mathbf{y}$.

B.4 EXTREME POINTS

Definition. A point $\mathbf{x}$ in a convex set C is said to be an *extreme point* of C if there are no two distinct points $\mathbf{x}_1$ and $\mathbf{x}_2$ in C such that $\mathbf{x} = \alpha\mathbf{x}_1 + (1 - \alpha)\mathbf{x}_2$ for some α, $0 < \alpha < 1$.

For example, in E^2 the extreme points of a square are its four corners; the extreme points of a circular disk are all points on the boundary. Note that a linear variety consisting of more than one point has no extreme points.

Lemma 1. *Let C be a convex set, H a supporting hyperplane of C, and T the intersection of H and C. Every extreme point of T is an extreme point of C.*

Proof. Suppose $\mathbf{x}_0 \in T$ is not an extreme point of C. Then $\mathbf{x}_0 = \alpha\mathbf{x}_1 + (1 - \alpha)\mathbf{x}_2$ for some $\mathbf{x}_1, \mathbf{x}_2 \in C$, $0 < \alpha < 1$. Let H be described as $H = \{\mathbf{x} : \mathbf{a}\mathbf{x} = c\}$ with C contained in its closed positive half space. Then

$$\mathbf{a}\mathbf{x}_1 \geqslant c, \qquad \mathbf{a}\mathbf{x}_2 \geqslant c.$$

But, since $\mathbf{x} \in H$,

$$c = \mathbf{a}\mathbf{x}_0 = \alpha\mathbf{a}\mathbf{x}_1 + (1 - \alpha)\mathbf{a}\mathbf{x}_2$$

and thus $\mathbf{x}_1$ and $\mathbf{x}_2 \in H$. Hence $\mathbf{x}_1, \mathbf{x}_2 \in T$ and $\mathbf{x}_0$ is not an extreme point of T. $\blacksquare$

Theorem 3. *A closed bounded convex set in E^n is equal to the closed convex hull of its extreme points.*

Proof. The proof is by induction on the dimension of the space E^n. The statement is easily seen to be true for $n = 1$. Suppose that it is true for $n - 1$. Let C be a closed bounded convex set in E^n, and let K be the closed convex hull of the extreme points of C. We wish to show that $K = C$.

Assume there is $\mathbf{y} \in C$, $\mathbf{y} \notin K$. Then by Theorem 1, Section B.3, there is a hyperplane separating $\mathbf{y}$ and K; that is, there is $\mathbf{a} \neq \mathbf{0}$, such that $\mathbf{ay} < \inf_{\mathbf{x} \in K} \mathbf{ax}$. Let $c_0 = \inf_{\mathbf{x} \in C} (\mathbf{ax})$. The number c_0 is finite and there is an $\mathbf{x}_0 \in C$ for which $\mathbf{ax}_0 = c_0$, because by Weierstrass' Theorem, the continuous function $\mathbf{ax}$ achieves its minimum over any closed bounded set. Thus the hyperplane $H = \{\mathbf{x} : \mathbf{ax} = c_0\}$ is a supporting hyperplane to C. It is disjoint from K since $c_0 < \inf_{\mathbf{x} \in K} (\mathbf{ax})$.

Let $T = H \cap C$. Then T is a bounded closed convex subset of H which can be regarded as a space of dimension $n - 1$. T is nonempty, since it contains $\mathbf{x}_0$. Thus by the induction hypothesis, T contains extreme points; and by Lemma 1 these are also extreme points of C. Thus we have found extreme points of C not in K, which is a contradiction. ∎

Let us investigate the implications of this theorem for convex polyhedra. We recall that a convex polyhedron is a bounded polytope. Being the intersection of closed half spaces, a convex polyhedron is also closed. Thus any convex polyhedron is the closed convex hull of its extreme points. It can be shown (see Section 2.5) that any polytope has at most a finite number of extreme points and hence a convex polyhedron is equal to the convex hull of a finite number of points. The converse can also be established, yielding the following two equivalent characterizations.

Theorem 4. *A convex polyhedron can be described either as a bounded intersection of a finite number of closed half spaces, or as the convex hull of a finite number of points.*

Appendix C GAUSSIAN ELIMINATION

This appendix describes the method for solving systems of linear equations that has proved to be, not only the most popular, but also the fastest and least susceptible to round-off error accumulation—the method of Gaussian elimination. Attention is directed toward explaining this classical elimination technique itself and its relation to the theory of **LU** decomposition of a nonsingular square matrix.

We first note how easily triangular systems of equations can be solved. Thus the system

$$\begin{aligned} a_{11}x_1 \qquad\qquad\qquad\qquad\qquad &= b_1 \\ a_{21}x_1 + a_{22}x_2 \qquad\qquad\qquad &= b_2 \\ &\ \ \vdots \\ a_{n1}x_1 + a_{n2}x_2 + \cdots + a_{nn}x_n &= b_n \end{aligned}$$

can be solved recursively as follows

$$\begin{aligned} x_1 &= b_1/a_{11} \\ x_2 &= (b_2 - a_{21}x_1)/a_{22} \\ &\ \ \vdots \\ x_n &= (b_n - a_{n1}x_1 - a_{n2}x_2 \ \dots \ - a_{nn-1}x_{n-1})/a_{nn} \end{aligned}$$

provided that each of the diagonal terms a_{ii}, $i = 1, 2, \ldots, n$ is nonzero (as they must be if the system is nonsingular). This observation motivates us to attempt to reduce an arbitrary system of equations to a triangular one.

Definition. A square matrix $\mathbf{C} = [c_{ij}]$ is said to be *lower triangular* if $c_{ij} = 0$ for $i < j$. Similarly, $\mathbf{C}$ is said to be *upper triangular* if $c_{ij} = 0$ for $i > j$.

In matrix notation, the idea of Gaussian elimination is to somehow find a decomposition of a given $n \times n$ matrix $\mathbf{A}$ in the form $\mathbf{A} = \mathbf{LU}$ where $\mathbf{L}$ is a lower triangular and $\mathbf{U}$ an upper triangular matrix. The system

$$\mathbf{Ax} = \mathbf{b} \tag{C.1}$$

can then be solved by solving the two triangular systems

$$\mathbf{Ly} = \mathbf{b}, \qquad \mathbf{Ux} = \mathbf{y}. \tag{C.2}$$

The calculation of $\mathbf{L}$ and $\mathbf{U}$ together with solution of the first of these systems is usually referred to as *forward elimination*, while solution of the second triangular system is called *back substitution.*

Every nonsingular square matrix $\mathbf{A}$ has an $\mathbf{LU}$ decomposition, provided that interchanges of rows of $\mathbf{A}$ are introduced if necessary. This interchange of rows corresponds to a simple reordering of the system of equations, and hence amounts to no loss of generality in the method. For simplicity of notation, however, we assume that no such interchanges are required.

We turn now to the problem of explicitly determining $\mathbf{L}$ and $\mathbf{U}$, by elimination, for a nonsingular matrix $\mathbf{A}$. Given the system, we attempt to transform it so that zeros appear below the main diagonal. Assuming that $a_{11} \neq 0$ we subtract multiples of the first equation from each of the others in order to get zeros in the first column below a_{11}. If we define $m_{k1} = a_{k1}/a_{11}$ and let

$$\mathbf{M}_1 = \begin{bmatrix} 1 & & & & \\ -m_{21} & 1 & & & \\ -m_{31} & & 1 & & \\ \cdot & & & \ddots & \\ \cdot & & & & \\ \cdot & & & & \\ -m_{n1} & & & & 1 \end{bmatrix}$$

the resulting new system of equations can be expressed as

$$\mathbf{A}^{(2)}\mathbf{x} = \mathbf{b}^{(2)}$$

with

$$\mathbf{A}^{(2)} = \mathbf{M}_1\mathbf{A}, \qquad \mathbf{b}^{(2)} = \mathbf{M}_1\mathbf{b}.$$

The matrix $\mathbf{A}^{(2)} = [a_{ij}^{(2)}]$ has $a_{k1}^{(2)} = 0, \quad k > 1$.

Next, assuming $a_{22}^{(2)} \neq 0$, multiples of the second equation of the new system are substracted from equations 3 through n to yield zeros below

$a_{22}^{(2)}$ in the second column. This is equivalent to premultiplying $\mathbf{A}^{(2)}$ and $\mathbf{b}^{(2)}$ by

$$\mathbf{M}_2 = \begin{bmatrix} 1 & 0 & & & & \\ 0 & 1 & & & & \\ \cdot & -m_{32} & 1 & & & \\ \cdot & -m_{42} & & & & \\ \cdot & \cdot & & & & \\ & \cdot & & & & \\ & \cdot & & & & \\ & -m_{n2} & & & & 1 \end{bmatrix}$$

where $m_{k2} = a_{k2}^{(2)}/a_{22}^{(2)}$. This yields $\mathbf{A}^{(3)} = \mathbf{M}_2\mathbf{A}^{(2)}$ and $\mathbf{b}^{(3)} = \mathbf{M}_2\mathbf{A}^{(2)}$.

Proceeding in this way we obtain $\mathbf{A}^{(n)} = \mathbf{M}_{n-1}\mathbf{M}_{n-2} \ldots \mathbf{M}_1\mathbf{A}$, an upper triangular matrix which we denote by $\mathbf{U}$. The matrix $\mathbf{M} = \mathbf{M}_{n-1}\mathbf{M}_{n-2} \ldots \mathbf{M}_1$ is a lower triangular matrix, and since $\mathbf{MA} = \mathbf{U}$ we have $\mathbf{A} = \mathbf{M}^{-1}\mathbf{U}$. The matrix $\mathbf{L} = \mathbf{M}^{-1}$ is also lower triangular and becomes the $\mathbf{L}$ of the desired $\mathbf{LU}$ decomposition for $\mathbf{A}$.

The representation for $\mathbf{L}$ can be made more explicit by noting that $\mathbf{M}_k^{-1}$ is the same as $\mathbf{M}_k$ except that the off-diagonal terms have the opposite sign. Furthermore, we have $\mathbf{L} = \mathbf{M}^{-1} = \mathbf{M}_1^{-1}\mathbf{M}_2{}^{-1} \ldots \mathbf{M}_{n-1}$ which is easily verified to be

$$\mathbf{L} = \begin{bmatrix} 1 & 0 & & & & \\ m_{21} & 1 & & & & \\ m_{31} & m_{32} & 1 & & & \\ \cdot & \cdot & & \cdot & & \\ \cdot & \cdot & & & \cdot & \\ \cdot & \cdot & & & & \cdot \\ m_{n1} & m_{n2} & \cdots & & & 1 \end{bmatrix}.$$

Hence, $\mathbf{L}$ can be evaluated directly in terms of the calculations required by the elimination process. Of course, an explicit representation for $\mathbf{M} = \mathbf{L}^{-1}$ would actually be more useful but a simple representation for $\mathbf{M}$ does not exist. Thus, we content ourselves with the explicit representation for $\mathbf{L}$ and use it in (C.2).

If the original system (C.1) is to be solved for a single $\mathbf{b}$ vector, the vector $\mathbf{y}$ satisfying $\mathbf{Ly} = \mathbf{b}$ is usually calculated simultaneously with $\mathbf{L}$ in the form $\mathbf{y} = \mathbf{b}^{(n)} = \mathbf{Mb}$. The final solution $\mathbf{x}$ is then found by a single back substitution, from $\mathbf{Ux} = \mathbf{y}$. Once the $\mathbf{LU}$ decomposition of $\mathbf{A}$ has been obtained, however, the solution corresponding to any right-hand side can be found by solving the two systems (C.2).

In practice, the diagonal element $a_{kk}^{(k)}$ of $\mathbf{A}^{(k)}$ may become zero or very close to zero. In this case it is important that the kth row be interchanged with a row that is below it. Indeed, for considerations of numerical accuracy, it is desirable to continuously introduce row interchanges of this type in such a way to insure $|m_{ij}| \leqslant 1$ for all i, j. If this is done, the Gaussian elimination procedure has exceptionally good stability properties.

BIBLIOGRAPHY

[A1] Abadie, J. and Carpentier, J., "Generalization of the Wolfe Reduced Gradient Method to the Case of Nonlinear Constraints," in *Optimization,* R. Fletcher (editor), Academic Press, London, 1969, pages 37–47

[A2] Akaike, Hirotugu, "On a successive Transformation of Probability Distribution and Its Application to the Analysis of the Optimum Gradient Method," *Ann. Inst. Statist. Math., Tokyo* **11,** 1959, pages 1–17

[A3] Antosiewicz, H. A., and Rheinboldt, W. C., "Numerical Analysis and Functional Analysis," Chapter 14 in *Survey of Numerical Analysis,* J. Todd (editor), McGraw-Hill, New York, 1962

[B1] Bartels, R. H., "A Numerical Investigation of the Simplex Method," Technical Report No. CS 104, July 31, 1968, Computer Science Dept., Stanford University, Stanford, Calif.

[B2] Bartels, R. H., and Golub, G. H., "The Simplex Method of Linear Programming Using LU Decomposition," *Comm. ACM* **12,** 5, May 1969, pages 266–268

[B3] Beckman, F. S., "The Solution of Linear Equations by the Conjugate Gradient Method," in *Mathematical Methods for Digital Computers,* Vol. 1, A. Ralston and H. S. Wilf (editors), John Wiley, New York, 1960

[B4] Broyden, C. G., "Quasi–Newton Methods and Their Application to Function Minimization," *Math. Comp.,* **21,** 1967, pages 368–381

[B5] Butler, T., and Martin, A. V., "On a Method of Courant for Minimizing Functionals," *J. Math. and Physics* **41,** 1962, pages 291–299

[C1] Carroll, C. W., "The Created Response Surface Technique for Optimizing Nonlinear Restrained Systems," *Operations Research* **9,** 12, 1961, pages 169–184

[C2] Charnes, A., "Optimality and Degeneracy in Linear Programming," *Econometrica* **20,** 1952, pages 160–170

[C3] Charnes, A., and Lemke, C. E., "The Bounded Variables Problem," ONR Research Memorandum 10, Graduate School of Industrial Administration, Carnegie Institute of Technology, Pittsburgh, Pa., 1954

[C4] Cohen, A., "Rate of Convergence for Root Finding and Optimization Algorithms," Ph.D. dissertation, University of California, Berkeley, 1970

[C5] Courant, R., "Calculus of Variations and Supplementary Notes and Exercises" (mimeographed notes), supplementary notes by M. Kruskal and H. Rubin, revised and amended by J. Moser, New York University, 1962

[C6] Crockett, J. B., and Chernoff, H., "Gradient Methods of Maximization," *Pacific J. Math.* **5,** 1955, pages 33–50

[C7] Curry, H., "The Method of Steepest Descent for Nonlinear Minimization Problems," *Quart. Appl. Math.* **2,** 1944, pages 258–261

[D1] Daniel, J. W., "The Conjugate Gradient Method for Linear and Nonlinear Operator Equations," *SIAM J. Numer. Anal.* **4,** 1, 1967, pages 10–26

[D2] Dantzig, G. B., "Maximization of a Linear Function of Variables Subject to Linear Inequalities," Chap. XXI of "Activity Analysis of Production and Allocation," *Cowles Commission Monograph* 13, T. C. Koopmans (editor), John Wiley, New York, 1951

[D3] Dantzig, G. B., "Computational Algorithm of the Revised Simplex Method," RAND Report RM–1266, The RAND Corporation, Santa Monica, Calif., 1953

[D4] Dantzig, G. B., "Variables with Upper Bounds in Linear Programming," RAND Report RM–1271, The RAND Corporation, Santa Monica, Calif., 1954

[D5] Dantzig, G. B., *Linear Programming and Extensions*, Princeton University Press, Princeton, N.J., 1963

[D6] Dantzig, G. B., Ford, L. R. Jr., and Fulkerson, D. R., "A Primal–Dual Algorithm," *Linear Inequalities and Related Systems*, Annals of Mathematics, Study 38, Princeton University Press, 1956, pages 171–181

[D7] Dantzig, G. B., Orden, A., and Wolfe, P., "Generalized Simplex Method for Minimizing a Linear Form under Linear Inequality Restraints," RAND Report RM–1264, The RAND Corporation, Santa Monica, Calif., 1954

[D8] Davidon, W. C., "Variable Metric Method for Minimization," Research and Development Report ANL–5990 (Ref.) U.S. Atomic Energy Commission, Argonne National Laboratories, 1959

[D9] Davidon, W. C., "Variance Algorithm for Minimization," *Computer J.* **10,** 1968, pages 406–410

[E1] Eaves, B. C., and Zangwill, W. I., "Generalized Cutting Plane Algorithms," Working Paper No. 274, Center for Research in Management Science, University of California, Berkeley, July 1969

[E2] Everett, H., III, "Generalized Lagrange Multiplier Method for Solving Problems of Optimum Allocation of Resources," *Operations Research* **11,** 1963, pages 399–417

[F1] Faddeev, D. K., and Faddeeva, V. N., *Computational Mtehods of Linear Algebra*, W. H. Freeman, San Francisco, Calif., 1963

[F2] Fenchel, W., "Convex Cones, Sets, and Functions," lecture notes, Dept. of Mathematics, Princeton University, Princeton, N.J., 1953

[F3] Fiacco, A. V., and McCormick, G. P., *Nonlinear Programming: Sequential Unconstrained Minimization Techniques,* John Wiley, New York, 1968

[F4] Fletcher, R., "A New Approach to Variable Metric Algorithms," *Computer J.* **13,** 13, 1970, pages 317–322

[F5] Fletcher, R., and Powell, M. J. D., "A rapidly convergent descent method for minimization," *Computer J.* **6,** 1963, pages 163–168

[F6] Fletcher, R., and Reeves, C. M., "Function Minimization by Conjugate Gradients," *Computer J.* **7,** 1964, pages 149–154

[F7] Ford, L. K. Jr., and Fulkerson, D. K., *Flows in Networks,* Princeton University Press, Princeton, New Jersey, 1962

[F8] Forsythe, G. E., "On the Asymptotic Directions of the *s*-Dimensional Optimum Gradient Method," *Numerische Mathematik* **11,** 1968, pages 57–76

[F9] Forsythe, G. E., and Moler, C. B., *Computer Solution of Linear Algebraic Systems,* Prentice-Hall, Englewood Cliffs, N.J., 1967

[F10] Forsythe, G. E., and Wasow, W. R., *Finite-Difference Methods for Partial Differential Equations,* John Wiley, New York, 1960

[F11] Fox, K., *An Introduction to Numerical Linear Algebra,* Clarendon Press, Oxford, 1964

[G1] Gale, D., *The Theory of Linear Economic Models,* McGraw-Hill, New York, 1960

[G2] Gass, S. I., *Linear Programming,* McGraw-Hill, third edition, New York, 1969

[G3] Gill, P. E., and Murray, W., "A Numerically Stable Form of the Simplex Algorithm," Report Maths 87, National Physical Laboratory (England), Division of Numerical and Applied Mathematics, August 1970

[G4] Goldfarb, D., "Sufficient Conditions for the Convergence of a Variable Metric Algorithm," Chapter 18 in *Optimization,* R. Fletcher (editor), Academic Press, London, 1969

[H1] Hadley, G., *Linear Programming,* Addison-Wesley, Reading, Mass., 1962

[H2] Hadley, G., *Nonlinear and Dynamic Programming,* Addison-Wesley, Reading, Mass., 1964

[H3] Hancock, H., *Theory of Maxima and Minima,* Ginn, Boston, 1917

[H4] Hestenes, M. R., "The Conjugate Gradient Method for Solving Linear Systems," *Proc. of Symposium in Applied Math.* **VI,** Num. Anal., 1956, pages 83–102

[H5] Hestenes, M. R., "Multiplier and Gradient Methods," *J. of Opt. Theory and Appl.* **4,** 5, 1969, pages 303–320

[H6] Hestenes, M. R., and Stiefel, E. L., "Methods of Conjugate Gradients for Solving Linear Systems," *J. Res. Nat. Bur. Standards,* Section B, 49, 1952, pages 409–436

[H7] Hurwicz, L., "Programming in Linear Spaces" in K. J. Arrow, L. Hurwicz, and H. Uzawa, *Studies in Linear and Nonlinear Programming,* Stanford University Press, 1958

[I1] Isaacson, E., and Keller, H. B., *Analysis of Numerical Methods,* John Wiley, New York, 1966

[K1] Karlin, S., *Mathematical Methods and Theory in Games, Programming, and Economics,* Vol. I, Addison-Wesley, Reading, Mass., 1959

[K2] Kelley, J. E., "The Cutting-Plane Method for Solving Convex Programs," *J. Soc. Indus. Appl. Math.* **VIII,** 4, 1960, pages 703–712

[K3] Kowalik, J., and Osborne, M. R., *Methods for Unconstrained Optimization Problems,* Elsevier, New York, 1968

[K4] Kuhn, H. W., and Tucker, A. W., "Nonlinear Programming," in *Proceedings of the Second Berkeley Symposium on Mathematical Statistics and Probability,* J. Neyman (editor), University of California Press, Berkeley and Los Angeles, Calif., 1961, pages 481–492

[L1] Lanczos, C., *Applied Analysis,* Prentice-Hall, Englewood Cliffs, N.J., 1956

[L2] Lemke, C. E., "The Dual Method of Solving the Linear Programming Problem," *Naval Research Logistics Quarterly* **1,** 1, 1954, pages 36–47

[L3] Levitin, E. S., and Polyak, B. T., "Constrained Minimization Methods," *Zh. vychisl. Mat. mat. Fiz.* **6,** 5, 1966, pages 787–823

[L4] Loewner, C., "Über monotone Matrixfunktionen," *Math. Zeir.* **38,** 1934, pages 177–216. Also see C. Loewner, "Advanced matrix theory," mimeo notes, Stanford University, 1957

[L5] Lootsma, F. A., *Boundary Properties of Penalty Functions for Constrained Minimization* doctoral dissertation, Technical University, Eindhoven, The Netherlands, May 1970

[L6] Luenberger, D. G., *Optimization by Vector Space Methods,* John Wiley, New York, 1969

[L7] Luenberger, D. G., "Hyperbolic Pairs in the Method of Conjugate Gradients," *SIAM J. Appl. Math.* **17,** 6, November 1969, pages 1263–1267

[L8] Luenberger, D. G., "A Combined Penalty Function and Gradient Projection Method for Nonlinear Programming," Internal Memo, Dept. of Engineering–Economic Systems, Stanford University, June 1970

[L9] Luenberger, D. G., "The Conjugate Residual Method for Constrained Minimization Problems," *SIAM J. Numer. Anal.* **7,** 3, Sept. 1970, pages 390–398

[L10] Luenberger, D. G., "Convergence Rate of a Penalty-Function Scheme," *J. Optimization Theory and Applications* **7,** 1, January 1971, pages 39–51

[L11] Luenberger, D. G., "The Gradient Projection Method Along Geodesics," *Management Science,* **18,** 11, July 1972, pages 620–631

[Mc1] McCormick, G. P., "Anti-Zig-Zagging by Bending," *Management Science,* (Theory), **15,** 5, January 1969, pages 315–320

[Mc2] McCormick, G. P., "The Variable Reduction Method for Nonlinear Programming," *Management Science* **17,** 3, November 1970, pages 146–160

[Mc3] McCormick, G. P., and Pearson, J. D., "Variable Metric Methods and Unconstrained Optimization," Chapter 21 in *Optimization,* R. Fletcher (editor), Academic Press, London, 1969

[M1] Morrison, D. D., "Optimization by Least Squares," *SIAM J., Numer. Anal.* **5,** 1968, pages 83–88

[M2] Murtagh, B. A., and Sargent, R. W. H., "A Constrained Minimization Method with Quadratic Convergence," Chapter 14 in *Optimization*, R. Fletcher (editor), Academic Press, London, 1969

[O1] Orchard-Hays, W., "Background Development and Extensions of the Revised Simplex Method," RAND Report RM–1433, The RAND Corporation, Santa Monica, Calif., 1954

[O2] Orden, A., Application of the Simplex Method to a Variety of Matrix Problems, pages 28–50 of Directorate of Management Analysis: "Symposium on Linear Inequalities and Programming," A. Orden and L. Goldstein (editors), DCS/Comptroller, Headquarters, U.S. Air Force, Washington, D.C., April 1952

[O3] Oren, S. S., "Self-Scaling Variable Metric Algorithms for Unconstrained Minimization," Ph.D. dissertation, Dept. of Engineering–Economic Systems, Stanford University, Stanford, Calif., 1972

[O4] Oren, S. S., and Luenberger, D. G., "The Self-Scaling Variable Metric Algorithm (SSVM)," Fifth Hawaiian International Conference on System Sciences, January, 1972

[O5] Ortega, J. M., and Rheinboldt, W. C., *Iterative Solution of Nonlinear Equations in Several Variables,* Academic Press, New York, 1970

[P1] Powell, M. J. D., "An Efficient Method for Finding the Minimum of a Function of Several Variables without Calculating Derivatives," *Computer J.* **7,** 1964, pages 155–162

[P2] Powell, M. J. D., "A Method for Nonlinear Constraints in Minimization Problems," in *Optimization,* R. Fletcher (editor), Academic Press, London, 1969, pages 283–298

[P3] Powell, M. J. D., "On the Convergence of the Variable Metric Algorithm," Mathematics Branch, Atomic Energy Research Establishment, Harwell, Berkshire, England, October 1969

[R1] Rosen, J., "The Gradient Projection Method for Nonlinear Programming, I. Linear Constraints," *J. Soc. Indust. Appl. Math.* **8,** 1960, pages 181–217

[R2] Rosen, J., "The Gradient Projection Method for Nonlinear Programming, II. Non-Linear Constraints," *J. Soc. Indust. Appl. Math.* **9,** 1961, pages 514–532

[S1] Shah, B., Buehler, R., and Kempthorne, O., "Some Algorithms for Minimizing a Function of Several Variables," *J. Soc. Indust. Appl. Math.* **12,** 1964, pages 74–92

[S2] Shefi, Adam, "Reduction of Linear Inequality Constraints and Determination of all Feasible Extreme Points," Ph.D. dissertation Department of Engineering–Economic Systems, Stanford University, Stanford, Calif., October 1969

[S3] Simonnard, M., *Linear Programming*, translated by William S. Jewell, Prentice-Hall, Englewood Cliffs, N.J., 1966

[S4] Stewart, G. W., "A Modification of Davidon's Minimization Method to Accept Difference Approximations of Derivatives," *J.A.C.M.* **14,** 1967, pages 72–83

[S5] Stiefel, E. L., "Kernel Polynomials in Linear Algebra and Their Numerical Applications," Nat. Bur. Standards, Appl. Math. Ser., 49, 1958, pages 1–22

[T1] Topkis, D. M., "A Note on Cutting–Plane Methods Without Nested Constraint Sets," ORC 69–36, Operations Research Center, College of Engineering, Berkeley, Calif., December 1969

[T2] Topkis, D. M., and Veinott, A. F. Jr., "On the Convergence of Some Feasible Direction Algorithms for Nonlinear Programming," *J. SIAM Control* **5,** 2, May 1967, pages 268–279

[T3] Traub, J. F., *Iterative Methods for the Solution of Equations*, Prentice-Hall, Englewood Cliffs, N.J., 1964

[V1] Veinott, A. F. Jr., "The Supporting Hyperplane Method for Unimodal Programming," *Operations Research* **XV,** 1, 1967, pages 147–152

[V2] Vorobyev, Y. V., *Methods of Moments in Applied Mathematics*, Gordon and Breach, New York, 1965

[W1] Wilde, D. J., and Beightler, C. S., *Foundations of Optimization*, Prentice-Hall, Englewood Cliffs, N.J., 1967

[W2] Wolfe, P., "A Duality Theorem for Nonlinear Programming," *Quar. Appl. Math.* **19,** 1961, pages 239–244

[W3] Wolfe, P., "On the Convergence of Gradient Methods Under Constraints," IBM Research Report RZ 204, Zurich, Switzerland, 1966

[W4] Wolfe, P., "Methods of Nonlinear Programming," Chapter 6 of *Nonlinear Programming*, J. Abadie (editor), Interscience, John Wiley, New York, 1967, pages 97–131

[W5] Wolfe, P., "Convergence Theory in Nonlinear Programming," Chapter 1 in *Integer and Nonlinear Programming*, J. Abadie (editor), North-Holland Publishing Company, Amsterdam, 1970

[Z1] Zangwill, W. I., "Nonlinear Programming via Penalty Functions," *Management Science* **13,** 5, 1967, pages 344–358

[Z2] Zangwill, W. I., *Nonlinear Programming: A Unified Approach*, Prentice-Hall, Englewood Cliffs, N.J., 1969

[Z3] Zeleznik, F. J., "Quasi-Newton Methods for Nonlinear Equations," *J.A.C.M.* **15,** 1968, pages 264–270

[Z4] Zoutendijk, G., *Methods of Feasible Directions*, Elsevier, Amsterdam, 1960

INDEX

INDEX

ABCDEFGH798765432